Reproductive Technologies in Farm Animals

Reproductive Technologies in Farm Animals

Ian Gordon

Professor Emeritus
Department of Animal Science and Production
University College Dublin
Ireland

CABI Publishing

CABI Publishing is a division of CAB International

CABI Publishing
CAB International
Wallingford
Oxfordshire OX10 8DE
UK

Tel: +44 (0)1491 832111
Fax: +44 (0)1491 833508
E-mail: cabi@cabi.org
Website: www.cabi-publishing.org

CABI Publishing
875 Massachusetts Avenue
7th Floor
Cambridge, MA 02139
USA

Tel: +1 617 395 4056
Fax: +1 617 354 6875
E-mail: cabi-nao@cabi.org

A catalogue record for this book is available from the British Library, London, UK.

Library of Congress Cataloging-in-Publication Data

Gordon, Ian R.
    Reproductive Technologies in Farm Animals / Ian R. Gordon.
        p. cm.
    Includes bibliographical references (p.   ).
    ISBN 0-85199-862-3 (alk. paper)
    1. Livestock--Reproduction.   2. Livestock--Breeding.   I. Title.

    SF871.G67 2005
    636.08′24--dc22

                                                        2004004673

ISBN 0 85199 862 3

Typeset by AMA DataSet Ltd, UK.
Printed and bound in the UK by Cromwell Press, Trowbridge.

Contents

_____

Preface

In the past half-century, great progress has been made in the reproductive management of farm animals, both mammals and birds. The present text aims to provide information on developments that have occurred over the years and to indicate areas in which reproductive technologies may usefully be employed, whether in commerce or research. Of the countless thousands of scientific publications dealing with reproductive technologies, only a small number are provided in the main text (essentially confined to reports appearing in the years 2000–2004). Tables dealing with landmark events in the development of the various technologies and pointing to those associated with them are provided in the various chapters. Those who wish to delve deeper into the subject matter can consult various of the texts mentioned in Appendix A. For a look at work at the research level and to find most of the papers dealt with in the text, Appendix B contains a list of journals that can be found in a good university library.

In the context of reproductive technologies emerging since the Second World War, the 1940s and 1950s saw the widespread establishment of artificial insemination and the introduction of frozen semen in cattle breeding; the end of the century saw the first moves to commercialize semen sexing technology. In the early 1960s, as well as being an ingredient in the human contraceptive pill, progestogens were shown to be highly effective when administered by the intravaginal route, leading to commercially acceptable techniques for the control of the oestrous cycle and breeding season in sheep and goats. In the late 1960s came a revolution in the measurement of hormones in body fluids with the introduction of the exquisitely sensitive radioimmunoassay (RIA) techniques, which provided a new approach to early pregnancy diagnosis; emerging around the same time were early forms of ultrasonics, which could also be employed in the detection of pregnancy.

The 1970s were to see embryo transfer and the freezing of embryos in cattle becoming commercial realities and their incorporation into increasingly effective breeding improvement programmes. With the availability of prostaglandins (PGs) and gonadotrophin-releasing hormone (GnRH), that decade also witnessed new possibilities for oestrus control in cattle and horses. The first steps towards mammalian cloning were taken in the mid-1980s and a few years later came the ability to produce cattle embryos in the laboratory in large numbers and at low cost. The 1990s were to witness the dramatic emergence of somatic cell cloning with the birth of Dolly the sheep, opening the way to new possibilities for the production of transgenic farm animals. It is perhaps worthy of a mention that there is an Irish connection, talking about recombinant DNA technology and its application in the production of transgenic animals. Dublin was where Erwin Schrodinger spent the years of the Second World War and wrote his seminal work *What is Life?*, which was published in

1944 and influenced the thinking of so many, including Watson and Crick, who, within a decade, were to lay the foundations of the revolutionary science of molecular biology.

The technologies that enable farmers to improve the biological efficiency of their livestock can also be valuable in other ways; farm animals are likely to play an important role in the production of high-value pharmaceutical proteins or, at some stage in the future, to be a source of organs and tissues for use in humans. Quite apart from the possible application of reproductive technologies in livestock production, saving endangered species and biomedicine, for those interested in a career in animal science or reproductive biology a knowledge of options available for the control and manipulation of reproduction may be of interest. In the matter of applying certain forms of technology, it should be noted that current and potential use of reproductive technologies in commercial practice is likely to vary according to the requirements of the regulatory authorities in different countries. What is permissible in North America may not be acceptable in a European Union (EU) country. Restriction on the use of certain techniques in assisted human reproduction is far from uncommon. In some EU countries, clinicians are unable to freeze or store fertilized eggs for use in *in vitro* fertilization nor are they able to test embryos for genetic diseases before transfer. It is not unexpected that similar restrictions apply to some of the reproductive technologies in the farm-animal world.

It should be emphasized that the provision of optimal nutritional management, maximizing animal comfort, minimizing stress and maintaining good animal health are all prerequisites for the successful application of reproductive technologies. As the ways in which farm animals are used become technically more complex, clearly the skills required by those working with them must be updated. Professionals need to have an extensive knowledge of animal science and the issues associated with animal care and welfare to ensure that emerging forms of reproductive technology are appropriately integrated into livestock production systems in a way that has the support of those who work in agriculture and society at large. It is certain that the demand for livestock products will greatly increase in the years ahead; the urgent need to sustain reliable food supplies at reasonable cost makes it important that the livestock industry is always serviced by a highly motivated and scientifically competent workforce.

Acknowledgements

_____

The author wishes to express grateful thanks to the authors and publishers who have given permission to reproduce illustrations and tables from books and original articles. Sincere thanks is due to the staff of CAB International for their excellent help in the preparation of the book.

The author is also most appreciative of the office and other facilities provided by University College Dublin at their Lyons Research Farm and to the staff of the Department of Animal Science and Production for their help on numerous occasions.

1

Introduction

1.1. The Changing Agricultural Scene

The present era is one of increasing public concern about food quality, animal welfare and environmental protection; food is no longer dissociated from its production system. Many see the need to question current food production and handling practices all the way through 'from plough to plate'. The emphasis of agricultural research is likely to shift increasingly from greater production efficiency to making farming more sustainable, not only in terms of the environment but socially and ethically.

The problems faced by the farming industry as it strives to deal with increasing concerns about the environment and other affairs on the land are likely to be all the greater because of the existing level of scientific literacy and the public's natural mistrust of many new technologies, which appears to be much greater on this side of the Atlantic than in the Americas. Part of the problem stemmed from the bovine spongiform encephalopathy (BSE) disaster. Three years ago (2001) there was a European Union (EU) surplus of 1 million tonnes of beef, which is only now being transformed into a deficit due to recovery in consumer demand and a decline in beef production.

Apart from the immense cost, the problem of BSE was to leave a legacy of mistrust among the consumers of the EU, which has implications for the uptake of many new technologies in European countries. While many countries around the world have moved forward with genetically modified (GM) foods, use of bovine somatotrophin (BST) in dairy cattle and hormonal growth promoters, all of which are capable of making valuable contributions to food production, the EU chooses a different path; there are those who see the current opposition to GM foods as being little different from the illogical opposition to growth promoters. In Ireland, there are those who point to the decline in the efficiency of beef production in the country over the past 15 years and note that the first direct attack on that efficiency came with the banning of growth hormones in the late 1980s, which was to have a real and measurable effect.

Considerable research effort in the USA and elsewhere was devoted to the development and testing of growth hormone preparations that may have applications in farm animals, particularly dairy cattle. Although BST has been employed to increase the yield of milking cows in various countries, principally the USA, there are other areas where it may have a useful part to play. It may, for example, be used to promote growth in the suckling young in species such as sheep and goats. For a variety of reasons, however, not the least being current ethical and consumer concerns, hormones in the EU are used much more in the control of reproductive diseases rather than in increasing the efficiency of lactation, growth and reproduction. Already, in some European countries,

©I.R. Gordon 2004. *Reproductive Technologies in Farm Animals* (I.R. Gordon)

restrictions have been placed on the use of hormonal techniques in dairy cattle fertility management. In Sweden, for example, oestrus synchronization as an aid to artificial insemination (AI) ceased after a decision by a farmers' association in 1996; this was based on the fear of consumer reactions which stemmed from ethical concerns arising from the replacement of management with hormones.

1.1.1. Species under consideration

During the 1970s, dairy farmers in Britain and Ireland enjoyed a market that was favourable to milk production and were paid a good price for the product. All this was to change in the 1980s with the introduction of milk quotas and other restrictions on production. None the less, in the countries of the EU, which currently produce 22% of world milk supplies, dairying remains one of the most profitable sectors of farming; milk yields per cow have increased in each member country over the past three decades and the trend within the dairy industry is towards increased intensification on a decreasing number of larger, more specialized production units (Van Arendonk and Liinamo, 2003). Figures on dairy cattle populations in the EU countries show a decline in cattle numbers over the 5-year period 1992–1997 (Table 1.1).

The main EU dairy production systems aim at high output of milk throughout the year and the primary objective of dairy farmers in most countries is to have cows producing a calf every 12 months without any particular calving pattern. If the cow does not show regular oestrous cycles, become pregnant at the appropriate time and deliver a live, healthy calf each year, then her other excellent qualities may be of little avail.

Longevity in dairy cows

It is not only a question of calving once a year; the dairy cow needs to keep this performance up over a much longer lifespan. Longevity is likely to be a trait of increasing importance in dairy cattle breeding programmes; it is important because feed costs in the first three lactations may not allow a cow to pay its way, the animal only becoming profitable when it reaches its fourth lactation. Farmers need cows that calve each year for at least six lactations, without problems from mastitis, lameness and subfertility. As matters stand in many countries, the rate of dairy replacements is seen to be far too high. French workers, for example, have reported a rate of 33%, almost twice the optimal replacement rate. Although ample scope remains to decrease culling rates by attention to traditional methods of breeding, feeding and management, there remains the possibility that

Table 1.1. Dairy cattle in the European Union (pre-2004). Number of dairy cows, average herd size, the proportion (%) of dairy farms with more than 50 cows and average annual milk yield in countries of the European Union.

Country	Dairy cows (× 1000 head)		Average herd size in 1997	Proportion of farms > 50 cows in 1995	Average yield (kg per head) in 1997
	1992	1997			
Belgium	751	642	32.3	18.7	5,005
Denmark	708	695	50.8	37.2	6,573
Germany	5,382	5,026	27.9	8.4	5,711
Spain	1,490	1,279	11.9	2.2	4,668
France	4,685	4,476	30.7	13.1	5,411
Ireland	1,262	1,268	32.4	16.9	4,232
Italy	2,443	2,088	20.5	9.5	4,988
Netherlands	1,821	1,674	44.0	37.9	6,524
UK	2,747	2,498	68.8	57.8	5,958
EU-15	23,460	21,760	24.0	10.8	5,513

valuable improvements can come from the appropriate application of reproductive technologies to improve cow fertility.

Zebu cattle and buffaloes

The majority of the world's zebu cattle (*Bos indicus*) and buffaloes (*Bubalus bubalis*) live in regions between the Tropics of Cancer and Capricorn, where nutrition, thermal balance, milk yield and reproduction are likely to be severely affected by high temperature and relative humidity. European cattle (*Bos taurus*) were introduced to the Tropics in the mid-1800s in an effort to increase the comparatively low levels of milk production in indigenous cattle; taurine cows, however, may not offer a viable option for milk production in tropical countries because of their poor survival rates. Zebu cattle are adapted to the tropical areas and generally possess a larger skin area and more numerous sweat glands to facilitate heat loss. They are also hardier and show greater resistance to the ectoparasites found in tropical areas; with their long legs and hard hooves, they are generally better fitted than taurine cattle as grazers of rough terrain (Fig. 1.1).

Looking at bulls, beef breeds are usually held to be less sexually active than dairy breeds and zebu bulls tend to be less sexually active than taurine bulls (Yates *et al.*, 2003). Success in the application of reproductive technologies in zebu cattle is more limited than in their European counterparts; part of the explanation lies in the fact that taurine cattle have been studied much more intensively over the years.

A major work published a quarter-century ago reviewed the information relating to the water buffalo (*B. bubalis*); the writings of Ross Cockrill in the 1970s stimulated much interest in the species as a producer of animal protein for human consumption. Since the early 1970s, there have been several world buffalo conferences at which numerous papers have been presented, dealing with all aspects of buffalo production. Globally, the world's domesticated buffalo population is classified, somewhat arbitrarily, into three types: swamp, river and Mediterranean. Domesticated swamp and river buffaloes are believed to have evolved from an ancestral swamp-like animal, the two types diverging many years ago in South-east Asia and spreading north to China and west to the Indian subcontinent.

Although India has less than 3% of the world's total land area, it supports 55% of the world's buffaloes; in that country cattle

Fig. 1.1.   Zebu bull at a Chinese AI centre.

and buffalo husbandry is a multipurpose system in which these animals are used mainly for milk, draught and manure. Worldwide, there is a growing awareness of the importance of the water buffalo, which has an estimated population of more than 150 million and plays a prominent part in rural livestock production, particularly in the countries of Asia. The productivity of the species clearly has a considerable bearing on the agricultural output in such regions. Ever-increasing demand for milk and meat from buffaloes has led to breeding programmes involving advanced reproductive technologies, but progress in this area has proved much slower and more difficult than in cattle.

Sheep, goats and deer

Although the major emphasis in the development of reproductive technologies has been on cattle, particularly dairy cattle, technologies such as AI and embryo transfer (ET) have also played a part in improving the biological and economic efficiency of sheep and goats. *In vitro* embryo production in small ruminants may also provide a low-cost source of embryos for research and for commercial applications in the emerging fields of cloning and the production of transgenic animals. One report in the late 1990s presented an overview of the sheep and goat industry worldwide. At that time, sheep (1096 million) and goat production (677 million) were the second and fourth largest livestock groups, respectively. In industrialized countries, sheep farming is more important (40%) than goat production (4.6%), although during the past two decades the number of sheep has decreased at world level. In general, in terms of meat, milk and fibre, production of small ruminants has grown substantially in developing countries but has either stagnated or declined in industrialized countries.

For 70 years, research workers around the world have been examining the possibility of employing hormones in the control of oestrus and ovulation in sheep; the fact that most ewes in the agriculturally productive countries are seasonal breeders and often produce smaller lamb crops than the farmer might wish was to make the sheep a rather obvious target for the reproductive physiologist's attention. It is now possible to control seasonal breeding activity in sheep and goats, and to some extent in deer, by way of hormones (steroids and gonadotrophins) and by the much fuller understanding of the reproductive processes in these animals.

Some draw attention to the fact that the European sheep industry is facing major challenges due to the changing nature of the retail market for lamb meat; one way of meeting this challenge may lie in manipulating the timing of supplies of lamb meat to the market. Manipulation of supply is likely to involve changes in the timing and spread of lambings, breed choice, nutritional regimes, optimization of the growth potential of lambs by controlling parasitism and the use of new meat storage technologies. There are many ways in which reproduction control in this species could have particular merit.

Goats are believed to be among the first ruminants to be domesticated; archaeological evidence suggests a long association between humans and goats, stretching back some 10,000 years. They are valued for milk and meat production as well as for providing the important fibres mohair and cashmere; 40% of the world's production of cashmere is in China, where transhumance (seasonal movement of stock to a different region) is a regular feature of the extensive management system employed. Farmers recognize that goats can look after themselves much better than some other forms of farm livestock; they can thrive in a variety of climates, ranging from extremes of tropical rainforests to dry deserts, where sheep cannot exist (see Fig. 1.2).

The deer is a recently domesticated farm animal, with the production of venison, with its low fat content, being one of its great virtues. Comparative data on carcass composition for red deer, cattle and sheep have shown lean meat percentages of 75.3%, 65.8% and 57.6%, respectively. Some countries use their deer for other purposes; in China, for example, deer are farmed almost entirely for the production of antler for medicinal use; velvet production is also

Fig. 1.2.   Tropical breed goats in China.

important in other Asian countries. Modern deer management practices have only been developed in recent decades; within a relatively short time frame, deer have been tamed but have undergone only limited genetic selection for improved productivity and domesticity. Deer, in fact, on average are only three to four generations removed from their wild progenitors; they are essentially wild animals habituated to the farm environment. The behaviour of deer can be in marked contrast to that of other ungulates such as sheep, goats and cattle, which have undergone profound physiological, morphological and behavioural changes as a result of thousands of years of artificial selection for domesticity, to such a degree that they often bear little resemblance to their wild progenitors. Clearly, the deer farming industry faces challenges which may be rather different from those encountered with the traditional farm species. In terms of reproductive physiology, the deer is of interest in losing few of its embryos during pregnancy, in contrast to conventional farm animals, such as the cow, which may lose 30% or more of its embryos.

As well as welfare aspects, disease and human health must also be kept in mind when dealing with deer on the farm. Since 1990, when deer farming became established in the Republic of Ireland, reports from that country showed the isolation of myco-bacteria from tuberculous lesions detected post-mortem in deer slaughtered in abattoirs in the 1990s; human health implications require that *Mycobacterium bovis* infection in deer is strictly controlled and where possible eradicated. As shown later in this chapter, there are ways in which reproduction control may be usefully employed in this species.

Pigs and poultry

During the past half-century, genetic selection has had a major impact on livestock production, with the pig and poultry industries at the forefront in applying genetic theory and nutritional wisdom towards improving their productivity. Some predict that the growth in demand for food will be greatest over the next 20 years in pig and poultry products (Kabanov, 2002). There are firm biological foundations for encouraging pig production; pigs can supply meat more rapidly and economically than most other farm animals. For horses, cattle, sheep and pigs, growth from birth to adulthood can see animal body weight increasing 10-, 14-, 10- and 208-fold, respectively. Of the farm-animal species, pigs have the potential to make a unique and valuable contribution in certain areas of human medicine. Good use has been made of miniature pigs in biomedical research, for example, especially in producing transgenic animals for use in human disease models.

As in pigs, artificial selection of domestic poultry over the years has led to the evolution of biological types differing considerably in their physiology and from their wild prototypes; in the UK and Ireland, 95% of the income from domestic poultry comes from chickens and turkeys, and reproductive techniques such as AI are of considerable importance in certain production systems.

Horses and camelids

Research is conducted in many countries on a wide range of problems relevant to the horse industry, including growth, racing performance, nutrition, housing, physiology, health, training, behaviour and reproduction. One author, commenting on the vitality of the horse industry in France, particularly in the field of equestrian competitions and leisure activities, suggested that there will be an increasing demand for more sophisticated knowledge of horse genetics and techniques by which it is possible to read the genome (Sellier, 2002). In terms of breeding improvement in racehorses, as reflected by their performance in races, it is said that little progress has been made over the years, presumably a reflection of the selection methods employed. Although there is ample scope for the application of assisted reproduction technologies in the horse world, all changes, whether in breeding or technology, have to take account of long-standing national and international regulations relating to these animals. In horse affairs, views are often sharply divided between those engaged in the wide range of sports and leisure activities and those dealing with Thoroughbreds and racehorses. While breed associations

such as the Thoroughbred Association and Jockey Club forbid intensive genetic selection made possible by AI and ET, others (e.g. the American Quarter Horse Association and the United States Polo Association) welcome assisted reproduction as a means of improving the genetic quality of their animals.

Old and New World camelids

It is believed that camels originated in North America, with some members of the family travelling north across the Bering Straits to give rise to the Old World camels (Bactrian and dromedary) and others moving south into South America and giving rise to the smaller New World camelids (llama, alpaca, guanaco and vicuña). In the UK and Ireland, there is a growing population of New World camels, which are currently kept as pets, as producers of fine fibre and as pack-animals for mountain-trekking holidays. In South America, researchers have given increasing attention to assisted reproduction techniques, such as ET.

Traditional and modern classification systems for dromedary camels in Africa, Asia and India have been described in a recent report. Traditional systems have classified camels according to their purpose (riding or pack-animals) or location (lowland or mountain) and descriptions are given for each category. In more recent times, attempts have been made to categorize camels into types comparable to those applied to cattle (e.g. meat, dairy, dual-purpose and racing types) but there appears to be no real justification for this at present. As a result of their nomadic way of life, there has been little differentiation of dromedaries into specialized types; they remain multi-purpose animals in most regions, with females being used primarily as milk producers and males for transport or draught, both sexes providing meat as a secondary product. According to one report, the global base on domestic livestock breeds held by the Food and Agriculture Organization (FAO) has information on fewer than 50 dromedary breeds, of which one is considered to be in danger of extinction.

The Bactrian camel (*Camelus bactrianus*) is a multipurpose animal, valuable for the production of milk, meat and work, as a riding or pack-animal and as a source of fuel from dried dung. The Arabian camel is seen by some to have the capability of being a better provider of food in the desert areas of the world than the cow, which can be severely affected by heat and a scarcity of feed and water. One of the most important factors affecting productivity, other than nutrition and disease, is the low reproductive performance of the camel; high fertility levels in the camel are essential, not only for profitable production, but also to provide opportunities for selection and genetic improvement.

Aquaculture

Quite apart from the potential applications of biotechnology to farm livestock, increasing consumer demand for seafood and the serious decrease in fish stocks around the world have led to a substantial commercial interest in aquaculture, which is seen to be a rapidly growing and extremely diverse form of animal production. Cod is one of the fish species, for example, thought to have the same potential for large-volume fish farming as salmon and rainbow trout; some believe there is potential to produce more than 500,000 tonnes of cod per year in Norway alone, far surpassing the wild catch (Adoff *et al.*, 2002). A recent article draws attention to the fish and shellfish that are currently genetically engineered; this is said to represent a global market of some US$50 billion, with an annual growth rate of about 10% (Young, 2003). There would appear to be great opportunities for research in aquaculure in helping to develop a sustainable and environmentally acceptable industry.

1.1.2. Consumer, social and ethical issues

What appears on the shelves of the supermarket must meet not only economic criteria, but also social and ethical criteria. A balanced judgement in arriving at such criteria must certainly mean having sound scientific information and a continuous and clear dialogue among all the parties involved. It has to be recognized that there are many consumers who question the safety of using advanced reproductive technologies such as *in vitro* embryo production, cloning and transgenesis in food animals. There is also the danger that a poorly informed public may often perceive risks that do not exist. Such concerns are to be seen at their most acute in Western Europe, where technologies developed and widely used in countries such as the USA to increase the efficiency of animal growth (growth promoters) and milk production (BST) in cattle have effectively been banned. In the EU, the BSE and foot-and-mouth outbreaks in the UK have focused consumer attention on food safety issues and have contributed to concerns over the production and use of GM crops.

In the context of reproductive technologies, some European countries have already placed restrictions on the use of hormonal techniques in dairy cattle fertility management. In Sweden, for example, oestrus synchronization as an aid to AI ceased after a decision by a farmers' association in 1996; this was stemming from the fear of consumer reactions based on ethical concerns arising from the replacement of management with hormones. In the context of ET technology, there is also evidence that farmers may already face a negative attitude towards their products. According to Galli *et al.* (2003a), it is not uncommon to find that retailers in Italy will not deal with beef produced by embryo-based technologies, even when produced by conventional multiple ovulation and ET techniques. According to some observers, the single-minded pursuit of animal productivity may no longer be appropriate; authors such as Hodges (2003) see a growing public unease with the direction being taken in livestock production. Forms of animal production that would be regarded as unacceptable by farmers and researchers alike are mentioned in Table 1.2.

It is important that scientists involved in the development of novel forms of

Table 1.2. Unacceptable livestock production.

1. The manipulation of body size, shape or reproductive capacity by breeding, nutrition, hormone treatment or gene insertion in such a way as to reduce mobility, increase the risk of injury, metabolic disease, skeletal or obstetric problems and perinatal mortality
2. The use of sedative drugs to overcome deficiencies in existing or planned husbandry systems
3. The use of potentially toxic drugs as an alternative to current husbandry practices (e.g. for chemical shearing)

breeding and the professionals and farmers involved in their application should take due note of the concern that is expressed (Van Arendonk and Bijma, 2003). While the introduction of AI into commercial farming brought few, if any, ethical problems or animal welfare issues, the same was not true for ET technology. A powerful incentive for the development of non-surgical embryo recovery and transfer techniques was the difficulty of dealing with donor and recipient cattle under farm conditions rather than in a carefully supervised research setting.

1.1.3. Human health considerations

Meat quality

Authors in Ireland and elsewhere have speculated on the likely future trends in the fat content of meat and meat products (Moloney, 2002). The fat content of meat has considerable implications for human health, and the combined and interacting effects of genotypes, gender, age at slaughter and nutrition before slaughter all play a part in this. There are many novel opportunities available to exploit the diet of meat animals to produce high-quality meat with increased concentrations of conjugated linoleic acid (with the potential to protect against obesity, neoplasms and heart diseases), a low fat concentration and a fatty acid profile more compatible with human dietary considerations. Such improvements are made all the more achievable because

of the many developments in reproductive technology.

Alternatives to conventional farming as a means of providing consumers with low-fat meat include deer farming for venison and crossbreeding between wild and domesticated animals. In Argentina, for example, the crossing of wild boars with domestic sows has produced meat with characteristics distinct from those of meat from purebred domestic pigs, making it suitable for a specialist market. Although the wild boar is an animal with pronounced reproductive seasonality, assisted reproduction, in the form of frozen semen, could permit the production of crossbred animals throughout the year.

Environmental pollutants

Researchers in recent years have drawn attention to the fact that environmental oestrogens and other potential hormone-disrupting compounds are widespread and persistent in the environment; they are likely to be present in drinking water, plastics, household products and food packaging and in the human food chain. To date, some 60 chemicals have been identified as endocrine disrupters (i.e. exogenous agents that interfere with various aspects of natural hormone physiology). Octylphenol (OP) is one of several compounds found in the environment that possesses oestrogen-mimicking action *in vivo*. The potential reproductive and health hazards posed by such environmental chemicals have generated concern among the scientific community, policy makers and the general public (Pocar *et al.*, 2003). There are those who claim that declining human male fertility may be due to global pollution with synthetic chemicals which have very weak oestrogenic and/or androgenic potency. There have also been suggestions that, during the last 20 years, puberty and the human menopause are occurring earlier in humans and that endocrine-disrupting compounds may be influencing the timing of adult reproductive transitions. Using sheep as their animal model, workers in Scotland found that administration of an environmental oestrogenic chemical (OP) would inhibit fetal follicle-stimulating hormone (FSH) secretion; this provided one explanation of how such chemicals may adversely affect adult reproductive potential. In Ireland, other workers concluded that maternal exposure to OP selectively inhibited FSH synthesis, with a subsequent decrease in testis size and number of Sertoli cells in the newborn lamb. Irish studies also found that exposure to OP *in utero* and in postnatal life may influence the onset of puberty and duration of the breeding season in ewes (Wright *et al.*, 2002).

Cloning and food products

Several commercial concerns, mainly in the USA, are currently developing strains of cattle by the use of somatic-cell nuclear transfer (see Chapter 12); there is an obvious need to ensure that any food product derived from such cattle (e.g. meat, milk, milk products) does not raise questions of public concern when it is marketed. In the USA, the Food and Drug Administration (FDA) has considered the various steps necessary in ensuring that the production of cloned animals is monitored and appropriately regulated. Cloning is also likely to be a crucial step in the production of GM animal products; mention of genetic modification of animals in many countries is guaranteed to raise public concerns about food safety. The thought of GM animal products is likely to generate the same reaction in the EU as that currently directed towards GM crops. It is essential to have in place appropriate regulations that meet the approval of consumer groups long before any attempt is made to market products from cloned animals. It is not only a matter of regulations but the need to have appropriately trained personnel to service the needs of cloned animals. In 1998, the UK's Farm Animal Welfare Council produced a report highlighting the need for the appropriate training of scientists who are capable of assessing and attending to the needs of farm animals produced by cloning; such special needs may simply be a consequence of the greater performance achieved by the cloned

animals. The same principle would also hold good for all GM farm-animal species. Few would disagree with the 'five animal freedoms' proposed by the Welfare Council some years ago (Table 1.3).

1.1.4. Organic livestock farming

There has been considerable growth in the number of organic farms, including those dealing with livestock, in the EU over the past decade (Younie, 2001; Hovi *et al.*, 2003; Vaarst *et al.*, 2003); reports dealing with organic farming systems come from countries near and far. In China, for example, where the organic movement started as recently as 1990, optimistic noises about its future are to be heard (Biao and Xiaorong, 2003). Although the early days of organic farming saw attention concentrated on crop production, animals now feature prominently and an appropriate set of rules to guide producers has been drawn up; since July 1999 (Boelling *et al.*, 2003), EU organic livestock producers have been expected to follow regulations laid down by legislators in Brussels. Organic livestock farming has set itself the goal of establishing environmentally friendly production, sustaining animals in good health, observing high animal welfare standards and producing products of high quality (Stocker, 2001; Sundrum, 2001; Zhu and Ji, 2002; Benoit and Veysset, 2003) – all noble aims.

No one can argue with the efforts of organic farmers to secure a niche market; where many would part company is with their contention that conventional

Table 1.3. The five animal freedoms.

1. Freedom from starvation or malnutrition
2. Freedom from thermal or physical discomfort
3. Freedom from pain, injury or disease
4. Freedom from fear or distress
5. Freedom to express most normal, socially acceptable behavioural patterns
And suggested additions:
6. Freedom from stress or suffering when transported
7. Freedom to die humanely

production methods are leading farmers to ruin. There is certainly the need, as pointed out by Hermansen (2003), for much greater interaction between mainstream agricultural research and research groups specializing in organic farming. One of the great advances in agricultural science of the last century was the discovery at the UK's Rothamsted and other research centres that elements essential for plant growth can be replaced by direct application to the soil from sources other than animal and crop waste. It may be a matter of taking nitrogen directly from the air, or it may be a matter of taking phosphate and potash from natural deposits or as the by-products of various industries. By their intelligent use, farmers have been able to produce greatly increased amounts of food while maintaining soil fertility at a high level.

Haughley experiment

Although it is said that organic agriculture first emerged in the Switzerland of the early 1920s under the inspiration of Rudolph Steiner, the organic farming concept in Britain is usually traced back to well-meaning but ill-informed opinion in the 1930s that food produced by intensive and agrochemical-based farming systems was fundamentally less wholesome than food produced in more 'natural' ways. In the UK, Lady Eve Balfour set up the Haughley experiment on her Suffolk farm in 1938 to produce scientific evidence for her belief that the health and future existence of humanity depend on the way in which humans treat the soil. In 1943 came the publication of her book *The Living Soil*, which was a factor that encouraged the founding of the Soil Association in 1946 as a forum for 'all those working for a fuller understanding of the vital relationships between soil, animal and man'.

Throughout the 1940s and 1950s, there was limited farmer interest in the activities of the Soil Association, widely regarded by most crop and livestock producers as irrelevant. By the 1960s, an increasingly productive British agricultural industry was witnessing a dramatic rise in the use of fertilizers, agricultural chemicals and machinery. Not surprisingly, government research funds did not find their way towards institutions that advocated organic production methods. With the coming of the 1980s, and the introduction in Europe of measures aimed at reducing rather than increasing animal and crop products, the British organic movement was to find a new respectability; in 1980, the Elm Farm Research Centre was founded as an educational charity at Newbury in Berkshire, conducting its own brand of research and operating an advisory service for those attempting to develop viable organic farming systems. The ever-increasing range of food products available in the supermarket led to consumer interest in organic products, which probably often arose from a genuine concern about food safety, considerations of animal welfare and a desire to lessen the environmental impact of agriculture.

However, the success of the expansion of organic systems is likely to depend on the production of animal products at a price which attracts the general run of consumers. The belief that there are real differences in quality between conventional and organic foods, certainly as projected by the Soil Association, is regarded by many as scientifically unsustainable. Where science does express an objective view, as in releases from the British Food Standards Agency, it finds no convincing evidence that the products of modern farming and intensive husbandry are harmful to the health of the consumer; there is also no convincing evidence which might differentiate organic from conventionally grown foods in terms of nutritional content or any other genuine measure of quality (Trewavas, 2001).

Animal health concerns

The welfare of animals in intensive and organic systems has been addressed by various workers; a report by Gade (2002), for example, dealt with welfare in the context of Danish organic pig production. Although organic meat production accounts for only a small proportion of the total meat production in the EU, it is increasing in size to take account of consumer demands. Organic production standards vary according to the organizations found within the different EU countries, with the regulations in some countries, such as the UK, being much more restrictive than in others. Although, in theory, there should be a greater opportunity for ensuring optimal animal welfare in organic systems compared with intensive systems, as management of the farm animals is expected to take account of their physiological, social and behavioural needs, there are also negative aspects. Organic farming comes at a cost, for the demands on management are greater in such systems, particularly those that are free range. Concern is expressed by some about animal health, since vaccinations, antibiotics and anthelmintics may only be used in a limited way in organic production and there is a tendency to use so-called 'natural' and homoeopathic products, which have little, if any, scientific credibility.

Integrating mainstream and organic research

What is needed is an integration of mainstream and organic research, certainly research that is government-sponsored, and a willingness to argue the merits of production systems without the emotional overtones that seemingly characterize the views of many committed to organic ideals. Some of the emerging technologies, such as genetic engineering, could possibly make life much easier on the organic farm (e.g. reducing the extent of pesticide and herbicide usage). A secure and sustainable agriculture, with adequate standards of animal welfare, should be the aim of all those engaged in livestock production.

1.2. Animal Welfare Considerations

Animal welfare is a complex issue, but one that the farmer and others who work with animals can ill afford to ignore. Animal production sytems on the farm, as the 21st century unfolds, are likely to constitute a rather different scene from those

experienced by farmers of 50 or even 25 years ago. There is a much clearer realization that farm animals are sentient beings (Table 1.4).

1.2.1. Education and training

Animal welfare has become a popular cause since the horrors of the Second World War because of its obvious appeal to human nature to prevent suffering and protect the weak. Apart from such natural instincts, however, it is very much in the farmer's best interests to take the initiative in improving the quality of animal life rather than having welfare dictated from outside the farm gate. Apart from taking account of the animal itself, improvements, in the form of appropriate education and training, should include the welfare of stockpersons and the staff who attend the animals; the welfare of the farm animal may well have implications for the psychological welfare of those who care for them. In the milking parlour, more refined milking systems are likely to prove valuable in monitoring the welfare as well as the health of dairy cows; the processing of sensor information should enable the early detection and treatment of many subclinical problems, resulting in healthier and more contented cows.

Table 1.4.   Animal and human feelings.

1. Humans are akin to animals by virtue of the process of evolution
2. Animal behaviour in response to traumatic events is similar to that of humans
3. The nervous system of animals is similar to that of humans
4. Animals adapt, as do humans, to meet the particular requirements of their environment
5. Many animals have a language of some form; gestural in bees, vocal as well as gestural in many vertebrates
6. Electroencephalograms of mice, rats, goats, sheep, cows and horses are broadly similar to those recorded in humans
7. Anxiety can cause psychosomatic symptoms (ulcers) in rats, etc.
8. Animals can learn or be conditioned

Increasingly, in the opening decade of the 21st century, there are those pointing to the need to re-evaluate stress in farm animals, particularly the need for a more integrated view of those stress reactions that are induced. Although, in the development of livestock production systems, humans were responsible for creating stress in farm animals, this effect can now be greatly reduced by their ability to provide the means of alleviating, removing or modifying potential stressors. In terms of modern livestock production systems, the need for such an approach is probably long overdue, since the demands of management systems designed to increase production have often had adverse effects, especially when measured against reproductive efficiency.

In all systems of management, there are two groups of stressors: those arising from environmental stress and those from management stress. The first group includes environmental temperature (worldwide, the most important stressor influencing reproductive performance), wind and humidity; the selection of animals for increased tolerance to high temperatures is regarded by many as the most important way of improving the productivity of farm animals in hot climates. Management-related stresses include animal density, handling procedures, movement of animals, social status, psychological distress, noise and physical trauma. Single factors from each group or combinations of such factors can serve as stressors and markedly challenge the homeostasis of the animal; the effect of the challenge will also depend on the duration of the stimulus and its intensity.

Workers such as Temple Grandin in the USA have provided a well-balanced view of the various manifestations of stress induced in animal handling. She has highlighted such stressful situations as health-care procedures, isolation of animals from herd-mates and rough handling during insemination, all factors well recognized in livestock production but not always put into correct perspective in assessing the causes of reduced fertility. In striving for optimum fertility, due attention should be given to the animal's ease. A recent paper by McInerney

(2002) in New Zealand provided a non-specialist overview of animal welfare issues to show a social science approach to what is often regarded as a matter for animal science.

1.2.2. Grazing and non-grazing animals

Over the past half-century genetic selection has had a major impact on livestock production, with the pig and poultry industries at the forefront in the application of genetic theory for improving their productivity. This improved productivity has often involved production systems which have raised welfare considerations. Beef cattle and sheep farming systems have largely escaped the attention of animal welfare pressure groups, probably because grazing animals appear happy and content; when contemplated by the casual observer, such cattle and sheep are apparently living a life not far removed from the days before domestication. Not so the dairy cow, which often finds itself in an environment far removed from green grass and open spaces. Dairy cattle production systems, however, are likely to be strongly influenced by climate and the temperature conditions under which they live. In the southern hemisphere, for example, New Zealand's dairy industry has been shaped by the country's climate, which permits year-round pasture growth; pasture provides an inexpensive and sustainable food for dairy cows, which they can efficiently convert into milk. For the consumer, pasture-based production systems, whether cattle, sheep or pigs, are generally held to be more natural; they can usually reconcile their thoughts on such systems with their views on protection of the environment, welfare of the animals and the general well-being of the farming communities involved.

Easy-care sheep

New Zealand is one of the countries where farmers and researchers have devoted thought to long-term improvements in animal welfare in grazing animals. Animal management systems in that country have laid emphasis on reduced human intervention; 'easy-care' sheep refer to animals that are able to successfully lamb and rear a lamb without human assistance in a difficult environment. Developing such 'ethically improved sheep' from polled sheep with a short tail that are devoid of wool on head, legs, belly and breech means that such animals require less handling than conventional sheep; they are thought to have a higher survival rate and lower lamb mortality and to require less shepherding at lambing than other sheep breeds or strains. 'Easy-care' sheep systems are more in keeping with the biology of the animal in an extensive environment, avoiding problems at lambing time in difficult terrain and reducing the need for skilled labour (Fisher, 2003). On the marketing front, there may also be the thought that such developments could help in pre-empting possible bans or trade barriers based on current sheep husbandry practices in New Zealand.

A pig's life

That modern production methods may often have a negative impact on the pig is well recognized. Before domestication, the pregnant sow lived in a matriarchal group and built a nest to protect its young; a firm bond was established between sow and piglets and suckling continued for at least 10–12 weeks. The sow's life nowadays is rather different, and does not usually permit nest-building activities; in most cases, the piglets are weaned at 4 weeks. However, in modern husbandry, it is important that the welfare of animals is respected and efforts are made to minimize anxiety-producing situations and to allow expression of normal behaviour as far as possible (see Orgeur *et al.*, 2002).

1.2.3. Biological measures of animal welfare

There are several biological measures that relate to an animal's welfare; these include behaviour, disease incidence and severity, many physiological indicators, life expectancy and reproductive rate. Physiological indicators include measures of immune status, stress hormones and homeostasis mechanisms.

Psychological stress in goats

An unusual response to stress may be found in goats, where it has been shown to prevent or terminate cyclical activity in females raised under extensive conditions; such stress would be best described as psychological, rather than stress arising from diet, disease or climate. There are instances in the USA which show that does, raised under range conditions, when moved to unfamiliar surroundings ceased to show cyclical breeding activity, even though they were in the accepted breeding season. Those with goats should be aware of the potential for stress-induced interference with reproduction in this species.

International standardization of measures

Although there are several ways of measuring animal welfare in a scientific manner, it is important that it should be measured on a scale that is recognized internationally, not just in the EU or the UK. Such standardization could be useful in assisting the consumer in seeking products from animals kept under the welfare requirements of their country of origin. In many studies of farm-animal welfare, workers have sought a biologically meaningful index of stress and have taken the activation of the hypothalamic–pituitary axis for this purpose. This activation may be inferred from a rise in cortisol concentration, which may be detected in body fluids, including blood plasma and saliva. Changes in other physiological variables may be causally linked in some way with elevation of cortisol concentration; there are, for example, studies in pigs suggesting that vasopressin may be secreted as a response to vehicular motion as perceived during road transport.

Various workers have demonstrated that stress-induced cortisol secretion can disrupt oestradiol-dependent events necessary for ovulation in the ewe; in the USA, studies have demonstrated that cortisol secretion blocked or delayed the pre-ovulatory surge of luteinizing hormone (LH) in sheep. In the UK, Liverpool workers reported that transport stress reduced LH pulsatility in sheep, the greatest effect being evident in the breeding season. In California, it was shown that stress-like concentrations of cortisol increased the negative-feedback potency of oestradiol; this was taken as evidence of serious endocrine changes that may be induced by stress in sheep. According to Smith and Dobson (2002), endocrine systems may be used as indicators of stress in two ways: (i) as part of the homeostatic response to a stimulus (e.g. adrenalin, corticosteroids); or (ii) as a hormone having a key role in a normal body function (e.g. reproduction), which stress alters to prevent normal function.

Although corticosteroids have been used as primary indicators of stress for many years, after an initial large response prolonged stimulation leads to gradually reducing plasma corticosteroid concentrations. None the less, the stress signal may still be operating and the animal feeling stress; thus, the welfare interpretation of a corticosteroid concentration may differ during the time course of a stress response. In interpreting hormone data for animal welfare purposes, it is important not to interpret a reduction in hormone concentration due to intrinsic hormone control mechanisms as a reduction due to a decrease in the stress stimulus.

1.2.4. Animal welfare and the consumer

A paper by McEachern and Tregear (2000) has reviewed the requirements of the main British farm animal welfare assurance schemes according to seven key criteria (origin and traceability, management and stockmanship, housing, health, nutrition, transport and slaughter). The authors draw attention to the differing requirements between the schemes currently operating and reflect on the extent to which such

assurance schemes are likely to result in improvements in farm-animal welfare on a large scale. There are those who believe that short-term improvements in animal welfare on livestock farms are unlikely unless consumer attitudes and behaviour change significantly; as they see it, achieving better animal welfare must start with the consumer, way of education and information, by price policies and appropriate marketing strategies.

1.2.5. Modification of management practices

Modern dairy cattle housing systems were developed primarily for economic and labour benefits and may unwittingly influence the behaviour and performance of the cow adversely or favourably; there has been a lack of information on the social and individual behaviour of cattle that are housed, although the behaviour of cows at pasture has been reported on at length. This consideration becomes all the more important with the trend in the industrialized countries for the size of the dairy herd to increase. In the space of a quarter-century, the average size of dairy herds in the USA has increased fourfold and similar increases have been evident in many other countries. Alongside the larger herd size, there is the trend towards more cows per stockperson and fewer man-hours per cow. Growth in the size of the average dairy herd and in the scale of buildings associated with the enterprise has brought the need to change conventional reproductive management practices to increase the efficiency of heat detection, AI, pregnancy diagnosis, cow handling and record keeping.

Animal–human interactions

There have been many studies showing that interactions between stockpersons and their animals can influence the productivity and welfare of the animals (Hemsworth, 2003). Frequent use of certain routines may result in farm animals becoming fearful of humans; such fear levels, through stress, may limit animal productivity and welfare.

The author emphasizes the need for research to identify the full range of stock-persons interactions that have implications for farm animals. According to Rushen (2003), the concepts of animal welfare used by researchers are often too limited and do not address many of the issues of concern to the public; too much reliance has been placed on physiological, immune and behavioural measures of welfare that have not been adequately validated. There has also been too much emphasis upon the type of housing used and less attention to other important sources of variability in animal welfare, especially the quality of stockmanship and nutritional needs. Animal welfare considerations may also involve appropriate contact between human and animal in the early period of the animal's life. In goats, for example, workers in the UK studied the effects of handling dairy goat replacements in early life on their subsequent milking-parlour behaviour; contact with humans in the first 5 weeks of life resulted in animals showing less fearful reactions when exposed to the novel experience of entering the milking parlour, being restrained by neck yoking and being milked.

Natural vs. artificial light

In domestic poultry, it has been noted that the visual system of birds evolved in natural light environments, which differ in many respects from the artificial light in poultry houses; a poor correlation between light provided and that required for effective vision may influence visually mediated behaviours such as feeding and social interaction, leading to distress and poor welfare. Although some systems require artificial lighting for production purposes, Prescott et al. (2003) have argued that it may be possible to rear birds more humanely in artificial environments that contain some features of natural light.

1.2.6. Altering animal temperament

As animal welfare considerations become increasingly important, particularly among consumer groups, it is essential that producers seriously consider all animal welfare aspects. Modification of management practices to reduce stress in farm animals is one way to improve animal welfare. A second option may be to improve the temperament of farm animals to reduce the amount of stress they experience during routine handling procedures – if such improvement can be readily achieved without reducing productivity. This may be achieved genetically (e.g. through selection of breeding stock for good temperament) or non-genetically (e.g. by modifying animal behaviour by way of training programmes).

1.2.7. Stress and slaughter

Care of deer

There have been many studies examining the effects of slaughter procedures on traditional farm animals, such as cattle, sheep and pigs, but relatively few studies on farmed red deer. Being a relatively newly farmed species, deer may be particularly sensitive to potentially stressful husbandry practices; their handling immediately before slaughter is potentially one of the most stressful experiences. A report by the UK Farm Animal Welfare Council proposed in the mid-1990s that deer should be rested for 1 h in lairage and for not more than 3 h. In New Zealand, where deer are farmed on a relatively large scale, there are dedicated deer slaughterhouses. In the UK, on the other hand, where deer may be shot in the field and processed locally, there is a need for more efficient and hygienic slaughter methods with adequate meat inspection facilities. The transport of deer to the abattoir is a further important welfare consideration; work in Scotland examined the effects of several factors on the behaviour of farmed deer during loading. Further studies by the same group led to the recommendation that deer should be stunned as soon as possible after entering the restraining pen at the abattoir; holding male deer overnight before slaughter may lead to problems with meat quality. The authors noted the need for further studies to devise handling systems which ease the movement of deer within an abattoir from the point of unloading to the point of slaughter.

1.3. Current Application of Reproductive Technologies

New reproductive technologies present farmers with many opportunities; such technologies include sperm sexing, oestrus control, multiple ovulation and embryo transfer (MOET), *in vitro* production (IVP) of embryos, cloning and cloning for the production of GM animals (Faber *et al.*, 2003). Such developments follow on from a long line of events, some of which are detailed in Table 1.5.

1.3.1. General considerations

Many of the current reproductive technologies are concerned with animal breeding programmes; most of these are concerned with the selection of animals based on an evaluation of their genetic quality and the dissemination of the superior genetic material from the nucleus to the commercial population. Genetic gain is generated in a small fraction of the population (the nucleus herd), and these animals are made available to the wider commercial population. In pigs and poultry, closed nucleus schemes are generally employed in which nucleus animals are kept on a small number of farms and only animals from such a nucleus contribute to genetic improvement of the nucleus population; a typical breeding programme usually involves a number of sire and dam lines.

In general, the nucleus breeding stock of such lines is centrally owned by breeding companies, which make crossbred breeding stock available to the commercial producers. In contrast to breeding programmes in dairy cattle, nucleus breeding stock in pigs and poultry are not used for commercial production. Crossbreeding in pigs and poultry is the rule due to the high reproductive rate and quick turnover in generations in such species, and breeding companies are able to protect their ownership of high-value genetic material by only selling crossbred animals to commercial producers. In cattle, particularly dairy animals, crossbreeding seldom occurs and élite female breeding stock are usually owned by the commercial producer.

The current commercial application of reproductive technologies varies in different countries. In New Zealand, for example, embryo-based reproductive technologies are usually not profitable other than in niche market situations where the returns from the resulting offspring are likely to be markedly greater than those achieved by way of natural matings or AI. Some authors contend that profitability can only be expected if the cost

Table 1.5. Milestone events in embryo transfer and related techniques.

Year	Species	Event	Researcher(s)
1890	Rabbit	Birth of young as a result of embryo transfer	Heape
1949	Sheep/goat	Birth of a lamb and kid from embryo transfer	Warwick and Berry
1951	Pig	Birth of piglets from embryo transfer	Kvansnicki
1951	Cattle	Birth of calf from embryo transfer	Willett *et al.*
1971	Cattle	First commercial cattle ET company formed	Alberta Livestock
1973	Cattle	Birth of calf after frozen stage (Frosty II)	Wilmut and Rowson
1974	Horse	Birth of a pony after embryo transfer	Oguri and Tsutsumi
1982	Cattle	Birth of calf after IVF	Brackett *et al.*
1983	Buffalo	Birth of calf after embryo transfer	Drost *et al.*
1986	Sheep	Lamb born via nuclear transfer	Willadsen
1988	Cattle	Cattle twins by IVP embryo transfer	Lu *et al.*
1997	Sheep	Birth of lamb cloned from adult cell	Wilmut *et al.*

IVF, *in vitro* fertilization.

of pregnancies in embryo-based reproductive technologies can occur at prices less than two to four times greater than AI or natural service (Smeaton *et al.*, 2003); the same authors note that two new uses for reproductive technologies in dairy cattle could be in the proliferation of novel or rare genotypes from gene discovery programmes and improving female reproductive rate for optimal marker-assisted selection (MAS).

1.3.2. Dairy and beef cattle

Reproductive technologies such as AI and ET have had far-reaching economic consequences in commercial dairy and beef herds. Although environmental factors may be manipulated in several ways, the cow's genotype is determined solely by its parents' genetic make-up; for that reason, reproduction plays a crucial role in determining the genetic progress made within the cattle enterprise. Embryo technologies are a combination of assisted reproduction, cellular and molecular biology and genomic techniques; their classical use in cattle breeding has been to increase the number of superior genotypes (Galli *et al.*, 2003a). However, with advances in biotechnology and genomics, they are likely to become tools in the production of transgenics and genotyping. Conventional forms of cattle ET, based on superovulation and the recovery and transfer of embryos, have been well established for many years and currently account for the majority of cattle embryos produced worldwide. However, little, if any, progress has been made in the last 20 years to increase the number of transferable embryos, despite the considerable additions to knowledge made during that period. *In vitro* embryo production, as an alternative to superovulation of donor cattle, is technically much more demanding than conventional cattle ET, requiring laboratory facilities and specialist equipment for its success. At the same time, the scope of this technology is seriously limited by the relatively small number of oocytes that can be obtained at any one time from donor animals.

The most widely utilized technologies in cattle are AI and ET. Although AI is widely used in the dairy industry for the production of herd replacements, it is not much used in the beef industry because of the practical problems in having such animals under observation in the way open to those dealing with dairy animals. For successful implementation of ET in cattle, the procedures of oestrus control, non-surgical embryo recovery and transfer and embryo cryopreservation are key factors; the absence of one or other of these factors in other farm species has been a factor limiting their application.

Optimum reproductive performance in beef suckler herds, in countries such as the USA, remains a key factor in ensuring that red meat retains a strong position as a source of protein in the human diet; applications of reproductive technologies should be viewed as potentially valuable in increasing the biological and economic efficiency of the beef enterprise and in improving the quality of the products. At one time, improvements in general living standards usually generated an increased demand for beef, whether for reason of social prestige or simply because people enjoyed eating it. Reproductive technologies applicable to beef herds can be broadly classified into those requiring a low, medium or high technical input. The low-technology category includes options such as pregnancy diagnosis and controlled calving; medium technology includes such options as oestrus control; high-technology options include twinning, sex control and emerging technologies such as cloning and the production of transgenics. Control and manipulation of reproduction in cattle covers several possibilities (Fig. 1.3).

One extremely important consideration in developing reproductive technologies is the likely cost to the farmer; to a great extent, cost is likely to be determined by the scale of operations and by the experience of the organization that brings them to the farm. It might also be mentioned that there is likely to be a close correlation between management expertise in a cattle enterprise and the successful adoption of a new procedure.

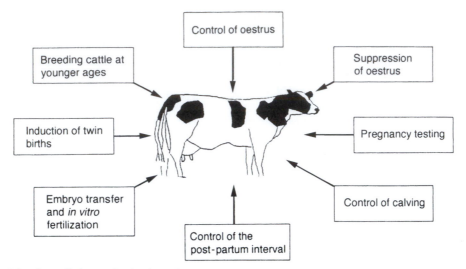

Fig. 1.3. Controlled reproduction in cattle.

1.3.3. Sheep and goats

From the farmer's viewpoint, it is fortunate that in the past four decades the ewe has become one of the animals favoured by the animal scientist in trying to understand endocrinological and physiological mechanisms used by mammals to control their reproductive activities; there have been several areas of interest in human medicine (e.g. fetal physiology) which have been greatly advanced using the ewe as the model. Current sheep statistics show that approximately 6% of ewes in England consistently fail to lamb, a figure little different from that recorded more than a century ago in Cambridge by Walter Heape; reducing that barren percentage to 2–3% would not seem an impossibility. In fact, sheep farmers in New Zealand, dealing with many thousands of ewes, would regard it as quite possible, a consequence of their careful selection and culling of animals.

Progress towards commercially acceptable controlled reproduction techniques was greatly accelerated in the early 1950s when Terry Robinson in Australia showed the crucial role of progesterone in the ewe's cycle and highly potent progesterone analogues designed for use in human fertility control became available for use in sheep. Controlled reproduction in sheep, as the

term might be applied to conditions in the UK and Ireland, could be expected to cover the full spectrum of lowland production systems (Fig. 1.4).

It may be a matter of breeding sheep towards the end of their normal anoestrus (early-lamb production); it may mean mean breeding ewes to top-quality rams by AI; or it may mean a rapid build-up of certain populations by ET. The scope for certain controlled reproduction applications is likely to vary with flock size and environment. A New Zealand farmer with an average flock size of 2600 breeding ewes or even a British farmer with an average of 200 would view compact lambings in quite a different light from an Irish farmer with fewer than 100 breeding ewes. It should also be emphasized that successful reproduction control in sheep is not only a matter of appropriate hormonal techniques but also one of ensuring that they are only employed in situations in which they are capable of achieving acceptable results. Difficulties in the past in some areas of reproduction control have arisen, not only from technical inadequacies, but in trying to pursue unnecessarily ambitious objectives, such as producing two lamb crops within the calendar year.

Despite recent improvements in embryo production and ET technologies made in

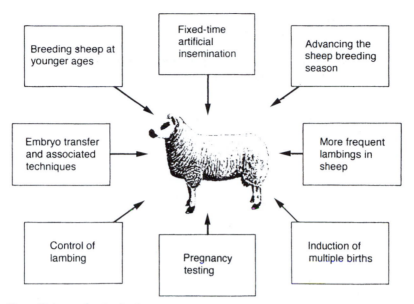

Fig. 1.4. Controlled reproduction in sheep.

recent decades, AI is the only reproductive technology used on any scale in countries where sheep and goats are important in farming. In some countries, France being an example, sheep and dairy goat production has followed much the same route as that taken in dairy cattle production, with milking parlours and similar husbandry methods. For normal breeding purposes, the use of ET in small ruminants is ruled out on a cost basis. As with cattle, the cost of a lamb or kid born after ET is likely to be ten times greater than one born after AI. On the other hand, it may occasionally be economical to use ET in small ruminants if it is a matter of health control or in salvaging germplasm from infected flocks.

The biology of early reproductive development, components of reproductive rate in the ewe (including genetic variation) and the biology and genetics of male reproduction in sheep have been dealt with in various reviews. In New Zealand, studies covered topics ranging from the advancement of the breeding season by manipulation of the photoperiod, or by the use of melatonin, the ram effect or hormones; the same authors also dealt with oestrus synchronization, AI and ET.

Lean meat sheep

If there is to be progress in producing types of sheep that are genetically lean and reach market weight in the shortest possible time, current understanding of the genetic control of lean tissue growth and development has to be improved; although there has been considerable research on the genetic control of growth in sheep, there has been much less on body composition and lean tissue growth rate. Provision of information in this area could enable researchers, in conjunction with farmers, to devise and implement appropriate and efficient breeding schemes; reproductive technologies, whether in the form of AI, ET or otherwise, are likely to be valuable in carrying through the necessary breeding programmes.

The use of selection indices in sheep has been adopted in various countries with the aim of reducing carcass fatness or increasing lean growth; this has often involved the ultrasonic measurement of fat and muscle depths as predictors of carcass composition. The purpose of sire reference schemes in sheep breeding programmes in the UK and elsewhere, showing how such schemes operate for the benefit of farmers, is dealt with in one recent report; the same work also discussed genetic trends in growth, body weight and carcass composition in the UK. It is believed that substantial responses to selection are achievable and the techniques have been employed in sheep breeding programmes in several countries; in Scotland, after 9 years of selection, Suffolk ram lambs were 6 kg heavier, with more muscle and less fat than their control-line contemporaries at 150 days of age. However, it appears there are much weaker genetic links between sheep flocks in most countries than those between cattle herds, because of the low use of AI in sheep. New Zealand workers note that selection within sheep breeds and crossbreeding using terminal sires often pay attention to total meat output rather than quality; they suggest that about half the selection pressure on males should be devoted to decreasing carcass fat content in order to meet present-day consumer demands.

AI used in conjunction with accurate progeny testing schemes could be one means of substantially increasing the rate of genetic progress in sheep. Unfortunately, AI in sheep has not been widely adopted, due to the poor fertility which can often follow intracervical insemination, particularly with frozen–thawed semen. This is a problem which should be soluble, given the required effort and research. In the meantime, laparoscopic intrauterine insemination for sheep has played an important role in the implementation of national sire reference schemes in the UK; this technology was developed at the same time as ultrasonic scanning for *in vitro* assessment of carcass traits and the use of best linear unbiased prediction (BLUP) for estimating genetic merit.

The goat is a seasonally polyoestrous animal with a breeding season in the UK and Ireland beginning about September and ending in the early months of the new year. However, most of the world's goats are located in tropical and subtropical regions and may show little response to photoperiod. Work conducted more than 60 years ago showed that in goats, as in sheep, the primary environmental cue used to regulate reproduction is day length. For those engaged in serious dairying, the extent of the breeding season is a matter of some importance, since it may be difficult to arrange matings in milking goats so that a herd will give a uniform yield of milk all the year round. Although there has been considerable research activity over recent years towards developing embryo production and ET technology in sheep and goats, AI is the only reproductive technology which is applied on any scale in countries where sheep and goat breeding has an important impact on the economy.

According to some workers discussing biotechnological advances in goat reproduction, developments in AI technology have enabled goat sperm to be stored successfully for several years before being used in cervical or laparoscopic AI (Fig. 1.5). Laparoscopic recovery of goat embryos to reduce adhesions resulting from repeated surgical interventions is seen to have considerable potential in improving embryo production for direct transfer or transfer after cryopreservation. The authors suggest that the diversified commercial value of goats and their convenient size make it easy to apply new technology for rapid genetic improvement. The domestic goat is also seen to be important in developments in the biotechnology industry in the USA; according to one recent report, the production of recombinant proteins in the milk of transgenic goats and other animals appeared to have an economic potential far beyond what researchers initially realized.

The control of goat reproduction involves the processing of semen, the induction of extraseasonal oestrus by manipulation of the daily photoperiod, the hormonal synchronization of oestrus and the production, collection, freezing and transfer of embryos. The use of new reproductive technologies in goat reproduction in France, as described in one recent review, covered topics such as the manipulation of light regimes to increase semen production in bucks and prevent the occurrence of a resting season. The same report dealt with the implications of the identification of a bulbourethral lipase in goat sperm for semen preservation. Out-of-season breeding, the

Fig. 1.5.   Semen collection in goats.

use of ultrasonography in reproductive management, AI, early pregnancy diagnosis, *in vitro* embryo production and cloning are among topics covered in goats by Bretzlaff and Romano (2001) in the USA.

Boer goat

The introduction of the Boer goat from South Africa into the USA led to interest in goat meat production in that country; it became evident that a lack of information among goat producers in the USA needed to be addressed. This led to agricultural colleges and other agencies attempting to rectify this lack of knowledge, especially in nutrition, health, genetics, reproduction and predator control. Discussing market trends and the potential for meat goat production, one author noted that the demand for goat meat in the USA has increased dramatically in recent decades. Although the number of goats slaughtered in the country is considerably less than that of cattle and sheep, the goat is the species that has increased significantly in number in recent times. The portion of the American population that prefers goat meat appears to be increasing, with the three largest goat-consuming ethnic populations driving the goat meat trade being Hispanics, Muslims and peoples from the Caribbean. There are moves in South America to explore the meat-producing capabilities of the Boer goat. In Chile, workers have reported the introduction of the breed using frozen–thawed embryos imported from New Zealand.

1.3.4. Pigs

Of all the farm mammals, the pig is the species which has been influenced to the greatest extent by modern intensive husbandry practices. In contrast to events in the permanently cyclic domestic sow, in the wild pig sexual activity usually only occurs in the autumn and farrowing at the end of the winter. In contrast to cattle and sheep, which have a relatively low reproductive output annually, the pig has the ability to produce a large number of young in a short space of time; given appropriate management, housing and feeding, pigs are capable of being among the most profitable of the meat-producing animals. Reproductive performance is usually based on the number of young produced each year by the sow. This characteristic comprises two important components: litter size and the number of litters produced each year by the sow.

As pig units grew larger and became more specialized, the needs of production placed increasing demands on reproductive performance; at the same time, larger-scale confinement production itself may well raise new problems which have to be solved if optimum reproductive performance is to be achieved. Those who have reviewed the performance of pigs in large production complexes in the former USSR noted that reproductive efficiency had declined markedly and that attempts to improve it had been ineffective. It should be kept in mind that reproductive efficiency in pigs can be influenced, in a way much less apparent in the other farm species, by the association of the sexes, whether this be in matters such as the elicitation of the full immobilization reflex in the oestrous sow or in triggering the onset of puberty in the gilt.

Biotechnology as it relates to pig production includes a wide range of reproductive technologies (Fig. 1.6) and an associated range of techniques in molecular genetics. Although techniques in molecular genetics are as yet poorly developed, reproductive techniques are more advanced and applied on the farm. In Germany, Niemann *et al.* (2003) described some of these techniques to illustrate the implications of biotechnology in future pig production. Sperm-sexing technology is likely to become available and be used to generate piglets of the desired sex by *in vitro* fertilization (IVF) and intracytoplasmic sperm injection (ICSI); initial studies also indicate reasonable success rates following intrauterine insemination with sexed sperm (Rath *et al.*, 2003a,b). Remarkable progress has been made in organ transplantation technology in recent years, driven by the increasing demand for organs in human medicine; the pig has featured prominently in this because donor organ anatomy and function are compatible. Some authors have predicted that pig heart transplants may be used in patients within the next 3–5 years (Niemann *et al.*, 2003).

Among major reproductive technologies applied to pigs, AI is the most notable, with semen from genetically superior boars disseminated to many females to produce high-quality breeding animals and fatteners. The use of deep uterine insemination with a low concentration of fresh semen or frozen or flow-sorted sperm is currently an active area of research. ET is now accepted as the safest way to introduce new genetics into pig herds; recent developments enable embryos to be collected and transferred by non-surgical techniques and pig embryos at the

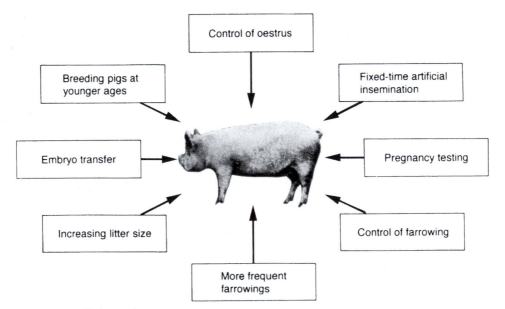

Fig. 1.6. Controlled reproduction in pigs.

expanded blastocyst stage can be frozen in liquid nitrogen. Reviewing progress in reproductive biotechnology in pigs, Techakumphu *et al.* (2002) in Thailand note that new advanced techniques, such as *in vitro* embryo production, somatic-cell nuclear transfer and transgenesis, are likely to prove useful both for pig production systems and human biomedical purposes.

Asia, with more than half the world's human population, has a thriving agriculture, a large part of which is devoted to pig production. The efficiency of production varies from country to country and is largely a legacy of whether the country relies on backyard farms using native breeds or intensive commercial farms using genetically improved breeds from Western Europe and North America. Reproductive efficiency in pig production units is usually below the levels found in European countries and this is the result of the hot and humid climatic conditions, diseases, substandard management, the difficulty of obtaining quality feeds and reliance on native breeds that are much less prolific than those in European countries (Kunavongkrit and Heard, 2000); clearly, there is considerable scope for improvement.

An article by Merks *et al.* (2000) in The Netherlands summarized current knowledge on genetic and management methods for improving fertility in sows; they note that fertilization rate, litter size and the interval between weaning and oestrus are traits that can be monitored on farms. A review of nutrition as it affects reproductive performance in pigs is provided by Prunier and Quesnel (2000); it is clear that, in female pigs, undernutrition may influence the growth of antral follicles, decrease ovulation rate and delay puberty and return to oestrus after weaning. Factors influencing the

reproductive performance of sows on commercial farms, including such measures of performance as the interval from weaning to first service, interval from weaning to conception, litter weight at weaning and subsequent litter size, have been covered by various authors.

1.3.5. Horses

The mare is unique among farm animals in several ways: in the hormonal characteristics of the oestrous cycle, having a long heat period (5–7 days), in the large size of the preovulatory follicle (40 mm in diameter) and in having ovulation triggered by a progressively increasing level of LH rather than a sudden surge. In contrast to the other farm mammals, in the horse, assisted reproductive techniques (e.g. *in vitro* embryo production) and manipulation of ovarian activity (e.g. superovulation) are generally less controllable.

Not so long ago, the horse in many ways was regarded as the most important domestic animal, being used in all types of farming, in transport, in war and, when necessary, as a source of meat. However, the many advances occurring in the developed countries in farm mechanization, especially from the time of the Second World War, have rendered the horse obsolete in many of its previous roles; for those in the Western world, the horse is now mainly associated with sport and recreation. One problem with the mare, now as in past days, lies in its low reproductive efficiency. Whereas conception rates to first service of the order of 85–95%, 80–90% and 55–65% are regarded as the norm in pigs, sheep and cattle, respectively, the equivalent value for the horse appears to be 40–50% (see Fig. 1.7).

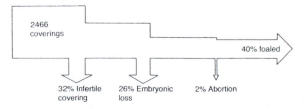

Fig. 1.7. Fertility in Thoroughbred mares.

Figures gathered in Ireland for Thoroughbred and non-Thoroughbred mares are matched by those gathered further afield. A study covering 22 stud farms in the Newmarket area in the UK by Morris and Allen (2002) found that 82.7% of the 1393 mares surveyed gave birth to a live foal at term in 1998, as compared with the proportion foaling in 1983 (77%); despite the evident improvement in foaling rates over the 15-year period, the overall rate of pregnancy failure remains high and clearly represents a major loss to the Thoroughbred breeding industry.

However, help may be on the way. As traditional production agriculture grapples with problems of oversupply and cut-backs in market prices and research funding, the horse is well placed to become the recipient of increasing attention from the reproductive biologist. Reproduction and reproductive technologies applicable to the horse are dealt with in many papers and reviews; one important source of information is a review by Allen and Antczak (2000). As previously mentioned, there may well be certain restrictions imposed by horse registries limiting the extent to which these reproductive technologies may be applied to full advantage in this species. There are often differences in the extent to which techniques such as ET and AI may be used. Even when AI is permitted by the breed society, the maximum time allowed from semen collection to insemination is likely to vary widely among breed associations, ranging from immediate insemination requirements to an indefinite storage period.

Although collection and transfer procedures developed in the 1970s are currently used commercially in various breeds (Fig. 1.8), techniques such as superovulation, embryo freezing, *in vitro* embryo production, oocyte transfer, gamete intra-Fallopian transfer (GIFT) and oocyte freezing are not yet commercially available. On the male side, the prediction of a stallion's fertility is one of the problems in horse breeding; despite many studies of the relationship between semen characters and fertility, it is not possible to predict with any great certainty the breeding capacity of a stallion.

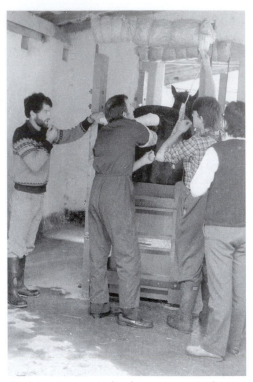

Fig. 1.8.   Non-surgical embryo recovery in the mare.

Stallion selection and fertility prediction are usually based on semen analysis and behavioural traits.

1.3.6. Buffaloes and yaks

The productivity of buffaloes is markedly influenced by inherent problems such as poor reproductive efficiency, often due to late maturity, poor expression of oestrus (particularly in summer), irregular oestrous cycle length, anoestrus, inactive ovaries and long post-partum intervals. For such reasons, information about basic reproductive patterns are of considerable interest in the context of attempts to improve the performance of animals. According to some authorities, dealing with buffaloes in Brazil, river- and swamp-buffalo females reach puberty at 16–20 and 20–24 months, respectively. Oestrous cycle length is 22–24 days. The gestation period is 300–310 days in river buffaloes and 310–330 days in swamp buffaloes. The same authors note that, in terms of meat production, buffaloes reach body weights of 400 kg at 2 years of age.

In worldwide terms, there has been a growing awareness of the importance of the water buffalo, which plays such a prominent role in Asian rural livestock production and to a limited extent in Mediterranean countries; factors affecting the productivity of the species have a considerable bearing on the agricultural output of such areas. The buffalo has probably been neglected for far too long as a valuable source of animal protein for human consumption in many countries. The well-recognized restrictions on reproductive efficiency in buffaloes include: (i) inherent late maturity; (ii) poor oestrus expression in summer; (iii) distinct seasonal reproductive patterns; and (iv) a prolonged inter-calving interval. However, growing demand for milk and meat has resulted in breeding programmes involving AI, oestrus control, ET and *in vitro* embryo production in this species (Fig. 1.9).

The importance of the water buffalo in a country such as India is well known;

currently, the buffalo contributes about 49% of milk supplies in that country, with cattle contributing 48% and goats 3%. The buffalo is recognized as having high fertility and being able to adapt better to heat stress than cattle under the same conditions; for that reason, buffaloes are seen to have greater potential as an animal protein source (meat and milk) than cattle in tropical regions. In many countries, buffaloes and cattle are managed in the same way. There are those, however, who point to the need for research to provide procedures, such as ET, specifically suited to the buffalo. Pioneering efforts by Martin Drost in Florida covered such topics as superovulation, ET, *in vitro* embryo production, embryo splitting, embryo sexing, cloning and gene transfer.

The yak (*Bos grunniens*) is a remarkable example of a domestic animal, a herbivore living in and around the Himalayas and, at altitudes of 2500–5500 m, usually above the tree line and at temperatures rarely climbing above zero; they provide food (meat and milk), transport, clothing and fuel where many other animals would fail to survive. A recent study was made in China of the scientific papers published worldwide on

Fig. 1.9.   River buffaloes (*Bubalus bubalis*) in China.

all aspects of yak science from 1950 to 1995; a total of 1452 papers was counted. The rate of publication increased rapidly from 0.6 per year in the 1950s to 99 per year in the early 1990s; 79% of papers were in Chinese, 15% in English and 4% in Russian. Many of the papers published in China are not readily accessible to readers in other countries; the authors draw attention to a book entitled *World Yak Literature Index* that has been produced with the aim of helping scientists to obtain information about this species.

The length of the oestrous cycle in the yak ranges from 18 to 21 days according to an Indian report; the duration of oestrus varies from 12 to 18 h. Behavioural signs of oestrus are similar to those observed in cattle, with a clear viscous vaginal discharge, vaginal swelling and reddening of the vaginal mucosa. According to some reports, oestrus is less obvious in the yak than in cattle and only about half the yaks in some regions show oestrus in the breeding season. Puberty occurs at 24–30 months of age and first calving occurs at 4 years; the gestation period is 255 days.

A low reproduction rate limits yak production and there is interest in methods of improving reproductive efficiency in the species (Zi, 2003). Unlike the situation in sheep, the physiology of seasonal breeding in the yak is poorly understood, but it appears that season is determined more by nutritional status and feeding levels than by photoperiodic sensitivity of the hypothalamic–pituitary axis. It is known that shortage of available pasture during the cold season is the most important constraint to reproductive efficiency and that food supplementation during the winter can improve fertility in the yak. Some amount of research has examined the induction of oestrus in the post-partum yak, using gonadotrophin-releasing hormone (GnRH) in combination with progesterone/progestogen-releasing devices.

Methods used in ET in the yak follow the lines of those developed in cattle. In AI programmes, yak bulls may have to be tamed for semen collection and processing. According to various reports, wild male yaks are ferocious; they live in groups of three to five males high in the mountains and descend to lower altitudes only to mate with female yaks. In work at the Tibetan Yak Research Centre, 50 wild male yaks were tamed so that they could be easily handled; they responded to simple verbal commands and semen could be collected with relative ease for producing frozen semen. The use of reproductive technologies such as oestrus synchronization and AI and of progesterone assays for pregnancy diagnosis have been described in some reports. Areas of research in yaks in China highlighted by workers in that country include: (i) hybridization of yaks with cattle and other species; (ii) selection and crossbreeding to improve productivity; (iii) AI using semen of Holstein cattle, wild yaks and semi-wild yaks; and (iv) development of the yak industry by improvements in pasture quality, feeding, management, housing and the processing of yak products.

1.3.7. Camelids

The reproductive activity of camelids has several unique features and information on oocyte physiology is limited, relative to that for the traditional farm animals. Techniques that may be employed to improve the reproductive performance of camels include AI, oestrus control, pregnancy diagnosis, ET and *in vitro* embryo production. Although, due to size differences, Old and New World camels are unable to mate naturally, with modern reproductive technology this need not be an insurmountable problem. Already a viable hybrid (dromedary × guanaco) has been born at the Dubai Camel Reproduction Centre; at the same Centre, a pregnancy has been established by way of a cryopreserved camel embryo. Recent literature contains information on sexual behaviour, conception, pregnancy, parturition and reproductive pathology of Bactrian and dromedary camels. The breeding season occurs within the period from October to May and varies from one region to another. Puberty occurs at 3–4 years. Ovulation is induced by copulation and the female is in sternal recumbency during mating. The duration of pregnancy is 370–406 days. The uterus has no caruncles and resembles that of the mare.

The development of techniques for assisted reproduction in camelids has been slow in comparison with that for farm species. Some progress has been made in the field of embryo freezing and ET in dromedary camels and in ET and oocyte recovery in llamas and alpacas. Much of the work in assisted reproduction in camelids has been focused on AI and storage of semen. There have also been attempts to hybridize Old and New World camelids using assisted reproduction techniques. The birth of a camel–llama hybrid, after numerous failed attempts, prompted Jones *et al.* (2002) in the UK to investigate the glycosylation of apposing fetal and maternal tissues of pregnant camels and alpacas; they found evidence suggesting that interspecific differences in glycans were factors that may account in part for the difficulty in producing a viable hybrid.

1.3.8. Deer

In New Zealand there are currently some 2.4 million deer farmed for venison and velvet production on some 4500 properties; many papers have been published on factors influencing the reproductive performance of farmed red deer (*Cervus elaphus*) in that country. There is also interest in the use of assisted reproduction to rescue endangered members of the deer family, bearing in mind that the survival of more than one-third of existing deer species may be threatened because of human activities. This growing interest in the preservation of various endangered species, which stretches all the way from the small Chinese water deer to the large North American wapiti, has seen the development of ET and other reproductive technologies suited to such animals.

One interesting difference between the deer and the cow lies in the apparent markedly lower incidence of embryo mortality that occurs in deer. Studies in New Zealand, for example, recorded a pregnancy rate to first service generally greater than 90%. In order to achieve a high pregnancy rate early in the mating season, it was recommended that farmers wean calves early, exclude hinds which had failed to rear a calf to weaning and hinds with a body score of 2.0 or less, join hinds early with one or more sire stags, use only experienced males for mating, limit the hind : stag ratio, use at least one backup sire after the peak of mating, keep mating mobs away from disturbance and avoid shifting or handling mating mobs. In a third paper, the same authors recommend the use of mating paddocks with limited gullies, hills and trees away from human disturbance; to increase the probability of yearling hinds conceiving, the deer should be in moderate to low body condition and diet during mating should be adjusted to ensure optimal growth rates.

In the USA, the captive breeding of wildlife, and in particular deer, is becoming big business (Long *et al.*, 2003). Unlike countries such as New Zealand, commercial developments in the USA in deer centre on shooting the animals for sport rather than on food production. The North American Deer Farmers Association claims to represent more than 75,000 deer, mainly in North America and Mexico. As well as showing high fertility, deer live much longer than regular farm animals. In New Zealand, studies have shown the percentage of red deer hinds weaning calves decreasing slowly from 90% at 6–7 years to 50% at 17 years. Considering that dairy cattle are lucky to see more than three calvings in a lifetime, the deer is clearly on a different plane in its lifetime reproductive performance.

Comprehensive accounts of red deer reproduction and approaches for improving the success of controlled breeding programmes are available (Berg and Asher, 2003), both for those in farming and for those who believe that such animals provide an ideal model for the development of *in vitro* technologies for endangered species conservation. Superovulation procedures are based on those used in sheep and cattle and embryo recovery is by surgical intervention;

ET is usually by way of a technique involving laparoscopy or by a Cassou pipette as used in AI. Although IVP of embryos is not currently practised commercially in deer, there are those who see an eventual place for such technology in the propagation of élite animals. The techniques of AI, ET and *in vitro* embryo production offer considerable advantages over natural mating for the propagation of deer species.

Artificial manipulation of reproductive seasonality in deer has been the focus of some research over the past two decades. Techniques ranging from the use of intravaginal progesterone-releasing devices, progestogen sponges and pregnant mare's serum gonadotrophin (PMSG) to exogenous melatonin delivery systems have been evaluated in red deer and to some extent in fallow deer. According to New Zealand reports, a highly effective technique involves the strategic administration of subcutaneous melatonin implants (Regulin: Schering, Australia) in spring/summer to synchronously advance ovulatory activity in females and spermatogenesis/rutting behaviour in males; such treatments have resulted in the calving season being advanced by some 6 weeks in red deer. An alternative approach is to treat females with progesterone/progestogen/PMSG and the males with melatonin, a combination offering a more effective control of the calving pattern. Treatment of the stags alone, making use of the 'male' effect, has also been found effective in that the females are induced into oestrus by the presence of the rutting stags, but the effect is less than when both sexes are treated. Apart from the cost in labour and materials and the need to repeat procedures annually, the use of melatonin may have consequences for other aspects of deer seasonal physiology, such as disruption of pelage moult cycles.

1.3.9. Poultry

The physiology of reproduction of chickens and turkeys differs from that of farm mammals in several ways, among them being the development of young outside the body of the dam, made possible by the inclusion of a large amount of yolk within the egg, the maturation and ovulation of a single egg each day, the development of only the left ovary and oviduct, the absence of a true penis and of counterparts to the accessory glands of the mammal and the capacity of chicken and turkey sperm to retain their fertilizing capacity within the female reproductive tract for weeks rather than days. There are reports of chicken sperm remaining viable for as long as a month after insemination and turkey semen for twice that period; in contrast to events in the mammalian oviduct, sperm are stored in sperm storage tubules in the lining of the oviduct.

Until predetermination of sex before hatching becomes technically feasible and economically acceptable, the routine killing of day-old chicks is likely to continue. According to Kagami (2003), the accumulation of new scientific knowledge and refinement of techniques could open a new vista for sex reversal and sex alteration in chickens. Sex reversal in chickens could have enormous commercial value in the poultry industry since broiler producers prefer males and egg producers prefer females. For that reason, there have been efforts, so far unsuccessful, dating back many years to achieve sex reversal in poultry.

1.4. Factors Affecting Male Fertility

Application of reproductive technologies can only be effective when made in the light of a reasonable understanding of reproduction in the various farm species. For that, there are several texts to which readers can turn (see Appendix A). It is clear from such texts that considerable species variation occurs and reproductive techniques that are highly effective in one species may be much less successful in the next. All the farm animals, however, share certain characteristics in some aspects of their reproduction, which deserve mention in any consideration of factors influencing fertility.

1.4.1. Sperm production

Spermatogenesis in the farm mammals is the process of division and differentiation by which sperm are produced in the seminiferous tubules of the testis and consists of spermatocytogenesis, meiosis and spermiogenesis (Fig. 1.10). Spermatocytogenesis involves mitotic cell division, which results in the production of stem cells and primary spermatocytes. Meiosis involves two cell divisions resulting in haploid spermatids. Spermiogenesis is the differentiation without division of spermatids, which are released as sperm; spermatids with spherical nuclei differentiate into sperm, which are released from the luminal free surface.

Spermiogenesis is a metamorphic process where no cell division is involved; a cascade of events results in the formation of the sperm tail and alterations can be seen in the male gamete nuclear proteins, cellular size, cellular shape and the position of proacrosomal granules and localization of the centrioles. The number of Sertoli and Leydig cells are related to sperm production, each Sertoli cell supporting a defined number of germ cells. The seminiferous epithelium is known to be sensitive to high temperature, dietary deficiencies, androgenic drugs (anabolic steroids), metals (cadmium and lead), X-rays, dioxin, alcohol and infectious diseases; such factors may elicit a temporary or permanent response, which can include an increase in the number of degenerating germ cells, reduction in the ratio of germ cells to Sertoli cells and a decline in sperm production. Various authors have dealt with the many factors affecting sperm production in farm mammals, especially in cattle, where extensive use of AI has been made, but in other species as well, including the horse.

Extremes of hot or cold environmental temperature are detrimental to sperm production; artificially increasing scrotal temperature to 41°C for 3 h in rams and boars causes rapid destruction of pachytene spermatocytes. Although increased scrotal temperature causes loss of round spermatids in the boar, pachytene spermatocytes in meiotic prophase appear to be the most thermosensitive in that species. In the stallion, increased testicular temperature induced by scrotal insulation can cause

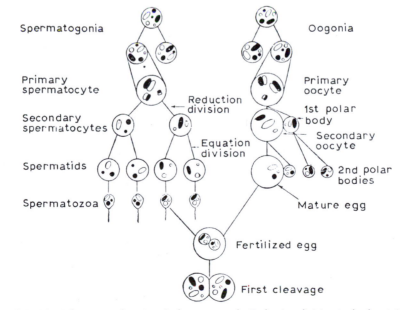

Fig. 1.10. Formation of sperm and oocytes in farm mammals. Reduction division in the formation of sperm and oocytes, showing how only one of each pair of chromosomes in the body cells of the farm animal pass to the germ cells (after Hammond *et al.*, 1983).

transient testicular degeneration as measured by semen characteristics; spermatogonia appear to be more resistant to elevated
temperature than are spermatocytes. As well
as high environmental temperature, other
conditions, such as fever and improper
scrotal descent, can interfere with the
thermoregulatory mechanisms involved in
keeping the testes cool and allowing normal
spermatogenesis.

The ability of mammalian sperm to penetrate and fertilize an oocyte is acquired during changes occurring in the female tract but
it is also dependent on maturational changes
within the epididymis (Fig. 1.11). As sperm
enter the caput epididymidis from the testes,
they are immature and unable to penetrate or
fertilize an oocyte. However, sperm from the
cauda epididymidis possess characteristics
and properties almost identical to those of

ejaculated sperm, clearly demonstrating
that modifications resulting from exposure
to specific factors within the epididymal
tract directly affect the fertilizing ability of
sperm. Ejaculated sperm must still undergo
several changes before being able to penetrate and fertilize an oocyte; these changes
include capacitation, hyperactivation and
the acrosome reaction.

1.4.2. Physiological and endocrinological factors

In the farm mammals, the testes descend
through the inguinal canal into the scrotal
sac. The purpose of this move away from
the body proper is to provide a suitably
cool environment for the testes to function;
whether this is to meet the needs of the
epididymis in the maturation and storage of
sperm cells or to protect sperm production
from temperature-induced errors remains
uncertain. It is well known that in the
farm mammals normal spermatogenesis is
dependent on a functional hypothalamic–
pituitary–testicular axis, which involves
the classic endocrine actions of gonadotrophins, feedback mechanisms of steroids
and proteins and the modulating effects of
various paracrine factors. Endocrine control
is in the form of changing secretory patterns
of the hypothalamic hormone GnRH, the
pituitary hormones LH and FSH and the
testicular hormones, such as androgens and
inhibin (Table 1.6).

In horses, it is also known that
testicular oestrogen plays a role, possibly as
a local paracrine factor in sperm production
and maturation. Testicular function is also
known to be affected by other, peripherally
produced hormones, such as prolactin,
growth hormone and insulin. Paracrine control mechanisms are known to coordinate
the various functions of the various
testicular cell types and possibly play a
role in modulating the testicular actions of
the pituitary gonadotrophins according to
local conditions and requirements. Potential
paracrine regulators of spermatogenesis of
proven physiological significance include

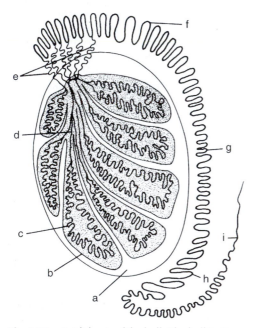

Fig. 1.11. Epididymis of the bull. The bull testis
and epididymis showing the pathway followed
by sperm on their journey from the seminiferous
tubules to the sperm duct that leads into the urethra.
(a) Tunica albuginea, the coat of the testis; (b)
septum of testis; (c) seminiferous tubule; (d) rete
testis (system of collecting ducts); (e) ducts leading
into the head (f) of the epididymis; (g) body of the
epididymis; (h) tail of the epididymis; (i) sperm duct
(after Hammond *et al.*, 1983).

Table 1.6.   Reproductive hormones in the male.

Hormone	Chemical description of molecule	Source of hormone	Target tissues	Main action in the male
Gonadotrophin-releasing hormone (GnRH)	Neuropeptide	Hypothalamus	Anterior pituitary	Release of FSH and LH from the anterior pituitary
Luteinizing hormone (LH)	Glycoprotein	Cells in the anterior pituitary (gonadotrophs)	Cells of Leydig (interstitial cells) in the testis	Stimulates testosterone production
Follicle-stimulating hormone (FSH)	Glycoprotein	Cells of the anterior pituitary (gonadotrophs)	Sertoli cells in the testis	Sertoli cell function
Prolactin (PRL)	Protein	Anterior pituitary	Testis	Influences behaviour
Oxytocin	Neuropeptide	Secreted by nerve cells in the hypothalamus, stored in the posterior pituitary: secreted by cells in the corpus luteum	Smooth muscle in the tail of the epididymis, the sperm ducts (vasa differentia) and the ampulla	Involved in PGF synthesis and the movement of sperm in the sperm ducts
Oestradiol	Steroid	Sertoli cells of the testis	Brain cells	Sexual behaviour
Testosterone	Steroid	Leydig cells of the testis	Accessory sex glands, various cells in the testis	Promotes sperm production, growth and secretions from the accessory glands
Inhibin	Glycoprotein	Sertoli cells in the testis	Gonadotrophs of the anterior pituitary	Inhibits FSH secretion
Prostaglandin $F_{2\alpha}$ ($PGF_{2\alpha}$)	Fatty acid	Epididymis	Affects activity of sperm and contractions of the epididymis	

factors such as testosterone, inhibin, activins, growth factors, oxytocin and vasopressin.

It is known that secretion of the gonadotrophins (LH and FSH) in males depends on the release of GnRH from the hypothalamus into the hypophyseal portal blood system; direct measurement of GnRH and LH in males has shown that GnRH and LH are secreted in a pulsatile manner, with a high degree of concordance between GnRH and LH pulses. Although the synthesis of FSH is clearly stimulated by GnRH, the extent to which its secretion depends on GnRH pulses is less clear.

Sperm production represents continuous proliferation and differentiation of germ cells occurring in a delicate balance with other testicular compartments, especially the supporting Sertoli cells. Although much remains to be elucidated about the roles of gonadotrophins and androgens in the initiation, regulation and maintenance of spermatogenesis in farm mammals, certain facts are well established. It appears that differentiation of A0 to A1 spermatogonia is sensitive to LH, while multiplication from intermediate to B2 spermatogonia, and consequently the rate of sperm production, depends on FSH. Meiotic divisions and spermiogenesis are maintained by testosterone. FSH is believed to be a key determinant of the rate of spermatogenesis and plays an important role in determining testicular size. It is known from various studies that, in the male animal, receptors for FSH are only expressed on testicular Sertoli cells. FSH is thought to play different roles during the life of the male; it functions as a growth factor during development and sustains sperm production in adults. There is evidence supporting the view that FSH plays a central role in the male at puberty through the control of testosterone production.

1.4.3. Genetic and environmental factors

Various forms of chromosomal abnormalities, although relatively uncommon, may be the cause of infertility in farm-animal males. Cattle, for example, normally have 60 chromosomes but occasionally two chromosomes fuse together in what is called a Robertsonian translocation (i.e. centric fusion), causing the total number to be reduced to 59. In most animals, including cattle and humans, such a Robertsonian translocation typically causes a reduction in fertility. One unusual form of subfertility occasionally encountered in bulls is a consequence of events which occurred in prenatal life. It is well known that there is a germ cell chimerism (XY/XX) in young bulls born in heterosexual twinning, due to exchange of primordial germ cells in embryonic life. Although it was thought at one time that these germ cells were eliminated in the young bull, more recent evidence suggests that even mature bulls (> 2 years old) may show evidence of spermatogonial chimerism (Rejduch *et al.*, 2000); such evidence may stimulate renewed interest in checking the possibility of survival and differentiation of germ cells from the female partner in the germ cell lines.

In terms of environmental effects, it has long been recognized in farm animals that high and low environmental temperatures may be responsible for reduced fertility. In the ram, for example, it appears that scrotal temperature is regulated independently of body core temperature by a feedback circuit involving scrotal thermoreceptors and effectors in the form of tunica dartos muscle activity and scrotum sweat-gland activity. These effector mechanisms, however, are insufficient to maintain scrotal temperature during extremes of heat and cold exposure.

Nutritional considerations

Although the classical view of the regulation of testicular function in the farm animals is that the effects of internal and external stimuli (e.g. puberty, daylight changes and social interactions) are relayed to the testis by the gonadotrophins, LH and FSH, it is also known that changes in nutrition can modulate testicular growth and the production of sperm. In rams, it appears that the effect of nutrition is mediated in a different way from that of other external stimuli because changes in testicular growth are not clearly associated with changes in gonadotrophin secretion.

1.4.4. Tests for predicting breeding potential of males

The importance of fertility in beef cattle in the USA is shown by the fact that fertility is regarded by farmers as five times more important economically than growth rate and ten times more important than carcass quality. For such reasons, many methods have been developed over the years to assess the fertility of young male farm animals that are to be used in natural service; details will vary according to species, but the general approach is much the same. The development of a fertility test for application in beef cattle under clinical field conditions that could be used on bulls of various ages and breeds has long been a goal; currently, such a test is not available. When a bull is selected for breeding, a number of factors are evaluated: breed, conformation, libido and mating ability. As well as that, the internal accessory sex organs and external genitalia are examined and semen quality is evaluated.

In the absence of a suitable fertility test, veterinarians use breeding soundness evaluation to predict the ability of a bull to achieve good pregnancy rates, an evaluation based on physical attributes and sperm characteristics. Tests that have been used to evaluate bull semen quality include examination of sperm morphology, sperm motility and motion analysis, cervical mucus and zona pellucida (ZP) penetration, supravital and fluorescence staining of sperm and flow cytometry. Many reports have dealt with the significance to bull fertility of morphologically abnormal sperm.

Using yearling bulls in beef production is an economic consideration in countries such as the USA; such animals, when well grown, can be as reproductively efficient as 2-year-olds. Selecting beef bulls as yearlings has become common in the USA because yearlings not selected for breeding can be slaughtered with little or no penalty for carcass quality. Many tests have been evaluated for predicting breeding potential of yearling beef bulls, but sperm morphology continues to be the most accurate single test available. Current guidelines for the evaluation of bulls recommend that at least 70% of ejaculated sperm be morphologically normal; the most common reason for rejection is immaturity, where the yearling ejaculates a high proportion of sperm with proximal cytoplasmic droplets or midpiece defects. Results of studies in Canada with beef bulls have shown that increased dietary energy may affect scrotal or testicular thermoregulation by reducing the amount of heat that can be radiated from the scrotal neck, thereby increasing the temperature of the testes and scrotum.

1.5. Factors Affecting Female Fertility

Fertility in female farm animals, as determined by average per cycle pregnancy rates, varies from 50 to 80% among the various traditional farm species, with most of the losses occurring after fertilization and prior to the third week of pregnancy. Such early losses are largely the result of defective embryo development; it is known that fertilization rates tend to approach 100%.

Poor fertility costs the dairy industry massive amounts of money each year, a large part of which is due to delayed pregnancy caused by early embryo loss. In the USA, it has been observed that, despite the virtual elimination of the specific infectious reproductive diseases prevalent when AI was introduced on a large scale in the 1940s, the dairyman's main problem remains low fertility in his cows. It has been shown, again talking about the USA, that the average dairy cow may only live for about 5 years, produce two calves and complete two lactations; reproductive failure is held to be a major reason for this short productive life.

1.5.1. Embryo mortality

Even in normal healthy cows, some proportion of embryos (25% or more) which pass through the oviduct into the uterus fail to continue development, generally during the first 3 weeks of pregnancy (Fig. 1.12). Embryonic mortality has long been recognized as a major source of loss in breeding cows and numerous studies have dealt with it. As described by various workers, the fertilization rate after the cow has been bred can be taken at about 90%, whereas the average calving rate to a single service may be some way below 50%; much of this loss is the result of embryonic mortality occurring between 1 and 3 weeks after breeding. From 3 weeks until 9 weeks into pregnancy, a further 10–15% of embryos die. When embryo death occurs before days 16–17, the cow can be expected to repeat after a normal oestrous cycle interval (i.e. 18–24 days); when embryo mortality occurs after days 16–17, the cow 'repeats' at long and irregular intervals. Between 7 weeks and full term, the incidence of fetal death is usually taken as 5–8%.

The incidence of embryo mortality in farm animals varies among the various species. The economic importance of cattle in the livestock industry has focused most attention on this species. The causes of embryo mortality in the cow can be divided into infectious and non-infectious categories. Specific uterine infections are caused by viruses, bacteria and protozoa that enter the uterus by way of the blood circulation or via the vagina; non-specific pathogens are mainly bacteria that enter the uterus by ascending infection. Uterine pathogens may cause embryo mortality by altering the

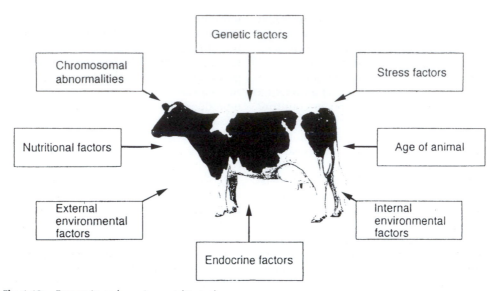

Fig. 1.12.   Factors in embryonic mortality in the cow.

uterine environment (endometritis) or by a direct cytolytic effect on the embryo.

Although attention has often been directed towards infectious agents, the evidence is that non-infectious causes probably account for 70% or more causes of embryonic death in cattle. In pigs, about 30% of oocytes shed at ovulation are not represented by piglets at birth, a source of considerable economic loss to the pig industry. It is known that most of the prenatal loss occurs in the first month of pregnancy, with some studies showing that only 75% of pig blastocysts present on day 9 of pregnancy remained viable at day 25. Much of this loss occurs around the time of maternal recognition of pregnancy and implantation (days 12–18 of gestation). It is known that, at the time of maternal recognition of pregnancy, there is an asynchronous array of developmental stages within the blastocyst population; it is believed that the more advanced embryos may bring about changes to the uterine environment that prove detrimental to less mature embryos.

Those who have investigated the timing of fetal mortality in pigs have shown that there are certain critical periods. Van Der Lende and Van Rens (2003) recorded an 8.7% fetal mortality rate, with critical periods occurring at the early fetal stage (days 35–40), shortly after mid-pregnancy (days 55–75) and after day 100 of gestation; these three periods corresponded with reported periods of change in porcine placental growth.

There are many reports showing that embryo mortality in sheep can account for one-third or more of losses occurring in pregnancy. In Germany, for example, workers have recorded an incidence of 33.5% embryo mortality by real-time ultrasonics in German mutton merino ewes. Many studies in sheep have shown the importance of nutrition in pregnancy. Results of work in Spain indicated that undernutrition can reduce the ability of the embryo to secrete interferon-tau (IFN-τ), with a consequent increase in the production of prostaglandin $F_{2\alpha}$ (PGF$_{2\alpha}$) from the endometrium, which can initiate luteolysis. Some workers have suggested that early onset of reproductive senescence in domesticated sheep, in comparison with a breed such as the Soay, may be the cost of selection for large litter size in this species. Workers in Norway claim to have found convincing evidence of early reproductive senescence in domesticated sheep, which occurs at 5–6 years of age, as measured on the basis of lamb weight and litter size (Mysterud *et al.*, 2002).

One remarkable fact in deer farming is the low incidence of embryo mortality in these animals. In contrast to events in cattle, sheep and pigs, red deer have a consistently high conception rate, markedly greater than that in cattle. In New Zealand, studies have examined post-blastocyst development in red deer to determine what features may contribute to this superior fertility; it was concluded that the superior fertility is a reflection of very early events in development and extremely low mortality. Understanding the causes of early embryonic losses is limited by the difficulty in accurately characterizing the events occurring during this period. In the USA, one recent report recorded the incidence of embryo mortality during the first 90 days after removal of the buck to be 2.8%, a value far below that recorded in cattle and markedly below that usually found in sheep. While deer may be regarded as 'wild' rather than domesticated, it is evident that a greater understanding of factors active in deer reproduction may help to shed light on the causes of embryo mortality in the traditional farm mammals.

1.5.2. Genetic and environmental factors

Progress in genetic selection of dairy cattle for increased milk yield has been partly achieved by selecting for differences in the animal's hormonal balance, especially growth hormone and insulin-like growth factors (IGFs); this has resulted in nutrients being partitioned towards the mammary gland and milk production at the expense of other organs. The dairy cow in the late stages of pregnancy is in a largely anabolic metabolic state. After calving, as lactation progresses, there is a shift towards catabolic metabolism. High-yielding cows typically mobilize body reserves (losing body condition) to maintain milk production until their feed intake matches or exceeds nutritional requirements. In the case of primiparous cows (usually 2 years old at calving), there is the additional problem of growth as well as lactation, which further

compromises their reproductive performance. In the UK, Taylor *et al.* (2003) showed that the high prevalence of delayed ovulation after calving in high-yielding primiparous cows had a detrimental impact on their fertility and was associated with marked physiological changes.

Intensive genetic selection for increased milk production over the past half-century, coupled with increased dry-matter intakes, has led to significant improvements in cow milk yield, estimated as about 40% in the past 20 years in the USA; this increase in milk output has unfortunately been at the expense of a marked decline in cow fertility. In the USA, the decline over the past 20 years has been estimated at about 0.5% per annum and in Britain about 1%. Workers at Nottingham University compared data for British Friesian cows in 1975–1982 and in 1995–1998; they looked at 20 commercial herds and found that conception rates had declined from 55.6% to 39.7% over the 20-year period.

In Ireland, throughout the 1960s and 1970s, herd fertility was high, with calving rates to first service between 60 and 69%; in the 1980s, data from research herds showed a significant decline in conception rate to first service between 1980 (67%) and 1988 (59%). Subsequent studies in commercial herds in the 1990s confirmed this trend, with calving rates to first service declining significantly by 0.7–0.9% per year. The most significant factors explaining such declines are believed to be the genetic changes associated with new strains of Holstein–Friesian cattle, increased herd size and possibly increased use of do-it-yourself (DIY)-AI. At this point in time, it is estimated that 48% of Irish dairy cows conceive to first service and 14% of cows remain non-pregnant at the end of a 15-week breeding season (Mee, 2003).

Fertility is a heritable trait in dairy cattle; for that reason, many urge that fertility traits should be incorporated into the breeding objectives of dairy cattle. According to Flint *et al.* (2002), bull proofs for daughter fertility have been available for more than 10 years in several European countries; however, despite the steady and well-documented decline in fertility in the

UK dairy herd, a fertility index has not yet been published in that country. The use of appropriate economic values for fertility in an overall index used in the dairy cattle industry would enable suitable emphasis to be placed on fertility to optimize herd profitability.

Stress effects

It is well established that stress in farm animals reduces fertility, although the precise mechanism by which stress affects fertility is not yet fully understood. It is believed that stressors activate the hypothalamic–pituitary–adrenal axis, resulting in the release of adrenocorticotrophic hormone (ACTH), which then stimulates glucocorticoid secretion from the adrenals. The release of ACTH and glucocorticoids interferes with the release of gonadotrophins through action on the hypothalamus and/or the pituitary. Some studies have investigated the potential influence of stress as a component of the 'repeat breeding syndrome' in cattle. The repeat breeding syndrome has been studied for some years in Sweden and is regarded as a multifactorial disorder. A survey of 57,616 dairy cows in 1541 randomly selected herds in Sweden showed the overall incidence of repeat breeding to be 10%. Workers there suggested from their data in that country that sustained adrenal stimulation associated with environment or social stress could be a factor in the syndrome, perhaps due to higher than normal progesterone levels during oestrus, which adversely affected fertilization. It is also possible that progesterone levels exceeding normal limits around the time of ovulation may lead to prolonged growth of the preovulatory follicle and delayed ovulation, with consequent detrimental effects on fertility.

Heat stress

Much work over the years has been devoted to improving the productivity of heat-stressed farm animals using different techniques. Researchers have examined physical, physiological, nutritional and management techniques for alleviating heat stress in all the farm animals and in buffaloes, camels and poultry; there are those who suggest that selection of animals for increased tolerance to high temperatures may be the best way forward in seeking to improve productivity in hot climates. In this context, genetic engineering may be able to make a useful contribution. The adverse effects of heat stress on reproduction in dairy cows have been well documented; they include a suppressed intensity of oestrus, a reduction in the strength of the preovulatory LH surge, a decreased secretion of progesterone, altered follicular development, decreased embryo development and reduced fertility.

It is clear that heat stress has many effects on the reproductive axis; some are direct effects on the hypothalamus, the anterior pituitary gland, the uterus, the follicle and its oocyte and the embryo itself; other effects are indirect, probably mediated by changes in the metabolic axis in response to reduced dry-matter intake. According to De Rensis and Scaramuzzi (2003), there is probably no single mechanism by which heat stress reduces post-partum fertility in dairy cows; the problem is due to the accumulation of the effects of several factors (Fig. 1.13).

Problems of heat stress in farm animals can be experienced over a wide geographical area, occurring in cows at air temperatures as low as 27°C. Summer infertility in Holstein dairy cows is a well-recognized phenomenon in Texas; workers in that state recorded that the marked seasonal decrease in Holstein fertility was less severe on farms that provided shade in the lounging or holding pen and dry cow areas and fans in lounging areas.

Heat stress decreases the intensity and duration of oestrus and is known to increase the incidence of silent ovulations (ovulation unaccompanied by oestrous symptoms). There is ample evidence showing that heat stress can alter concentrations of circulating reproductive hormones by increasing circulating concentrations of corticosteroids. Heat stress also influences follicle dynamics, the quality of follicles and their

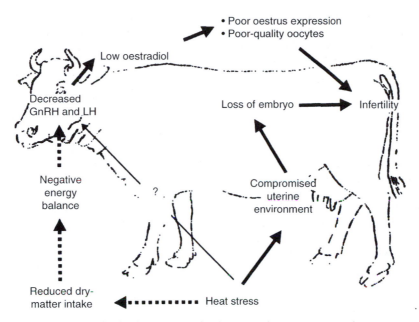

- Poor oestrus expression
- Poor-quality oocytes

Low oestradiol

Decreased GnRH and LH

Loss of embryo ➝ Infertility

Negative energy balance

?

Compromised uterine environment

Reduced dry-matter intake ◀ ••••••••• Heat stress

Fig. 1.13. Mechanisms involved in heat stress in the dairy cow (from De Rensis and Scaramuzzi, 2003).

steroid-producing ability. It is also clear that high ambient temperatures affect the early developing cattle embryo, but with decreasing effect as the embryo develops. The use of shade, fans or evaporative cooling reduces but does not eliminate the fertility problems associated with heat stress; additional reproductive strategies are required to counteract the adverse effect of heat stress. It is likely that heat stress adversely affects reproductive performance in dairy cattle indirectly by influencing energy balance and metabolism as well as having direct effects on the reproductive tract and the early embryo. An increase in maternal body temperature is likely to result in an increase in the ambient temperature of oocytes, zygotes or embryos in the oviduct or uterus of the cow. An *in vitro* study by Sugiyama *et al.* (2003) in Australia sought to mimic the *in vivo* scenario of a heat-stressed cow or heifer, examining the effects of increased ambient temperature on zygote and early embryo development; exposure of insemi-nated oocytes to high ambient temperature in the first 2 days of *in vitro* culture had a significant adverse effect on embryonic development and survival.

Overcoming heat stress

Although infertility in the bull caused by heat stress can be eliminated by using AI with semen collected and frozen from males in cool environments, the female poses more difficult problems. None the less, it could be argued that ET represents a method analogous to AI in that embryos could be collected from non-stressed donors under appropriate temperature conditions and transferred to heat-stressed recipients. Workers have shown that the effects of maternal hyperthermia on bovine embry-onic survival are most pronounced in the very early stages of development. It appears that the ability of the embryo to withstand elevations in temperature increases as it develops; surprisingly, oocytes are more resistant than two-cell embryos, suggesting a thermo-protective role of cumulus cells or oocyte-derived factors.

Seasonal changes

In hot countries, conception rates of lactat-ing cows can decrease from about 50% in the winter to less than 20% in the summer.

However, fertility in the autumn (30%) is lower than that in winter, although ambient temperatures decrease and animals are no longer under heat stress. The fact that it takes 40–50 days for small antral follicles to develop to the ovulatory stage means they are exposed to heat stress in their early stages of development; this can result in a defective oocyte at time of ovulation. According to some studies, both the immediate and the delayed responses to heat stress may be involved in the low fertility experienced by cattle during the summer and autumn.

One area of research in heat-stressed dairy cattle for increasing embryonic survival is likely to involve the manipulation of heat-shock protein (HSP) synthesis; it is known that the bovine embryo can produce increased amounts of HSP70 in response to elevated temperature as early as the two-cell stage. It remains to be seen whether manipulation of HSP synthesis can increase embryonic survival following exposure to maternal hyperthermia or other shocks. In Spain, Lopez-Gatius (2003) presented a 10-year (1991–2000) survey of factors influencing dairy cattle fertility in that country; cyclicity and pregnancy rates of all cows bred by AI in the warm period of the year (May–September) declined significantly but remained practically constant in the cool period (October–April).

Seasonal effects on reproduction in pigs in Finland were examined in a report dealing with 1081 herds; this work identified clear seasonal effects on various aspects of fertility in the sow. It was found that the poorest reproductive performance was consistenly observed in late summer and autumn. Recent advances in research on seasonal infertility in pigs have been reviewed by Peltoniemi *et al.* (2000), with special emphasis on the implications of the generally recommended restricted post-mating feeding strategy of the early pregnant gilt and sow. It is believed that, after an initial progesterone-mediated beneficial effect on embryonic survival, a restricted post-mating feeding strategy may have a negative effect on the maintenance of early pregnancy in the sow and gilt in the summer–autumn period.

Although the endocrinological mechanisms involved in the seasonal disruption of pregnancy have yet to be determined, there is evidence that LH is reduced in the summer–autumn period and this reduction is amplified by the commonly applied restricted post-mating feeding strategy. These changes in LH secretion, although not leading to luteal regression, may exert a progesterone-mediated adverse effect on the capability of embryos to produce adequate embryonic signalling, leading to a seasonal disruption of pregnancy.

Heat stress in sheep and pigs

There are reports of heat stress affecting reproduction and pregnancy rates in sheep. In Australia, for example, some studies led workers to conclude that placental growth in sheep is restricted in animals that do not thermoregulate well when exposed to a hot environment. The authors note that the ability of flock managers to identify high body temperature status, which will subsequently lead to restricted placental growth, can be used to identify low-producing individuals early in their commercial life. In Argentina, workers used shade during the breeding season in a subtropical climate to reduce heat stress in ewes and to increase the overall flock pregnancy rate; ewes with shade had a significantly higher conception rate to first service than controls.

Lactating sows are often exposed to ambient temperatures above the upper limit (22–25°C) of the thermoneutral zone; under such conditions, sows increase heat loss and ultimately decrease heat production. The increased heat loss is mainly achieved by way of increased evaporative losses by polypnia and behavioural adjustments whenever possible (such as watering of skin and change in posture). A decrease in heat production is likely to involve a reduction in feed intake, which becomes acute above 27°C; decreases in basal metabolic rate, physical activity and lactogenesis also contribute towards heat reduction. The reproductive ability of the sow is inhibited as a result of such changes, as shown in reduced LH pulsatility during lactation and delayed return to oestrus after weaning.

1.5.3. Management and nutritional considerations

Reduced fertility is more evident in high-yielding dairy cattle and is believed to be due to several management factors, including increased milk production, food intake and fluctuations in body condition and body reserves. The decline in fertility is known to be the result of embryonic loss and it is clear from many studies that nutrition plays a role in determining the extent of this loss. Nutrition can influence reproductive parameters, although the pathways by which nutrition exerts its action are not fully understood. Metabolic hormones, such as insulin and IGFs, and their binding proteins appear to synchronize with gonadotrophins at the level of the gonads. Nutrition is known to have a direct effect at the level of both the ovaries and the reproductive tract.

A recent study of 1000 dairy cows on Irish farms showed an embryonic loss rate of 7% between 4 and 12 weeks of pregnancy, a loss that was significantly greater in cows losing body condition between days 28 and 56 of gestation than in cows that either maintained or improved in body condition (Silke *et al.*, 2002). Data discussed by Boland (2002) suggest that high dietary intake or high metabolic load associated with management of high-yielding dairy cows adversely affects normal oocyte development and the establishment of pregnancy; such effects are often evident within a few days of fertilization. A much fuller understanding of the role of nutrition in influencing oocyte quality and the extent of embryonic mortality is essential; nutrition is one factor that can be manipulated by the farmer in many ways, but only effectively if applications are soundly evidence-based.

Assessing body condition

Body condition score (BCS) has proved to be a useful management tool in assessing

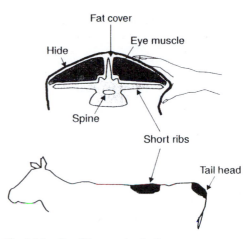

Fig. 1.14.   Condition scoring in the cow.

the nutritional status of dairy cows (Fig. 1.14). Cows are usually scored on a 5-point scale from 1, indicating thin, to 5, indicating fat. During the last 30 years, the traditional subjective appraisal of the body fat stores in farm animals, made by eye and touch, has been rationalized by the introduction of these numerical systems of rating specific points. Palpation of the lumbar vertebrae, the pin and hook bones (tail head) and occasionally width behind the shoulders provides an assessment of the fatness of the animal. There is ample evidence to show that body condition at calving can have a direct effect on the health, milk yield and fertility of cows; it represents the cumulative effects of the dry period, the BCS at drying off and the loss of condition during the dry period.

In horses, too, body condition scoring has been used as a means of maintaining the maximum reproductive performance of mares. In Croatia, Cacis and Ivankovic (2001) used body condition scoring as a guide for the efficient feeding and reproductive management of mares; this is on the basis of evidence that the amount of body fat, as an indicator of stored body energy, has a definite influence on fertility. A body condition scoring method has been described, based on visual assessments of fatness and the prominence of vertebrae, ribs, hips and pin bones.

Diet and fertility

In dairy cattle, the more intensive production systems and associated higher
milk yields generally involve a decreased
dependence on grass and an increase in the
energy content of the diet; clearly, it is
important to understand how the changing
composition of the dairy cow's diet affects
fertility. Some studies suggest that the
polyunsaturated fatty acid content of
the diet may influence ovarian and uterine
function. Other work suggests that the
spacing of feeds may influence fertility. It
is believed that high dry-matter intake associated with high milk production reduces
progesterone levels by increased clearance
of progesterone by the liver; feeding cows
four times a day rather than once or twice
can help in maintaining high progesterone
levels which may improve reproduction.

Well recognized in dairy cattle management is a marked drop in fertility when cows
first go out to grass in the spring. It is now
known that the impact of spring turnout on
fertility mainly affects ovulation, fertilization and/or early embryonic development;
there appear to be no adverse effects from
20 days of pregnancy onwards (Laven et al.,
2002). The practical implication of such
information for farmers is to avoid breeding
in the first 3 weeks after turnout.

On a worldwide basis, there are several
factors other than milk yield which may
influence cow fertility. In Africa, for example, cows have to deal with problems other
than milk production. In Zimbabwe, where
they are often used as draught animals, it is
known that nutritional stress may be more
important in suppressing ovarian activity
than work stress and that dietary supplementation can reduce the negative effects of
draught on ovarian activity.

Pigs and sheep

In pigs, some studies have shown that
feeding diets high in fibre to breeding sows
can have a positive effect on reproductive
performance. Workers in Scotland, for
example, found no increase in ovulation
rate in gilts fed high-fibre diets but did find

evidence that such diets fed over the rearing
period improved early embryo mortality.
The nutritional influences on the hormonal
control of reproduction in the female pig
were reviewed by Prunier and Quesnel
(2000); it was concluded that decreased
metabolic clearance of progesterone, resulting in increased plasma concentrations
of this hormone, could be involved in
the inhibition of gonadotrophin release and
reduction of the ovulation rate occurring in
feed-restricted cyclic gilts. A study reported
by Ferguson et al. (2003) suggested that a
pre-mating nutritional regimen previously
shown to improve embryo survival in
Meishan gilts may have reproductive
benefits in European breeds as well; they
showed that increased feed intake for 19
days increased oocyte nuclear maturation
by increasing the percentage of oocytes
reaching metaphase II and by changing
follicular fluid composition. Such findings
contribute to a growing body of evidence
that altered nutritional regimens before
mating can influence oocyte and follicle
characteristics in pigs and in farm
ruminants.

Many studies have examined the
influence of nutrition on reproductive efficiency in sheep. In Aberdeen, researchers
have shown that excess rumen-degradable
nitrogen in the diet of ewes increased urea
and ammonia levels in plasma and the
uterus, with an associated increase in
embryo mortality. Other authors have
produced evidence showing that maternal
undernutrition may result in retarded
embryonic development at 8–11 days after
mating and reduced pregnancy rates after
2 weeks of pregnancy. Among reasons
advanced for such effects is variation
in peripheral progesterone concentrations
due to nutrional effects; this may induce
asynchrony between embryo and uterus.
One recent study found that progesterone
concentrations in blood collected ipsilateral
to ovaries bearing a corpus luteum (CL) were
higher than those in contralateral samples;
it was concluded that undernutrition can
reduce the endometrial content in the
first week after mating and hence influence
embryo survival.

Plant oestrogens

The effect of plant oestrogens on ruminant reproduction has been the subject of much research since the mid-1940s; phyto-oestrogens in herbage can have a very deleterious effect on reproduction, particularly in sheep and to a limited extent in cattle. Plants that contain these compounds include some of the economically important members of the Leguminosae family, and, while such plants play a valuable role in improving soil fertility, they can also adversely affect sheep fertility. With breeding ewes grazing oestrogenic clover prior to and during mating, the animals suffer temporary infertility from which they can recover within a few weeks after removal to non-oestrogenic pasture. Prolonged grazing on oestrogenic clover for several years can result in a permanent and progressive infertility in ewes.

The phyto-oestrogen formononetin, the main isoflavone present in subterranean clover and red clover, is implicated in such problems; although the compound is not oestrogenic itself, it is metabolized in the rumen, mainly to equol, which is oestrogenic. Other oestrogenic substances may be associated with fungal growth on plants, which also reduce reproductive performance, especially ovarian function and early events in embryonic development. Although various researchers have suggested methods that may protect the ewe against phyto-oestrogens, the long-term solution lies in using forage plants with low levels of these compounds. In the meantime, pasture management plans can be devised to avoid feeding oestrogenic herbage at critical times to breeding ewes.

1.5.4. Disease and metabolic disorders

Uterine infections have long been recognized as a major cause of reduced fertility in mares; it is also evident that resistance to antibiotics is a common feature among bacteria isolated from the uteri of mares with fertility problems. Careful use of antibiotics in stud farm practices is indicated, preferably together with a bacteriological diagnosis to reduce the risk of development of further antibiotic resistance. Persistent endometritis is held to be the third most serious clinical problem in horses after colic and respiratory-tract disorders. In the normal healthy mare, the uterus is well protected from external contamination by physical barriers that include the vulva, the vestibule, the vagina and the cervix; any compromise of these barriers may predispose the mare to a chronic uterine infection. The mating process itself is another source of uterine contamination; intrauterine deposition of semen causes an inflammatory reaction due to bacterial contamination of the ejaculate and to the sperm themselves. There are reports suggesting that some 15% of a normal population of Thoroughbred brood mares may develop persistent endometritis after mating. Endometritis can have serious effects on the fertility of affected mares; a persistent inflammation can often lead to premature luteolysis and embryonic loss because of increased $PGF_{2\alpha}$ concentrations. Uterine inflammation may also interfere directly with the survival of the embryo; it is believed that the uterus in the normal mare should be capable of spontaneously clearing inflammation within the first 5 days after breeding.

Lameness

In cattle, one of the serious animal welfare problems is the high incidence of lameness to which dairy animals are predisposed when walking and standing on wet concrete in cubicle houses. It has been estimated that the incidence of lameness in modern dairy farming runs somewhere between 2 and 20% and occurs during the first 2–3 months of lactation rather than at other times. Lameness can seriously affect the profitability of the dairy operation by reducing milk yield and fertility as well as leading to higher culling rates. There are many reports in the literature showing an association between lameness and fertility in dairy cattle; in Korea, for example, workers have reported calving intervals (CIs) increasing from 109.6 to 150.6 days in lame cows.

Ovarian cysts

1.6. Enhancing Female Fertility

Cystic ovarian disease (COD) is the most commonly diagnosed reproductive disorder of dairy cows; the mechanisms involved in COD are not well understood, but it is believed that abnormally large follicles may involve high levels of LH secretion. Ovarian cysts (structures having a diameter > 2.5 cm in the absence of a CL in either ovary) may be classified as follicular or luteal on the basis of their histological and morphological appearance. Cows with follicular cysts are usually characterized by high LH secretion. Ovarian cysts are estimated as occurring in 10–30% of milking cows and usually present a problem in the first 3 months after calving. There are several risk factors associated with cysts, including genetic disposition, nutrition, season, milk production and management system. Some reports suggest an association between lameness and ovarian cysts. A study in the USA, based on historical records from a 3000 Holstein farm, although it failed to establish a cause–effect relationship, showed that cows that became lame within the first month of calving were associated with a higher incidence of ovarian cysts, a lower likelihood of pregnancy and lower fertility than control animals (Melendez *et al.*, 2003).

A considerable volume of literature is available on attempts to enhance pregnancy rate in farm animals, particularly dairy cattle, by either modifying diet or using hormones (Fig. 1.15).

1.6.1. Nutritional approaches to enhanced fertility

The development of diets to keep pace with the demands of ever-increasing milk yields in dairy cattle is a major challenge to nutritionists; regimes are required that will support milk production at an economic level and maintain the cow in good health and fertility. Much nutrition work concentrates on energy requirements, particularly since cows in early lactation are likely to be in negative energy balance. There are also nutritional experiments showing that high levels of dietary protein are associated with low pregnancy rates, the problem apparently being due to rumen-degradable protein, with older cows being more susceptible than younger animals. Turnout in the spring, a time at which various observers have recorded a marked decline in fertility, apparently results from embryo loss,

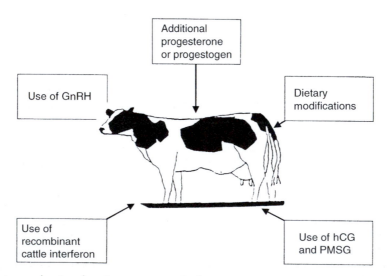

Fig. 1.15.   Approaches to enhancing pregnancy rates in cows.

particularly where nitrogenous fertilizer has been applied.

Some experiments have examined the effect of feeding rumen-protected fatty acids to post-partum primiparous beef heifers, showing that this lipid feeding increased plasma levels of linoleic acid and PGF metabolite, but did not improve fertility in their subsequent breeding performance. Elsewhere, however, there have been strong indications that linoleic acid, a member of the n-6 family of unsaturated fatty acids, a specific inhibitor of prostaglandin synthetase, may be involved in the action of IFN in preventing luteolysis at the time of maternal recognition of pregnancy. There are also fatty acids belonging to the n-3 family, present in fish oil, which may help to explain some studies showing a 20% improvement in conception rates in cattle fed on diets containing fishmeal. There are those who believe that there are many exciting possibilities to be explored in nutritional management which may help to improve dairy cow fertility. Less work has dealt with nutritional influences that may improve the quality of sperm in male farm animals. In Scotland, Rooke et al. (2001) showed that feeding tuna oil increased the proportion of boar sperm with progressive motility and reduced the proportion of sperm with abnormal morphologies.

1.6.2. Hormones to enhance fertility

Hormones play a vital part in the reproductive processes of the female (Table 1.7) and numerous attempts have been made to reduce the incidence of embryo mortality with exogenous hormones, including progesterone, IFNs, GnRH and human chorionic gonadotrophin (hCG), all of which have produced variable results. Despite the importance of progesterone during early pregnancy, many studies with cattle fail to show any improvement in pregnancy rate following progesterone supplementation. This is thought to be due to several factors, including the potential down-regulation of high levels of exogenous progesterone on endogenous production and the fact that many of the treated cows may not require additional progesterone. Stimulating the production of progesterone by the CL may take various forms. Studies in England led workers to conclude that melatonin can act directly on the CL to increase progesterone production and that this action may be related to the reported improved luteal function late in the breeding season after the prolonged exposure of ewes to melatonin.

GnRH treatment

Much interest in commercial dairy herds has centred around the use of GnRH between days 11 and 13 after breeding, which has increased pregnancy rates in many trials. It is believed that GnRH treatment suppresses follicular oestradiol-17β and $PGF_{2\alpha}$ secretion, possibly representing a physiological mechanism for an indirect anti-luteolytic effect of GnRH. Studies in horses have also shown an increase in pregnancy rates after buserelin (GnRH) administration 8–11 days after ovulation and AI, although there has been no evidence of secondary ovulations or luteinization of follicles after GnRH administration. Given during the luteal phase (day 10 after AI), Kanitz et al. (2003) recorded pregnancy rates of 48.4% for GnRH-treated animals and 36.4% for controls, the inseminations including frozen–thawed as well as fresh semen and lactating as well as maiden mares.

In sheep, there is some evidence suggesting that treatment with hCG or GnRH on day 12 of pregnancy can improve embryo survival. In Wales, workers found that hCG (200 iu) may improve embryo survival by stimulating luteal and embryonic growth, but there was less indication of buserelin (GnRH analogue) acting through a similar mechanism.

In pigs, a study conducted by Peters et al. (2000) on sows kept outdoors employed GnRH given at 24 h after first service or on days 11 or 12 after first service; they found no significant effect of treatment on farrowing rate to first service or on litter size. Elsewhere, however, a study reported on the

Table 1.7. Reproductive hormones in the female.

Name of hormone	Chemical description of molecule	Source of hormone	Target tissues	Main actions in the female
Gonadotrophin-releasing hormone (GnRH)	Glycoprotein	Hypothalamus	Gonadotrophs of the anterior pituitary	Release of FSH and LH from the anterior pituitary
Luteinizing hormone (LH)	Glycoprotein	Anterior pituitary	Ovarian cells (theca interna cells and luteal cells)	Induces ovulation, formation of corpus luteum and secretion of progesterone
Follicle-stimulating hormone (FSH)	Glycoprotein	Anterior pituitary	Ovarian cells (granulosa cells)	Follicle development and the synthesis of oestradiol
Prolactin (PRL)	Protein	Anterior pituitary	Mammary cells	Lactation, maternal behaviour
Oxytocin	Neuropeptide	Synthesized in the hypothalamus and stored in the posterior pituitary; synthesized by cells of the corpus luteum	Myoepithelial (smooth-muscle cells) in the uterus and mammary gland	Motility of the uterus, promotes the synthesis of prostaglandin $F_{2\alpha}$; ejection of milk
Oestradiol	Steroid	Granulosa cells of the ovarian follicle, placental cells	Hypothalamus, reproductive tract and mammary gland	Sexual receptivity, release of GnRH, secretory activity of the reproductive tract; uterine motility
Progesterone	Steroid	Corpus luteum and placenta	Endometrium of the uterus; mammary gland, myometrium; hypothalamus	Inhibits GnRH secretion, promotes the maintenance of pregnancy; involved in endometrial secretions; inhibits reproductive behaviour
Testosterone	Steroid	Cells of the theca interna	Granulosa cells; brain	Substrate for oestradiol synthesis
Inhibin	Glycoprotein	Granulosa cells	Gonadotrophs of the anterior pituitary	Inhibits FSH secretion
Prostaglandin $F_{2\alpha}$	Fatty acid	Uterine endometrium	Corpus luteum; myometrium of the uterus	Luteolysis; ovulation
Prostaglandin E_2	Fatty acid	Ovary, uterus and embryonic membranes	Early corpus luteum; motility in equine oviduct	Ovulation
Human chorionic gonadotrophin (hCG)	Glycoprotein	Chorion of the developing embryo	Ovary	Influences production of progesterone by the ovary
Equine chorionic gonadotrophin (eCG; PMSG)	Glycoprotein	Chorionic girdle cells	Ovary	Influences production of progesterone by the ovary
Placental lactogen	Protein	Placenta	Mammary gland	Mammary gland activity

effect of a GnRH analogue (rismorelin) given to gilts in the 12–18 day period after their pubertal oestrus; mated at the second oestrus, gilts showed an increased ovulation rate (18.6 vs. 15.7) and a higher total number of embryos and viable embryos than controls. Other work in the same American laboratory showed that rismorelin increased litter size only in high-prolificacy gilts; it was evident that genetic as well as other factors interact to affect the number of fetuses when ovulation rate is altered.

Heat-stressed animals

Some have attempted to use hormone treatments to assist conception in heat-stressed dairy animals. In the USA, Willard *et al.* (2003) found that supplemental administration of GnRH after AI (day 5 or 11) resulted in an improved endocrine status and ultimately greater pregnancy rates for cows synchronized and inseminated during summer heat stress. Although GnRH administration apparently provides a protective effect within the uterine environment to improve embryo survival, the precise mechanism by which the hormone exerts its effect is not fully understood. As a management tool for summer breeding, the use of GnRH appears to be beneficial and is similar to the results achieved elsewhere under typical production management conditions (Peters *et al.*, 2000). There have been other situations in which a programme of oestrus control and fixed-time insemination improved the fertility of dairy cows suffering heat stress. In Italy, working with Holstein cows during the summer, De Rensis *et al.* (2003) demonstrated that a protocol of fixed-time AI without oestrus detection was able to reduce summer infertility; it was believed that improvement in fertility was the result of enhanced folliculogenesis and oocyte quality after hormonal treatment.

Growth hormone and insulin

Some studies have sought to evaluate the effect of recombinant BST at the time of oestrus on progesterone concentrations and conception rates of repeat-breeding Holstein cows; it was shown that this treatment improved conception rate, the effect being greatest in cattle that had repeatedly failed to become pregnant to previous breedings. The effect was associated with an increase in circulating progesterone concentrations on day 18. The influence of exogenous insulin given before breeding on litter size has also been the subject of some studies; these have shown the total number of piglets born to be significantly higher after insulin treatment for 4 days, starting on the day of weaning.

Supplemental progesterone

Some workers have examined the effect of exogenous progesterone following mating on embryo survival in various of the farm animals, including pigs. It is known that high ovulation rates, such as may follow increased feed intake prior to mating (flush-feeding), are associated with an increase in embryo mortality. Although the reason for this negative association is not well understood, an inverse relationship between the level of feed intake and circulating progesterone concentrations has been demonstrated in pigs by several workers. It has also been demonstrated that supplemental progesterone administered during early pregnancy can enhance embryo survival in flush-fed gilts.

Oxytocin

Impaired uterine clearance of inflammatory debris is a significant contributor to persistent endometritis in the mare; persistent post-breeding endometritis, in turn, is a major obstacle to pregnancy. Enhancing mechanical clearance post-breeding by the use of oxytocin and prostaglandin has been used as a method of improving pregnancy outcome in mares susceptible to endometritis. A study by Nie *et al.* (2003) in the USA sought to determine if luteal function or pregnancy outcome differed between mares given oxytocin or prostaglandin (cloprostenol); results indicated that cloprostenol could be used to treat mares until the second day after ovulation without the treatment decreasing pregnancy outcome.

1.6.3. Trophoblastic vesicles

Much research attention in recent decades has been directed towards elucidating signalling mechanisms between the conceptus and the mother before embryo attachment (Goff, 2002). It is evident that interactions between the embryo and uterus are complex and essential for normal embryonic development and survival; it is believed that problems in the signalling mechanisms play a significant role in early embryonic mortality in view of the high loss of embryos that occurs during this period. During early pregnancy, embryos must inhibit the development of luteolytic mechanisms to maintain the secretion of progesterone necessary for continuing development; thus, embryos secrete IFN-τ, which acts locally within the uterus to inhibit luteolytic PGF secretion by inhibiting the development of oxytocin receptors on the luminal epithelium and by inducing development of a PG synthesis inhibitor. In terms of embryo mortality, there is evidence showing that poorly developed bovine embryos fail to inhibit oxytocin-induced PGF release, whereas well-developed embryos inhibit luteolytic PGF release, the presumption being that the poor embryos failed to produce sufficient IFN-τ. Nottingham workers have concluded that successful recognition of pregnancy in cows relies on an adequate degree of embryonic development and production of IFN-τ, which in turn depends on an appropriate hormonal environment, particularly in terms of an appropriate pattern of progesterone secretion after ovulation. It should also be remembered that the level of progesterone may be influenced by the LH surge and the quality of the CL which is established.

Trophoblastic vesicles (TVs) have been used in attempts to increase conception rate in cattle. It has also been shown that removal of the inner cell mass of bovine blastocysts by laser irradiation on day 7 is an effective method to produce TVs; such vesicles may be used in co-transfer (ET + AI) to increase conception rate. Japanese workers have shown that cattle TVs can be obtained from *in vitro*-produced embryos on a large scale and used for cryopreservation – such TVs are reported to support pregnancies when co-transferred with embryos.

Low-fertility cows represent a severe financial liability for the dairy farmer; where there is no mechanical obstruction in the reproductive tract, repeated failure to conceive may be due to problems in the maternal recognition of pregnancy. Work in Canada has shown that the use of *in vitro*-derived embryos may be an effective means of getting valuable repeat breeders in calf and ultimately in milk. They were able to demonstrate a pregnancy rate of 34% in cows bred four times without becoming pregnant; surprisingly, most calves were born to the AI rather than the IVP embryo, suggesting that the IVP embryo may increase the rate of AI-derived embryo implantation.

1.7. The Years Ahead

1.7.1. New opportunities and fresh challenges

Farmers have been effectively manipulating the genome of their animals for centuries, although their efforts were more an art than a science; they did, however, select superior individuals as progenitors for the next generations and thereby increased the population frequency of favourable alleles and allelic combinations at numerous quantitative trait loci (QTL). Recent decades have seen the emergence of a new scientific discipline, genomics, which results from the convergence of genetics, molecular biology, informatics and robotics, with the objective of analysing complex genomes.

Long-established research institutes, such as the Roslin in Scotland, set up to deal primarily with farm animals and improve livestock production, in the last two decades have seen a change in direction away from the farm and increasingly towards human

health care. In the UK, there has been a gradual but serious erosion of public spending
on research in agriculture and much of
the funding of institutes now finds itself
directed towards the biomedical area, much
of it focused on the use of transgenic technology to produce human therapeutic proteins in the milk of genetically modified
sheep and cattle. Inevitably, those engaged
in carrying out the research and in directing
research may not be expected to have the
same commitment to farming and farm livestock as once was the case and this at a time
when agriculture and those who make their
living from the land need all the help they
can get, not in increasing production but in
increasing the efficiency of farm systems.

Disease control needs appropriate support

Until 1996, the diseases of animals known
as transmissible spongiform encephalo-
pathies (TSEs), such as BSE, or 'mad cow
disease', and scrapie, were regarded as
farming problems with no known medical
implications for humans; this illusion was
rudely shattered when it became apparent
that BSE, in the form of new variant
Creutzfeldt–Jakob disease (nvCJD), could be
transmitted to humans. The occurrence of
'mad cow disease' (BSE), first diagnosed in
the UK in 1986, was a devastating blow to
beef consumption in that country, which
fell by 20% in a matter of months. The
disastrous effect of BSE on the image of
British farming and on the reputation of scientists working on problems in the farming
industry and on science generally, followed
by the devastating outbreak of foot-and-
mouth disease in the UK, serves to illustrate
the critical need to focus research on problems of animal health and to devise ways
and means of protecting the farmer and the
consumer from the impact of such diseases.

1.7.2. Maternal recognition of pregnancy

Much fertility research in farm animals has
been directed towards elucidating factors
involved in the maternal recognition of

pregnancy. In cattle, for example, it is estimated that 13–15% of pregnancies in dairy
cows are lost around this time, i.e. 14–19
days into pregnancy, probably due to
failure of the anti-luteolytic IFN secretory
mechanism. The availability of a recombinant IFN molecule may make possible the
development of commercial products to
improve pregnancy rates on farm. It is,
however, essential to understand why some
cattle embryos die because they fail to produce sufficient IFN; it is clearly necessary
to understand the genetic control of IFN
production in the embryo. If this is not
taken into account, there may be a risk that,
in assisting the survival of IFN-deficient
embryos, future dairy cows may have even
lower fertility.

The ruminant conceptus is free-living
in the uterine lumen until day 16 (ewe) or 20
(cow) of pregnancy. Maternal recognition
of pregnancy is achieved by secretion of
IFN-τ by the conceptus, inhibiting the
up-regulation of uterine oxytocin receptors,
which initiate luteolysis; inadequate early
growth of the embryo will result in the failure of this mechanism. Growth of conceptus
and placenta is influenced by the IGF system, which in turn is regulated by maternal
nutrition. While IGF-I within the uterine
compartment is apparently derived mainly
from the maternal circulation, IGF-II is produced by the maternal caruncles and the
fetal allantochorion. It is believed that, in
the later stages of gestation, the IGF system
provides a mechanism whereby the fetus
can alter placental growth to suit its own
nutritional requirements.

In red deer, experiments were undertaken to investigate the role of anti-luteolytic
IFN as a means of increasing the calving rate
after asynchronous ET; workers in the UK
were able to show the benefits of exogenous
IFN treatment in overcoming the failure to
establish pregnancy due to a lack of synchrony in embryo–maternal signalling. The
same group also reported that the IFNs were
closely related to the IFN-τ and IFN-ω found
in bovines and giraffes, showing > 85%
nucleotide sequence homology and > 74%
predicted amino acid similarity. There have
also been reports suggesting that the red deer

conceptus secretes an anti-luteolytic IFN for which the endometrium expresses a receptor during early pregnancy; the presence of IFN receptors in the hypothalamus and posterior pituitary would suggest the involvement of the central nervous system in the maternal recognition of pregnancy in deer.

1.7.3. Nutrition and reproduction

The development of diets to keep pace with the demands of increasing milk yields in dairy cattle is a major challenge to researchers in nutrition; the need is for regimes that will support milk production at an economic level and maintain the cow in good health and fertility. Much nutrition work concentrates on energy requirements, particularly since cows in early lactation are likely to be in negative energy balance. Reproduction can clearly be influenced by changes in energy balance, this can be caused by undernutrition occurring in extensive husbandry or when nutritional requirements are greatly increased in intensive husbandry. The main factors that link metabolism and reproduction, such as insulin, IGFs, glucose, leptin and neuropeptide-Y, act on the hypothalamic–pituitary axis by altering gonadotrophin secretion, as well as on the gonads by directly altering gametogenesis. Nutritional experiments also show that high levels of dietary protein are associated with low pregnancy rates, the problem apparently being due to rumen-degradable protein, with older cows being more susceptible than younger animals.

2

Artificial Insemination

Artificial insemination (AI) is used in animals ranging all the way from the honeybee to the elephant. Some of the milestone events in the development of AI technology are detailed in Table 2.1. The earliest reference to any form of AI is in 13th-century Arabic scriptures featuring the horse. The first systematic exploitation of this technology was also in horses, with the work of the Russian physiologist Ivanov at a government stud farm more than a century ago. Without doubt, AI has been the most

Table 2.1. Milestone events in the development of AI technology.

Year	Event	Researcher(s)
1677	Discovery of sperm by the use of a magnifying lens	Anton van Leeuwenhoek
1780	Artificial insemination of a dog bitch and the subsequent birth of pups 62 days later	Spallanzani
1803	Freezing of stallion sperm in the snow and motility recovered after warming	Spallanzani
1890	AI in horses first attempted in France	Repiquet
1899	Started work on horse AI at Moscow State University	Ivanov
1912	Demonstrated AI in horses, achieved results comparable to those obtained by natural service. Achieved success in cattle and sheep AI and trained hundreds of inseminators	Ivanov
1914	Start of work in Italy which led to artificial vagina for semen collection in the dog	Amantea
1920s and 1930s	Development in Russia of artificial vaginas for use in bulls, stallions and rams; development of simple diluents	Milovanov
1936	Shipment of ram semen from Cambridge in the UK to Poland; birth of lamb after AI	Arthur Walton
1937	Development in Denmark of the rectovaginal method of AI in cattle	Various Danish workers
1941	Development of egg-yolk citrate semen diluent for cattle	Glenn Salisbury
1946	Antibiotics (penicillin and streptomycin) used to control pathogenic microorganisms in semen used for AI	Almquist
1949	Method of freezing sperm of several species discovered	Chris Polge
1952	First calf born (Frosty I) after use of frozen–thawed bull semen in Cambridge	Chris Polge and Tim Rowson
1960	Liquid nitrogen became the refrigerant of choice for preserving bull semen. Most countries used 100% frozen bull semen	Many researchers in various countries

important reproductive technology applied
during the 20th century to cattle; unlike
technologies such as embryo transfer,
which in the cow calls for considerable
expertise on the farm and in the laboratory
to be successful, AI is relatively cheap and
simple to apply. Some reliable authorities
estimate that the contribution made by AI
to improved dairy production worldwide
since the Second World War was equal to
the combined contributions of better health,
husbandry and nutrition; the technique was
to greatly accelerate genetic selection, most
notably with dairy cattle.

Current usage in cattle

Current world statistics for AI in cattle
stand at 232 million doses of semen produced as a frozen product and 11.6 million
as liquid (Vishwanath, 2003). Fresh rather

than frozen semen is primarily restricted
to New Zealand, with limited amounts
used in Africa, Australia, France, Germany
and Eastern Europe (Table 2.2). Problems of
overproduction and lower profitability in
European Union (EU) countries have seen a
decrease in cattle AI usage in the recent
decade. In Ireland, for example, statistics
show AI covering only 37% of breeding
cows, markedly down from the figure
in the 1980s. On the other hand, in North
America, the use of AI by pig producers has
increased dramatically; estimates given by
Lamberson and Safranski (2000) speak of
increased usage from less than 5% in 1986,
to 30% in 1996 and to 50% in 1998. Such
increased usage is a reflection of increased
numbers of sows under confinement
conditions, the need to minimize disease
introduction into herds and the desire
of producers to use genetically superior
boars. However, it is also clear from many
reports that successful AI programmes in
pigs require skilled management to attain
optimal conception and farrowing rates.

The discovery by Chris Polge of the
cryoprotective properties of glycerol in the
late 1940s led to the widespread commercial
exploitation of frozen semen in farm animals, particularly in dairy cattle. Figures
given in Table 2.3 for 1998 show that
Europe produced the highest percentage
of all frozen bull/buffalo semen (48%),
followed by the Far East (28%) and North
America (19%). Although several potential
cryoprotective substances have been examined (e.g. dimethyl sulphoxide (DMSO),

Table 2.2. Overall impact of AI in cattle and
buffaloes (from Thibier and Wagner, 2000).

Regions	Total females of breeding age (40% of total cattle and buffaloes)	Total first-service AI
Africa	69,121,454	870,892
North America	45,206,360	11,203,880
South America	140,755,113	1,366,678
Far East	240,860,059	58,181,005
Near East	32,600,776	1,068,991
Europe	67,628,246	33,872,942
Total	596,172,008	106,564,388

Table 2.3. Fresh and frozen semen doses in cattle and buffaloes (from Thibier and Wagner, 2000).

Regions	No. of SCC	No. of semen banks	No. of bulls	No. of doses produced Fresh	No. of doses produced Frozen
Africa	18	161	646	55,204	1,484,850
North America	69	73	9,267	0	43,270,500
South America	71	138	530	0	5,917,269
Far East	188	644	9,228	8,874,920	63,938,027
Near East	17	124	268	16,794	2,559,640
Europe	239	455	19,803	2,694,903	115,176,785
Total	602	1,595	40,102	11,641,821	232,347,071

SCC, semen collection centres.

propanediol), glycerol was to remain the cryoprotectant of choice for sperm in all farm species; even so, the basis of its cryoprotective properties still remains somewhat unclear. In addition to the cryoprotectant glycerol, the basic composition of diluents for the freezing of semen are: (i) ionic and non-ionic substances that maintain osmolarity and provide buffering capacity; (ii) a source of lipoprotein or high-molecular-weight material to prevent cold shock, such as egg yolk, milk or soy lecithin; (iii) glucose or fructose as an energy source; and (iv) other additives, such as enzymes, bacteriostats, fungistats and antibiotics.

Much has been written on AI; the literature of the late 1990s contains accounts of the application of the technology in species ranging from cattle to camelids. Recent contributions to the literature include a review by Hopkins and Evans (2003), covering many AI topics: disease control through AI; semen collection; collection techniques; maintenance of the collection equipment; analysis of semen; semen dilution and cryopreservation; insemination procedures; and the ageing of gametes.

2.1. Advantages of Artificial Insemination

2.1.1. Cattle and buffaloes

AI was to have a major impact on cattle breeding schemes after the Second World War. The start of the 1950s saw the introduction in the UK of progeny testing to make efficient use of the possibilities offered by AI; since then, AI has enabled the large-scale progeny testing of bulls and the subsequent widespread use of those identified as being of superior genetic merit. In Nordic and Western European countries, AI in dairy cattle has been used extensively. On the technical front, the discovery of cryoprotectants, the freezing of semen using liquid nitrogen as the refrigerant and the introduction of the plastic straw as a semen container have been noteworthy

milestones. Freezing of semen opened the way to developments in international trade; in many countries, for animal health reasons, it resulted in the disappearance of fresh semen from the commercial scene. Improvements in the insemination procedure permitted AI sperm doses to be decreased and enabled the maximum number of inseminations to be performed with a single ejaculate.

It can be claimed, with justification, that the use of frozen semen revolutionized dairy cattle breeding; where once one bull was kept to breed 30–40 cows, it became possible to think in terms of an outstanding dairy bull siring thousands of calves in a year, with his semen being used in several countries simultaneously and for years after his demise. In economic terms, the widespread application of AI in countries such as the USA has resulted in a steady improvement in the genetic quality of dairy animals and a doubling of milk yields during the past 30 years. Throughout the dairy cattle populations of the Western world, the availability of AI was to lead to the virtual replacement of natural service by the technique. In countries such as India, state governments were able to support crossbreeding programmes with the semen of exotic breeds like Holstein–Friesian, Brown Swiss and Jersey. In many developing countries, however, the appropriate combination of factors to make AI widely acceptable in cattle has been much less evident.

Although, in most countries, frozen semen is exclusively employed in cattle AI (see Table 2.4), fresh semen continues to play a part in some places. Fresh semen has been appropriate for breeding cattle in New Zealand because of the very marked seasonal pattern in semen demand; nearly all cows are bred to calve within a 2-month period around mid-August. To meet such seasonal demands, New Zealand researchers have developed semen technology that enabled them to achieve acceptable conception rates with insemination doses as low as 2 million sperm per dose, in contrast to the conventional 12–20 million live sperm per dose used in most countries (Fig. 2.1).

Table 2.4.   Frozen semen usage in cattle (data refer to the year 1998).

Country	No. inseminations	% frozen semen	Sperm dose
Australia	1,600,000	100	25 million
Brazil	2,861,852	100	12–15 million
Canada	1,500,000	100	15 million
Denmark	787,828	100	15 million
France	4,800,000	90	20 million
Italy	2,450,000	100	18 million
Japan	2,173,456	99	20 million
Holland	1,659,496	100	Varies with bull
New Zealand[a]	3,800,000	37	1 to 2 million
Spain	1,800,000	95	30 million
USA	10,466,000	100	10–30 million

[a]New Zealand uses 1–2 million sperm for liquid semen in Caprogen during their restricted breeding season; the sperm dose for frozen sperm in that country is 10–30 million (from Vishwanath, 2003).

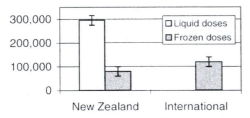
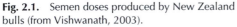

Fig. 2.1.   Semen doses produced by New Zealand bulls (from Vishwanath, 2003).

Beef cattle

The use of AI in commercial beef suckler cattle can be valuable, for example, in allowing the use of best linear unbiased prediction (BLUP)-tested sires to maximize the quality of the calf crop. The use of synchronized AI can also enable all cows to be bred on day 1 of the breeding season, thus permitting a more compact calving pattern than that possible with natural service. The use of AI in beef herds can also eliminate the risks of venereal diseases such as *Campylobacter fetus venerealis*. The problems associated with accurate heat detection in beef cows, especially with cows at pasture during the summer months, means that the application of AI is often limited to one synchronized service with fixed-time AI, followed by natural service for the cows that return to the AI. This system calls for the use of 'sweeper' bulls, which adds to costs and may carry disease risks. To eliminate the use of 'sweepers', it

is possible to resynchronize oestrus in cows that fail to conceive to first service, as demonstrated in Scotland by Lowman and associates at the Scottish Agricultural College. The practical outcome on one farm, where a triple oestrus synchronization protocol was applied, saw 75% of the suckler cows calving over a 4-week period and 90% over 46 days (see Chapter 5, Section 5.5.2).

Tropical cattle

The use of AI in cattle in developing countries, many of them in the tropics, can be particularly valuable in facilitating crossbreeding of zebu and taurine cattle. Some point to the need for more information on the optimum time for insemination relative to the onset of oestrus, bearing in mind that the duration of oestrus appears to be shorter in tropical cattle than in those raised under temperate conditions.

Buffaloes

AI is a prerequisite for any efficient breeding improvement programme in buffaloes. In contrast to cattle, however, returns to service in these animals are difficult for the herdsman to detect due to the less obvious external signs of oestrus. For that reason, the average number of services per conception reported for the buffalo may be higher than that recorded for cattle. Some workers have also recorded a significantly lower conception rate in heifers than in adults; this difference may be explained by the greater technical difficulty of passing the inseminating instrument through the cervix of the heifers. Although the principles and precautions taken in handling frozen buffalo semen and bull semen are similar, many reports suggest that frozen semen has not always met with the same success as in cattle; conception rates have been quoted as 50–60% with fresh semen and 25–45% with frozen semen.

The importance of AI in buffalo breeding has been discussed in a recent report from Italy, the author dealing with frozen semen production, milk recording, progeny testing, crossbreeding programmes and breeding schemes in which partners in Asia and Europe might cooperate. In Argentina, workers scored buffalo cows on the difficulty to pass the insemination gun through the cervix; buffaloes with high, medium and low difficulty scores had pregnancy rates of 0, 80 and 73.1%, respectively. Of interest is a new AI device (Ghent device) for uterotubal junction insemination developed in Ghent and assessed in breeding buffalo cows (Verberckmoes *et al.*, 2003); a preliminary trial with one-quarter of the standard insemination dose was conducted without a reduction in conception rate.

2.1.2. Sheep and goats

A comprehensive training manual on AI in goats and sheep by Chemineau and associates in France in the early 1990s covered the reproductive physiology of male and female animals and dealt with various factors (seasonal, environmental, genetic) that influence reproductive traits. Mention is also made of the housing and management of animals in AI centres, control of oestrus and ovulation, AI methods and methods of pregnancy diagnosis. A recent account from Brazil has dealt with the history of sheep AI in that country and refers to the currrent position and future prospects for the application of the technique in that country.

Artificial insemination plays an important part in sheep breeding in some regions of the world, particularly when used in conjunction with accurate progeny testing. The first serious efforts with sheep AI were in the former USSR in the 1920s and the number of ewes bred by this method increased to the point where 42–44 million were inseminated annually, representing 72–76% of all animals; in certain regions, 90–95% of ewes were bred by AI. Development of AI in Soviet Russia was part of a massive upgrading programme which started after the First World War, when Australian merinos were imported into the country. The AI procedures adopted at that time were uncomplicated and remained so over the years; ram semen was collected and used immediately, occasionally with a limited degree of dilution. Although Russia was the first to apply sheep AI on the farm, in terms of subsequent developments in ram semen technology it has not been associated with any notable advances. The artificial vagina for the collection of ram semen, first used in Russia, was one of the contributions to AI technology made in that country (Fig. 2.2).

After the Second World War, the sheep industry in the USSR made considerable progress, not only in expanding sheep numbers, but in moving rapidly towards a predominance of merino types by the extensive use of AI. Sheep were maintained on large farms carrying anything from 3000 to 60,000 sheep. The usual arrangement was for breeding ewes to be kept in flocks of 600–800; for AI, heat periods among ewes were checked by running eight to ten colour-marked 'aproned' rams with the flock and selecting out those marked each morning. About 70% of ewes were inseminated using freshly collected, undiluted semen, the insemination being carried out within 20–30 min of collection, using 0.05 ml volumes and sperm doses estimated at 120–150 million. The 'workload' of any one ram was usually no more than 400–600 ewes in a breeding season; conception rates to first service of 75–80% were claimed. In passing, it may be noted that the first successful insemination in sheep in Poland took place in 1935 using semen imported from Arthur

Fig. 2.2.   Collecting ram semen by artificial vagina.

Walton at Cambridge's Animal Research Station in England. This constituted the first successful instance of long-range breeding in farm livestock and was the result of Walton's pioneering efforts in AI over a period of many years.

With the exception of France, sheep AI has not been a common practice outside Russia in the other sheep-producing regions of the world, including Australia, New Zealand and America, North and South. This lack of application is due to several problems relating to the management of the ewe flock, the costs involved in AI and the handling procedures necessary for ram semen. At the same time, there are those in countries with small sheep populations, such as Ireland, who maintain that the availability of an efficient sheep AI service might yield similar benefits to those achieved in cattle and would greatly enhance the scope for pedigree sheep breeders to respond to consumer demands (less fat in the carcass, for a start). However, the fact that most sheep AI, as currently practised in Russia, is still based on fresh undiluted semen, as it was 80 years ago, serves to show that methods of dilution and frozen storage of semen

applicable in cattle do not necessarily hold true for sheep.

AI can be the means of accurately comparing the genetic merit of rams from different flocks, enabling genetically superior males to be identified. This can be achieved by sire-reference schemes, in which a small number of high-merit rams are mated via AI to a proportion of ewes in each of a number of ram-breeding flocks. This could be a very useful practical application of AI, but is dependent on achieving acceptable conception rates. In Australia, for example, AI has an important role in sheep breeding, particularly in conjunction with accurate progeny testing, such as sire evaluation schemes. The growth of sheep AI in that country has paralleled the development of technology for the use of frozen semen, but has remained limited by the low fertility achieved by intracervical insemination.

Although the introduction of laparoscopy for intrauterine insemination in the early 1980s has resulted in a steady increase in the number of ewes inseminated by this approach, the number remains small compared with the total number of sheep. Even so, the number of ewes inseminated has been a significant proportion of the genetically influential 'stud' ewe population for wool growers in that country. Finally, it should be kept in mind that there is much more to the introduction of sheep AI than the technical problems associated with the technique. There needs to be, for example, an organization within the country to offer a widespread and continuing service; in that regard, sheep farmers in France have been fortunate in having a national agency such as the Institut National de la Recherche Agronomique (INRA) to implement their AI programmes.

Goats

In France, AI in goats has an important role in the improvement of milk production systems, used in conjunction with progeny testing. The majority of goats are inseminated in advance of the breeding season, using deep-frozen semen after oestrus control with progestogen and pregnant mare's serum gonadotrophin (PMSG). AI is done on the understanding that it permits higher selection intensities for males and unbiased genetic evaluation by separating genetic (sire) and environmental (herd) effects and allows the monitoring of bucks for the selection for major genes, such as the alpha-sl-casein gene; the programme in France also aims at decreasing adverse effects from embryonic mortality and pseudopregnancy. As described in one report, since the mid-1980s AI has been carried out on a considerable scale in France; the use of fresh and

frozen semen, oestrus control and improvements in the insemination technique are covered by these authors. In 1997, the average conception rate recorded in 49,533 inseminated does was 63.4%; in very young females, however, the conception rates were regarded as too low and warranted further studies. In Argentina, one report described goat production systems, noting that proposals for the future of the goat industry in that country included a plan for an AI programme and possible embryo imports from Australia. Worldwide activity in sheep and goat AI has been described by Wagner and Thibier (2000), who conducted a survey in 1998; the overall number of semen doses produced, by regions, in sheep, goats and pigs, is given in Table 2.5.

2.1.3. Pigs

AI can be used in several ways in pig herds. It may be a matter of using fresh or occasionally frozen semen from an AI centre or it may be a matter of using AI with semen collected on the farm; it may also be a matter of using AI in combination with natural service.

Research in recent decades has established reproductive technologies which have become an integral part of the global pig industry; AI is an example of a technology that has continued to expand from its early use in European countries such as Denmark and Holland to the USA and

Table 2.5. Semen doses produced in sheep, goats and pigs (1998) (from Wagner and Thibier, 2000).

| Regions | Number of doses produced within the region in 1998 | | | | | |
| | Ovine | | Caprine | | Porcine | |
	Fresh	Frozen	Fresh	Frozen	Fresh	Frozen
Africa	800	65,000	0	1,147	0	0
North America	3,500	316	3,500	500	25,555,000	10,000
South America	0	0	0	0	765,630	0
Far East	2,000,000	13,590	300,000	312,494	3,500,300	0
Near East	3,100	8,446	0	5,313	400	0
Europe	4,679,184	348,200	14,910	66,100	35,158,320	5,900
Total	6,686,594	435,552	318,410	385,554	64,974,650	15,900

Canada, where it has been estimated by Day (2000) that a majority of the 6–7 million sows bred are artificially inseminated. AI has had considerable impact worldwide in its application to commercial pig production. In meeting demand for semen in North America, boar stud facilities have dramatically increased in number within recent years, with boar holding capacities approaching 800 at some studs. The purpose of such studs is to provide farms with a dependable, high-quality, freshly diluted semen product for the insemination of sows and gilts.

The application of AI in commercial pig production has varied from country to country according to factors influencing the market value of the end-product. In Greece, for example, AI in pigs is not widely used (this is apparently due to lack of knowledge and inability to eliminate the weaknesses of AI practice); workers in that country noted that boar semen processing is critical in influencing successful use of the technology. A paper by Leiding (2002) dealt with the current position and future for pig AI in Germany; in the year 2000, 4.5 million inseminations were carried out in the country, mainly by farmers and breeders, compared with 1.2 million in 1990. The same author considered that the future of pig AI in Germany was dependent on the quality of the resultant progeny.

In the USA, as noted by Seidel (2003), the use of pig AI has quadrupled in recent years due to the work of pig breeding companies that have developed much leaner pigs with superior carcass quality compared with the traditional breeding stock available from the purebred breeders. In the same country, Kelley (2003) described intrauterine insemination in which a special catheter is manipulated 10–14 inches deep into the sow's cervix, until it reaches the uterine cavity. The attraction of deep insemination is that it reduces the amount of sperm per dose. Disadvantages include the equipment and labour costs, effects in genetic pool and possible decline in reproductive performance and it cannot be used with gilts. New strategies for low-dose AI in sows, including deep intrauterine AI to optimize the use of

ejaculates and to introduce new sperm-based technologies into pig reproduction, have also been reviewed by Rath (2002).

Current methods of freezing boar semen generally involve cumbersome processing procedures, and inseminations are often followed by low fecundity rates and small litter sizes. The main advantage of using fresh liquid boar semen is that fertility is maintained even with low numbers of sperm in the inseminate. Using fewer than 1 million sperm cells per breeding unit can achieve conception rates with liquid semen similar to those with frozen–thawed semen containing approximately 15 million sperm; reduced fertility makes cryopreservation of boar sperm an uneconomical option for commercial usage. The high return rates associated with frozen semen are believed to be due to disrupted sperm function and structural integrity due to the freeze–thaw process; there is also the requirement for large sperm numbers and inseminate volume, which adds to processing and storage difficulties. Figures for pig AI conducted throughout the world in 1998 show that very few doses of frozen semen were used in that species (Table 2.6).

2.1.4. Horses

There are various advantages to using AI in horses, some of which are outlined in Table 2.7. The use of fresh or frozen semen, for example, can be one means of avoiding the need to transport mares over long distances to visit the stallion of choice. The technique can obviously be the means of markedly increasing the number of mares covered by a stallion in the stud season. Numerous factors are known to influence the pregnancy rate in horses bred by AI; these include the inherent fertility of the mare and stallion, the type of semen used for insemination (i.e. fresh, cooled–transported or frozen–thawed), the number of sperm in the insemination dose, the concentration of the diluted semen and the length of time liquid semen is stored prior to AI. In general, the equine AI industry has evolved a system in

Table 2.6. Fresh and frozen semen usage in sheep, goats and pigs (from Wagner and Thibier, 2000).

| | Number of first inseminations in 1998 | | | | | |
| | Ovine | | Caprine | | Porcine | |
Regions	Fresh	Frozen	Fresh	Frozen	Fresh	Frozen
Africa	400	20,400	0	920	16	0
North America	3,500	250	3,500	300	20,500,000	1,000
South America	0	0	0	0	308,030	306
Far East	1,307,900	458	195,000	196,154	1,997,300	1,700
Near East	2,392	1,400	0	1,900	400	0
Europe	1,854,068	136,705	12,510	57,308	17,521,547	3,804
Total	3,168,260	159,213	211,010	256,942	40,327,293	6,810

Table 2.7. Advantages of artificial insemination in horses.

Applications and advantages of a fresh semen AI service for horses
- Permits hygienic service with better disease control
- Allows better and more efficient use of the stallion, even in old age
- Avoids the need to transport mares, thereby saving time, expense and inconvenience, especially when mares have a foal at foot
- There is less risk of injury to mare and/or stallion
- Each time AI is used on a mare, her reproductive tract is examined, thereby helping the early detection and treatment of problem mares
- AI permits the breeding of abnormal mares that could not be used in natural service
- Semen quality can be assessed prior to each insemination
- May achieve an improvement in the pregnancy rates
- Opens the way to developing an international trade in stallion semen

Advantages of frozen semen
- Allows the indefinite preservation of stallion semen
- Would enable the establishment of a national semen bank
- Enables stallion to be used even several years after death
- Permits greater utilization of stallions in training by collecting and freezing their semen when they are temporarily inactive
- Would enable a major expansion in international trade

which mares are inseminated with fresh or cooled semen every 48 h until oestrus is no longer detected (usually 1–2 days after ovulation) or until ovulation is detected by transrectal palpation and/or ultrasonographic examination of the ovaries.

The use of AI in horses and trade in stallion semen increased worldwide in the 1970s and 1980s, although there have been several breed registries refusing to register foals born from inseminations with transported semen. The acceptance in recent years of frozen semen as a method of producing registered foals by two of the world's largest breed associations (American Quarter Horse and American Paint Horse) stimulated new interest in frozen semen technology; pregnancy rates achieved with frozen and liquid cooled semen in a commercial setting in the USA are regarded as acceptable. In the German riding-horse population, it was estimated that more than 60% of all coverings were performed by AI in 1999; reports from that country indicate that the correct application of AI can result in similar, and sometimes better, fertility compared with natural service.

There have been reports describing the practice of AI in French national studs; these noted that all AI techniques are more expensive than natural service; costs are given as 30 euros for immediate AI (< 5 min after collection), 120 euros for transported semen and 300 euros for frozen semen (including fees for scanning the ovaries). Suggested guidelines for utilizing stallions are as follows: one-third of mares served at collection place; one-third of mares within the regional area (with fresh semen < 12 h after collection); and one-third outside the region (using frozen semen).

The advantages and disadvantages of breeding mares with cooled transported semen are discussed by workers in the USA; their reports have provided a list of American breed organizations that permit use of this technology. The same authors also dealt with the fertility achieved by breeding with transported equine semen and the factors that may affect fertility when this technology is used. Common practice in the horse breeding industry has been to send two insemination doses (in separate bags within the same shipping container) for breeding with transported cooled semen, one to be used for an initial insemination upon arrival and the other held for a repeat insemination on the following day; however, there may be no advantage to this procedure and some workers have recommended placing all the semen in the mare's uterus as soon as possible after semen collection. Use has been made in the USA and Australia of a bioabsorbable implant of the gonadotrophin-releasing hormone (GnRH) agonist deslorelin (Ovuplant) to induce ovulation of large (> 35 mm) follicles during oestrus in mares. Accurate control of ovulation is regarded as vital when frozen semen is used; the Ovuplant device has been registered for use in the USA and Australia.

2.1.5. Poultry

AI was first applied to chickens but was later to have the greatest impact in the turkey industry, where all breeding birds are usually inseminated. Commercial broiler chicken breeders have yet to incorporate AI into their breeding programme. The advantages of breeding poultry by AI include: (i) accelerating genetic progress by allowing greater selection pressures; (ii) reducing the number of males used in breeding programmes; (iii) reducing the spread of reproductive pathogens; (iv) overcoming periods of low semen production due to disease or environmental stress; and (v) overcoming problems associated with natural mating (physical incompatibility in turkey strains).

Companies that develop and sell specific genetic strains of chickens (primary breeders) rely on natural matings and restrict the use of AI to pedigree breeding. Secondary breeders of broiler chickens rely almost entirely on natural mating. The lack of interest in AI by some primary and almost all secondary chicken breeders is probably due to the economic disadvantages of using the technique; these include the high cost of the skilled labour required for carrying out AI and the need to use caged rather than floor-reared breeders.

AI involves the collection and deposition of semen within the bird's reproductive tract; semen is prepared either for short-term use (24 h or less) or long-term (frozen) storage. Although some turkey breeders inseminate undiluted semen, most dilute the semen with an appropriate medium. Excellent sperm survival and fertility can be obtained when diluted turkey semen is stored for 6 h in a cool liquid state; however, liquid storage beyond 6 h or freezing of semen is unlikely to preserve the viability of turkey semen at the level acceptable for commercial use. The improvement of long-term liquid storage procedures for turkey semen is commercially important because production of this bird relies almost entirely on AI.

Long-term storage of avian semen at −196°C (liquid nitrogen) is achieved by freezing semen in a diluent containing a suitable cryoprotectant. Subsequent fertility in the chicken following AI of thawed sperm is reported to be about 60–70% whereas that of the turkey is generally much less (about 30%). Although the potential for worldwide distribution of genetic material exists, the commercial application of frozen semen technology has been limited primarily to the freezing of semen from selected genetic lines of birds for subsequent pedigree breeding. Frozen semen technology, however, has been used to preserve semen from endangered avian species. It is also important in preserving a suitably diverse genetic pool of avian semen which should help to counteract problems with inbreeding that may follow from commercial breeding programmes.

Many factors affect semen production in chickens and turkeys; there are wide differences in the onset of semen production and in semen quality between species, strains within species and individuals within strains. In general terms, chicken and turkey ejaculates obtained by abdominal massage average about 0.25 ml in volume and contain about 5000 million sperm/ml and 9000 million sperm/ml, respectively. Insemination involves the deposition of a predetermined volume (usually less than 0.1 ml) of semen with a minimum of 100–200 million viable sperm within the hen's vagina. Commercial AI operators inseminate 1000 turkey hens per man hour, with fertility maintained above 90% during a 22-week egg production season. In AI programmes, inseminations are repeated at weekly intervals in chickens and every 2–3 weeks in the turkey. When laying hens are inseminated with good-quality sperm, they lay their first fertile egg 48 h after the insemination.

Regardless of natural mating or artificial insemination, it should be noted that the mechanism of sperm storage in the chicken oviduct has remained a mystery since the 1960s, when sperm storage tubules were discovered between the shell gland and the vagina; it is these tubules which enable the hen to lay a series of fertilized eggs following a single insemination. Prior to the 1960s, it was known that only motile sperm could ascend the vagina and enter these tubules; how they remained viable for many days in that location remained a mystery. A study by Froman (2003) involved a synthesis of previous knowledge of sperm storage tubules with new knowledge gained from computer-assisted sperm motion analysis; this technique enabled a greater understanding of how chicken sperm motility is maintained. This synthesis of information represents the first coherent explanation of sperm storage in birds.

2.1.6. Deer and camelids

The cryopreservation of sperm combined with AI is the reproductive technology that has been most extensively applied to deer species. The frozen storage of semen permits semen samples of high genetic value to be stored for many years; this can be important commercially in the farming of red and fallow deer and in the breeding of rare or endangered species such as Père David's deer. An obvious advantage of frozen semen is as a method of moving genetic material across international boundaries and over long distances without the risks and costs of live animal shipments. For such reasons, the use of AI with frozen semen is expanding in deer species. Most procedures that are employed to freeze and thaw deer sperm have been modified from those for domestic ungulates (Soler et al., 2003).

There has been interest in countries such as New Zealand in the application of controlled breeding and AI in deer; this is usually aimed at increasing the utilization of superior sires. Deer hinds have been treated with controlled internal drug release (CIDR) intravaginal devices over a 14-day period and bred by intrauterine insemination, using frozen–thawed semen, at 65 h after progesterone withdrawal.

Old World camelids

Interest has been increasing in recent decades in the application of reproductive technologies to camels. According to Hafez and Hafez (2001), the functional anatomy of male reproductive organs and the reproductive physiology of dromedary and Bactrian camels are similar other than in the seasonal pattern of reproductive events; the same authors note that the control of sexually transmitted diseases is an important component of camel stud management to improve reproductive performance in this species. The Bactrian camel is a seasonally induced ovulator; factors in seminal plasma are believed to be responsible for triggering ovulation and the formation of a functional corpus luteum. In AI practice, ovulation can be successfully induced with frozen–thawed semen supplemented with GnRH and acceptable fertility rates achieved.

2.2. Growth and Development of AI Technology

2.2.1. Natural matings

During mating in cattle, the bull deposits millions of sperm in the anterior vagina of the cow; the cervix, however, is a major obstacle and the number of sperm eventually reaching the uterine body rarely exceeds 1%. This is the reason why, in cattle AI, semen is generally deposited directly into the uterine body, bypassing the cervix and permitting the use of a markedly reduced number of sperm. Vital aspects of Fallopian tube physiology in pigs have been discussed by Hunter (2002), who reviewed aspects of oviduct physiology that relate to successful fertilization and

early development of the zygote; other areas covered by the same author included glycoprotein secretions in the caudal isthmus, functional sperm reservoirs and controlled release of viable sperm.

Of the many millions of sperm deposited in the female reproductive tract at mating, few are destined to reach the ampulla at the time of ovulation. In most animals, the sperm reservoir in the caudal isthmus of the oviduct is created by the binding of sperm to oviductal epithelium; the sperm reservoir is believed to control sperm transport, maintain sperm viability and modulate capacitation (Fig. 2.3). It seems likely that the binding and modulation of sperm attached to the oviductal epithelium represent a mechanism for selecting functionally competent sperm and prolonging their lifespan by delaying capacitation.

It is known that transient inflammation is a normal occurrence after mating in the mare, serving to remove excess sperm, seminal plasma and contaminants from the uterus; resolution of this inflammation prior

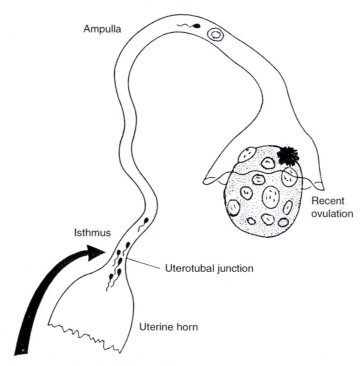

Fig. 2.3. Keeping sperm numbers low in the oviduct.

to the descent of the embryo into the uterine lumen is necessary for the embryo to survive. Mares that fail to clear the semen-induced inflammation from the uterus may develop a persistent mating-induced endometritis, which results in infertility. The cause of a delay in uterine clearance is not fully understood. According to one recent report, delayed uterine clearance is seen most commonly in pluriparous mares greater than 14 years of age; in such animals, the uterus may be located more ventrally in the abdomen, making them more predisposed to fluid retention.

Rapid sperm transport is important in the mare, for the post-mating uterus soon becomes a hostile environment for sperm; they need to reach the oviduct within 4 h of mating in order to survive and eventually fertilize the oocyte. In the mare, as in the other farm species, the dramatic reduction in the number of sperm reaching the site of fertilization is due to various checkpoints along the way. It is known that uterine contractions carry sperm towards the oviduct and at the same time eliminate excessive sperm; seminal plasma contains factors which cause an immediate increase in uterine blood flow. Sperm and seminal plasma are believed to provoke uterine contractions; sperm induce leucocytosis in the equine uterus by activating complement, and seminal plasma is known to have immunosuppressive effects in the uterus (Katila, 2001). Sperm in the uterus trigger an influx of polymorphonuclear neutrophils into the uterine lumen by activation of complement; seminal plasma appears to have a modulatory effect on the chemotaxis and migration of neutrophils.

After natural insemination or AI, a reservoir of sperm is established in the lower part of the oviduct (see review by Suarez, 2002); this reservoir serves to ensure successful fertilization by providing the appropriate number of sperm in the proper physiological state for fertilizing the oocyte after it enters the Fallopian tube. It appears that sperm are retained in the reservoir by binding to specific carbohydrate moieties on the surface of the mucosal epithelium of the oviduct; it is known that a bovine seminal plasma protein associates with sperm and confers on them the capacity to bind to the carbohydrate moiety. It is now clear that only a small, highly selected population of morphologically normal, motile bull sperm reach the site of fertilization, in the ampulla region of the cow's oviduct. This reduction in sperm numbers is important, since it represents the first mechanism to reduce/prevent polyspermia (penetration of the oocyte by more than one sperm). This sperm selection occurs regardless of how the sperm population is deposited, whether by natural service at the mouth of the cervix or in the uterine common body by conventional AI.

Lifespan of sperm in Fallopian tubes

In the cow or the ewe, there may be a day or more from the time of mating and the time of ovulation; in the mare it is a matter of several days and in chickens and turkeys it is a week and more. Clearly, sperm must have the capability of surviving for such periods and still be capable of achieving fertilization; several mechanisms are known to operate in the farm animal which help to maintain sperm in a fertile condition over a period of hours, days or even weeks. Attention has been drawn by Italian workers to the possible role of oviductal secretions, specific glycoproteins or cell contacts in maintaining the fertilization competence of sperm (Gualtieri and Talevi, 2003); it seems possible, for example, that the bull's ejaculate contains at least three different subpopulations: the first is made up of sperm still uncapacitated and unable to bind to oviductal cells; the second is composed of sperm able to bind, having undergone the early stages of capacitation; and, thirdly, there is a category of sperm in an advanced state of capacitation and are unable to bind. The Italian workers, exposing bovine sperm to an oviductal monolayer for 1 h, were able to split a sperm suspension into two subgroups; sperm that bound to oviductal cells showed a threefold greater competence to effect fertilization than those that did not bind.

2.2.2. Collection and processing of semen

Collection methods used in the farm animals vary according to the ejaculation patterns of the different species (Fig. 2.4). In the stallion, for example, before effective ejaculation occurs, a clear fluid originating from the accessory glands, principally from the prostate, is emitted as a pre-ejaculatory secretion. The ejaculate itself consists of several successive jets, the first of which contain high numbers of sperm and mainly epididymal secretions; this is followed by fractions containing decreasing sperm numbers and increasing proportions of seminal plasma. In some stallions, gel, derived from the vesicular glands, is secreted towards the end of the ejaculate with low sperm-count fractions. Semen can be collected either by closed artificial vagina models (for recovering the entire ejaculate) or open-ended models, which enable the operator to decide which fractions will be collected.

The open-model has several advantages, including the ability to avoid collecting bacteria-rich non-sperm ejaculatory fluid and gel; there is also evidence that sperm-rich semen fractions may tolerate cooled storage better than total ejaculates. Results of recent work in Finland demonstrated that semen of better hygenic quality and higher sperm concentration could be collected by the Equidame phantom than by the conventional artificial vagina; they recommended this device for semen collection when fractionation of semen is desired or when it is important to have low bacteriological contamination or to avoid a high proportion of seminal plasma (Fig. 2.5).

The observance of stringent animal health precautions is important at all stages, from semen collection until the time sperm is safely deposited in the female tract. Sources of contamination and recommendations to reduce the bacteriological content of boar semen have been discussed by Pinto *et al.* (2000) in Portugal; it is known that the presence of high levels of bacteria in boar semen has a harmful effect on sperm, reducing their viability and introducing a risk of infection for inseminated animals. Preputial washing and antibiotic treatment (e.g. streptomycin + penicillin + gentamycin) are routinely used in AI bulls to reduce bacterial load and increase the viable lifespan of sperm. It is well known that even short-term forms of stress can influence the quality of semen produced by bulls; studies have shown that the incidence of sperm abnormalities may increase significantly in bulls after dehorning and treatment with

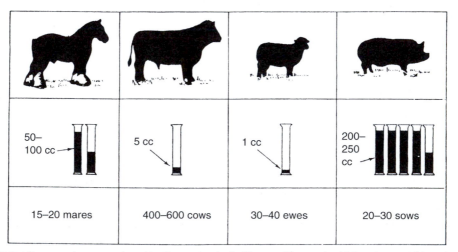

Fig. 2.4.   Semen volumes produced by farm animals. Stallions and boars give high-volume and low-density semen; bulls and rams give low-volume and high-density semen. The number of females that might be inseminated with a single semen dose is also indicated (after Hammond *et al.*, 1983).

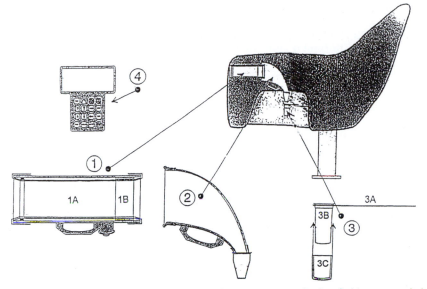

Fig. 2.5. The Equidame® phantom. The Equidame® phantom consists of a detachable open-ended artificial vagina (1A) and a rubber ring (1B) inside the vagina, a detachable metallic funnel (2), five detachable metallic holders (3A) and five plastic cups (3B). Small styrofoam shields (3C) envelop the plastic cups to isolate semen. Near the phantom there is a computer control panel (4) with a keyboard and a display for programming the phantom for semen collection (from Lindeberg *et al.*, 1999).

double injections of antibiotics compared with the pretreatment period.

Male management

Many strategies have been employed in the management of males in the semen collection process; management is important in maintaining the sex drive (libido) of the male and can also influence the quality of the ejaculate (Prado *et al.*, 2002, 2003). Studies in France in the 1990s dealt with data for 771 rams and 55,360 ewes which showed that ejaculate volume was significantly affected by technician, time of collection (a.m. or p.m.), pattern of collection frequency, week of collection and ram. The conception rates in ewes were significantly affected by AI technician, ram, age of ewe, fertility in previous breeding season and interval from lambing to AI. In Ireland, studies showed that the presence of an oestrous rather than a non-oestrous teaser ewe significantly increased total sperm number collected, by 17% (Table 2.8); similar data are available for bulls.

Table 2.8. Effect of teaser on sperm output in rams.

Ram breed	No.	Oestrous teaser Total sperm output (sperm/ ejaculate × 10)	Non-oestrous teaser Total sperm output (sperm/ ejaculate × 10)
Suffolk	5	2.10	1.85
Texel	5	1.63	1.23
Dorset Horn	5	2.29	2.02
Total	15	2.01	1.70

Collection in camels

Methods used in conventional farm mammals for semen collection are not suitable for recovering semen from camels. In Saudi Arabia, Al-Eknah *et al.* (2001) adopted a new approach in which a semen collection area was constructed to permit semen collection from underneath the male camel; their approach facilitated safe collection and provided full view of the ejaculation process.

Hormonal stimulants

Hormonal therapy may be used on some occasions as an aid to semen collection In the USA, for example, Estienne and Harper (2000) examined the effect of prostaglandin $F_{2\alpha}$ ($PGF_{2\alpha}$) on the training of sexually active boars (i.e. boars experienced in natural mating) to mount an artificial sow for semen collection; use of PG facilitated the training of such boars to mount and significantly reduced their reaction time (elapsed time after entering collection pen until the start of ejaculation).

Ambient temperature diluents for pigs

The difficulties encountered in the cryopreservation of boar semen in the early 1950s led to the examination of alternative procedures. The short-term preservation of boar semen in various diluents increased during the 1960s; this was a period marked by the development of several fresh semen diluents that were adapted for the commercial use of pig AI. The pig industry subsequently made extensive use of fresh semen stored for 2–4 days at ambient temperature. As with diluents employed in other farm animals, pig semen diluents are designed to afford protection against temperature and pH changes, increase semen volume, provide a source of nutrients for the sperm and inhibit bacterial growth. The compositions of some fresh semen diluents are given in Table 2.9. A sperm dose of 2000–3000 million per insemination is commonly used, in a diluent volume of some 70 ml. Of the various diluents employed in commercial pig AI, one that has often been used is based on ethylenediamine tetra-acetic acid

(EDTA) and usually referred to as the Kiev diluent; this diluent proved popular because it was simple to prepare and sperm could be stored for up to 3 days without any appreciable loss of fertility. Problems and procedures in the liquid storage and freeze–thawing of boar semen are dealt with in several reports (Johnson *et al.*, 2000; Paulenz *et al.*, 2000).

Assessing male fertility

The assessment of male fertility is usually based primarily on semen evaluation, using conventional parameters such as sperm motility, semen density and the incidence of sperm abnormalities. The extensive use of AI technology makes it essential to use semen assessment parameters which correlate with sperm fertility. The relationship between the percentage of morphologically normal sperm and fertility has been widely described in farm animals and humans. However, finding a laboratory test reliable enough to predict the potential fertility of a given semen sample or a given bull for AI is not possible and not considered as likely to be possible in the future.

Semen evaluation has long been used by animal scientists in dealing with the selection of males for both natural and assisted reproduction; the development of a field test to accurately predict the performance of semen has been the objective of many researchers. In recent decades, the introduction of automated sperm morphology analysis systems has permitted objective measurements of sperm dimensions to be made. In Spain, workers have reported a computer-assisted approach which has shown a high degree of objectivity and

Table 2.9. Composition of some boar semen diluents.

Diluent	Semen life (days)	Ingredients
Kiev	3	Glucose, sodium citrate, sodium bicarbonate, EDTA
BTS	4–5	Glucose, sodium citrate, sodium bicarbonate, EDTA, potassium chloride
Zorpva	5	Glucose, trisodium citrate, sodium bicarbonate, EDTA, TRIS, citric acid, cysteine, PVA
Reading	4–6	Glucose, trisodium citrate, sodium bicarbonate, EDTA, potassium chloride, TRIS, citric acid, cysteine, PVA, trehalose

repeatability in accurately and precisely measuring sperm heads; they suggest that the application of computer-assisted sperm morphometry could be useful in improving assisted reproduction techniques in farm animals.

In New Zealand, workers also reported an accurate method of assessing ram sperm head morphology, using computer-assisted morphometry analysis. Fertility prediction has economic merit, as observed by Roudebush and Diehl (2001) in the context of pig production systems; these authors demonstrated that platelet-activating factor (PAF) is present in boar sperm and that levels are significantly higher in individuals with a high farrowing rate status and high number of piglets born; it was suggested that the mean level of PAF in the ejaculate may be a useful parameter for predicting fertility.

Predicting male fertility

Sperm motility and morphology are two of the criteria used to evaluate bulls for breeding soundness; present methods of bull evaluation are reliable for detecting bulls that have the potential for high fertility and those that do not. There is a well-established relationship between age and semen quality in bulls; reports show that the concentration and motility of sperm may be lower in intermediate and old bulls than in young Holstein bulls (young = 2–3 years; intermediate = 5–6 years; old = > 9 years). However, many bulls may be difficult to categorize because of the unknown importance of certain sperm aberrations. Defects such as the knobbed acrosome defect have been reported in many breeds, including Friesian, Charolais, Simmental, Maine Anjou, Salers, Hereford and Normande; bulls severely affected may be virtually sterile and semen with a high proportion of knobbed acrosome defects is likely to be of low fertility. In this context, it may be noted that *in vitro* fertilization (IVF) may be a useful research tool. It has been shown that knobbed sperm completely fail to penetrate the zona pellucida, and that normal sperm coexisting with knobbed

sperm may also be functionally deficient, often giving rise to poor-quality embryos. For such reasons, it is probably not wise to compensate for the knobbed sperm defect by increasing the sperm dose in the belief that more normal sperm are available for fertilization.

The cattle AI industry has long been in need of an objective, rapid, but inexpensive method of evaluating frozen–thawed bull semen ejaculates; much work has been devoted to developing stains for assessing sperm and measuring the integrity of the sperm plasma membrane and the status of the acrosomal membrane. Fluorescence staining has been suggested by several workers. In the context of stallion fertility, Colenbrander *et al.* (2003) see flow-cytometric analysis of sperm with appropriate stains as offering considerable promise. Finnish researchers reported a new fluorescence method that used an automatized fluorometer and fluorophore stain, propidium iodide, which stained cells with damaged membranes. In a study involving more than 92,000 inseminations, they showed that there was a 3.9% higher non-return rate per cent (NRR%) for Holstein–Friesian cattle than for Ayrshires. They also showed that there was clear seasonality in the NRR%, which was highest in summer/autumn and lowest in winter.

Diluents

Studies over many years have been directed towards formulating diluents which can improve the fertility of sperm used in AI. A wide range of media has been used, some of them unusual; one such was coconut milk, which turned out to be surprisingly effective in tropical countries. Most of the early diluents were developed for use in cattle AI. Two important considerations in the cattle AI industry relate to the control of non-pathogenic, environmental contamination of semen and total quality management; the preparation and composition of semen diluent affect both of these. Complete preparation of semen diluents is usually carried out in the AI laboratory; the preparation is labour-intensive and occasionally human errors may result in diluents of variable quality. There is also a desire to eliminate any possibility of contaminants from animal products; the bovine spongiform encephalopathy (BSE) problem in cattle served to emphasize the need for caution in this area. For such reasons, ready-to-use, manufacturer-guaranteed diluents available on the market may minimize certain of these problems; as an example, a diluent concentrate containing soybean extract as a substitute for egg yolk is currently used in some semen diluents.

2.2.3. Storage and cryopreservation of semen

Cattle

With increasing technical efficiency in the cattle AI industry came the development of an international trade in semen and genetically superior animals. The internationalization of dairy cattle breeding led to substantial changes in many commercial dairy cattle populations into which foreign genes have been imported; examples of rapid progress in the breeding of high-quality dairy cattle are usually found in countries that have looked beyond their borders to find the best genetic material available. Various improvements in semen diluents and in the freezing and thawing of semen, together with better AI routines, particularly in the timing of insemination in relation to expected ovulation, have made it possible to decrease the number of sperm in the AI dose, thereby permitting the maximal number of inseminations to be performed from a single bull ejaculate. It is known, however, that the freeze–thaw process, even with optimal protocols, reduces the functional bovine sperm population by 50% or more – and that there is a close relationship between the number of viable sperm and fertility after AI.

In cattle, there is evidence that certain steps in the cryopreservation of bull semen result in the production of toxic reactive oxygen species (ROS); the freeze–thawing of bull semen has been shown to cause a strong reduction of sperm intracellular antioxidants such as glutathione (GSH) and superoxide dismutase (SOD). Control of the level of ROS to promote sperm motility and survival of sperm by adding antioxidants or by using conditions that reduce oxidation have been successfully tested for the preservation of unfrozen bull semen; adding antioxidants such as α-tocopherol and ascorbate has a protective effect on metabolic activity and the cellular viability of cryopreserved bull sperm. According to Bilodeau *et al.* (2001), the addition of thiols to diluted semen should be considered. It is thought that, when the concentration of sperm is low and high post-thaw motility is critical, dilution of semen by a diluent leads to a low level of GSH and other antioxidants, which may partly explain the lower fertility of frozen–thawed semen in comparison with fresh semen. A report by Foote *et al.* (2002) dealt with 17 laboratory studies and two field trials that were conducted with 174 semen collections from bulls in an AI station; they found that certain combinations of antioxidants may be useful for those processing bull semen using an egg-yolk-based diluent. It is known that bull sperm, before they are frozen, are exposed to aerobic conditions during processing, but have little endogenous antioxidant to protect them against ROS.

The quality of frozen semen is obviously an important factor determining conception rate in cattle under field conditions. Some of the factors involved in the successful freezing and thawing of bull semen are detailed in Table 2.10. It is believed that the release of toxic substances by dead and abnormal sperm may seriously affect the fertilizing capacity of their companion cells. It may be noted that semen cryopreservation procedures (dilution, cooling, freezing/thawing) induce capacitation-like changes in sperm. This cryocapacitation is believed to be partly responsible for the reduced fertility of frozen–thawed bull semen. A better understanding of the events involved in both capacitation and cryopreservation could lead to a substantial improvement in the quality of frozen–thawed semen (Cormier and Bailey, 2003).

Table 2.10. Successful freezing and thawing of bull semen.

Factor important in cryopreservation	Commonly used conditions or substances
Processing semen promptly	Immediately following collection
Macromolecules to protect sperm against prefreeze cold shock	Buffered egg yolk or heated milk
Cooling rate and prefreeze time	Cool in 1–2 h; prefreeze time varies with extender and species
Special cryoprotectant	Usually glycerol
Freezing rate	Varies: about 10 min from +5°C to −100°C
Storage temperature	−196°C in liquid nitrogen
Thawing rate	Usually thaw at 30–37°C

Filtration of sperm

It has been found that the surface charge is changed after the death of a sperm and this has led to the use of various filtration procedures based on this fact. Using a Sephadex ion-exchange filter, for example, it has been possible to improve the initial and post-thaw motility of Holstein bull sperm as well as increasing the number of sperm with intact plasma membranes and normal acrosomes. The effect of Sephadex filtration has also been favourably reported by those dealing with buffalo semen; workers in India found sperm motility and survival to be higher and the percentage of abnormalities to be lower in Sephadex-filtered semen than in untreated semen or semen from which seminal plasma had been removed.

Reducing sperm damage during freezing

Although damage to sperm can occur at any stage during the freeze–thawing process, it is known that sperm are particularly vulnerable when undergoing freezing itself and when cryoprotectants are removed. In Colorado, Graham and Purdy (2002) found that increasing the cholesterol content of bull sperm membranes altered sperm membrane physiology, permitting greater numbers of cells to survive the freeze–thaw process; the same authors suggest that such treatment may improve the fertilizing potential of frozen sperm and may permit the use of fewer sperm in an insemination dose.

Novel freezing techniques

Workers in Israel have developed new techniques to improve freezing and vitrification of sperm, oocytes and embryos (Arav et al., 2002); they described a novel freezing technology based on their 'multi-thermal-gradient' freezing apparatus, which they claim has the ability to control ice-crystal propagation. By changing the thermal gradient or the liquid–ice interface velocity, it is possible to optimize ice-crystal morphology during freezing of cells and tissue; using such apparatus the Israeli workers have been able to freeze bull, stallion, boar, ram, fowl and human sperm with normal post-thaw motility/pre-freezing motility of 70–100%.

Buffaloes

In buffaloes, there is ample evidence that the composition of the diluent in which semen is diluted before freezing plays a major role in the successful cryopreservation of buffalo sperm. Substances of high osmolarity, such as glycerol, protect sperm during the freezing process and energy-rich compounds, such as pyruvate, provide extra energy during capacitation and fertilization (Fabbrocini et al., 2000). A study by Ahmad et al. (2003) showed that a semen filtration system containing Sephadex ion-exchange columns effectively removed immotile, dead and abnormal sperm from a diluted low-quality buffalo semen ejaculate; the filtered semen showed improved freezability characteristics.

Sheep

Although AI with frozen ram semen by cervical insemination gives unacceptably low fertility rates, frozen semen in 0.25 ml straws (100 million/ml) has given good fertility results using laparoscopic intra-uterine insemination (Anel *et al.*, 2003). Although techniques for the insemination of sheep with freshly collected semen have been available for many years, the development of freeze–thawing methods has been slow and procedures still leave much to be desired. In many comparative trials, the fertility of frozen–thawed ram semen after cervical insemination has been much below that of fresh semen; this reduced fertility has been attributed to impaired sperm transport through the cervix, resulting in failure to establish adequate numbers in the sperm reservoirs that are known to exist in the ewe's reproductive tract. It is, however, now well established that the insemination of frozen–thawed semen directly into the uterus can result in a fertilization rate much the same as that of fresh semen; such evidence is usually taken as supporting the view that the basic problem is one of establishing adequate numbers in the sperm reservoirs (cervical and isthmic).

Workers in Australia found that the treatment of frozen–thawed ram sperm with seminal plasma significantly improved fertility after AI; it was surmised that this may have been due to its decapacitation effect, combined with a substantial improvement in sperm transport. Further reports from the same Australian group suggested that seminal plasma might help to protect ram sperm from the adverse effects of the freeze–thawing process (Maxwell and Evans, 2000).

Integrity of sperm plasma membrane

The integrity of the sperm plasma membrane is of crucial importance for the functioning of the sperm cell; freezing and thawing are detrimental to this plasma membrane. There are many different methods for freezing and thawing, which may affect the sperm membrane in various ways. In Sweden, workers froze stallion semen in 0.5 ml straws and showed a significant difference (percentage live sperm) after thawing at 37°C compared with that at

75°C. It was concluded that stud farms using frozen semen should thaw the straws at 37°C rather than 75°C; the lower temperature was easier to work with, as thawing at the higher temperature requires special equipment and has to be timed very carefully to avoid damage to the sperm.

There is ample evidence from several farm mammal species to show that the freezing process destabilizes the membranes of sperm and transforms them into a more advanced capacitation-like state than those freshly ejaculated. In Australia, workers reported data indicating that freezing causes membrane changes in ram sperm that are functionally equivalent to capacitation. It is probable that such functionally capacitated sperm may react differently from fresh sperm in the female reproductive tract. There is also evidence indicating that fresh and frozen–thawed ram sperm behave differently within the ewe tract, suggesting that this may have implications for the timing of AI when frozen–thawed sperm are used in intrauterine insemination. Many studies have focused on the effect of the freeze–thawing process on the membrane integrity of sperm. In Sweden, for example, studies reported on the effect of different thawing procedures on ram sperm integrity.

Freezing goat sperm

Freezing procedures for use in goats have been described by French workers; these procedures involve removal of seminal plasma by washing sperm as soon as they are collected and before dilution and freezing. It was earlier found that an enzyme produced by the bulbo-urethral (Cowper's) glands catalysed the hydrolysis of lecithins in egg yolk to fatty acids and lysolecithins, which are toxic to sperm. The implication of this discovery was that egg-yolk diluent should not be used for semen conservation unless sperm are first washed. There are reports showing that successful preservation of goat semen requires removal of seminal plasma and that dilution with skimmed milk could result in higher conception rates than dilution with egg-yolk buffers.

Sensitivity of boar sperm

After the successful use of glycerol in the freeze–thawing of bull sperm, the same approach was attempted with boar semen. It soon became evident that boar sperm responded to cooling and freezing quite differently from bull sperm; the addition of glycerol or cooling to temperatures below 15°C led to a marked reduction in sperm survival. It is now clear to all that differences in the sensitivity of mammalian sperm to cryopreservation make it almost impossible to extrapolate well-established techniques from one animal species to another. Although, by the mid-1970s, commercially applicable freezing procedures had been developed in several countries, it was clear that frozen semen had much reduced fertility, even when much larger sperm numbers were inseminated compared with fresh semen. For such reasons, freeze–thaw procedures are currently used commercially on a very limited scale, usually for the international movement of boar semen.

The current procedure for AI in pigs using frozen–thawed semen involves the infusion of a high number of thawed sperm (5000–6000 million) in a large volume of fluid (80–100 ml) into the cervical canal. Despite using such high sperm numbers, there appears to be insufficient colonization of the oviductal sperm reservoirs in the sow's tract, resulting in inadequate numbers of functionally intact sperm in the oviducts to ensure a high fertilization rate. For that reason, AI in pigs with frozen–thawed semen often results in conception rates below 20%, far below the rate achieved with fresh semen; this clearly influences the use of frozen pig semen in commercial pig AI programmes.

A paper by Thurston et al. (2003) examined various freezers and reported on a controlled-rate freezing machine developed in their laboratory; they showed that preserving boar sperm using one freezer (Watson) improved post-thaw semen quality. One way of avoiding the need for high sperm numbers is to place appropriate numbers of thawed sperm near the utero-tubal junction; in recent years, several research groups have described methods for non-surgical deep intrauterine catheterization in sows, which allows the transcervical deposition of semen into the uterine horns. A study by Roca et al. (2003) in Spain has indicated that application of deep intrauterine insemination can provide acceptable fertility in weaned sows using a relatively low number of frozen–thawed sperm.

Longevity of stallion sperm

There are those who believe that the longevity of sperm in the oviduct may be particularly important in the horse; there was one report from the 1940s suggesting that fertilization may occur several days after breeding the mare. As with other farm species, it is believed that contact of stallion sperm with oviductal epithelial cells maintains their viability in the oviduct. There have been various reports dealing with the evaluation of stallion semen before and after freezing; these suggest that semen characteristics after freezing/thawing, the hypo-osmotic swelling (HOS) test and sperm motility after storage at 4° or 37°C are the best indicators of semen quality. The HOS test, applied just after semen collection, can be a useful predictive test for the freezability of stallion semen. Further work in the same laboratory showed that the addition of glutamine to the INRA82 diluent improved post-thaw motility and may provide the means of dealing with those stallions considered to give 'unfreezable' semen (post-thaw motility < 35%). According to French workers in a recent report, current research investigations on chilled semen tend to favour storage at temperatures above 4°C (15 and 20°C) and the use of antioxidants in semen diluents. For semen freeze–thawing, the use of an amino acid (glutamine) in addition to glycerol in the diluent and the modification of the lipid content of sperm membranes by addition of liposomes or cholesterol carrier are recommended.

2.2.4. Insemination procedures

In the 1940s, vaginal or shallow cervical insemination in cattle, performed with the aid of a vaginal speculum, was replaced by deep cervical or intrauterine insemination, involving the technique of cervical fixation *per rectum*; this method proved more efficient and was rapidly adopted as standard by the cattle AI industry (Fig. 2.6). Based on the assumption that the deposition of semen nearer to the site of fertilization (ampulla of the oviduct), the next logical step in cattle AI was to attempt deep AI; however, when this was done, it was found to have little effect on the success of AI. Deep intracornual deposition of semen in cattle was first proposed in the late 1950s to increase the efficiency of AI, to reduce the required number of sperm per insemination dose and to enhance the use of sperm from genetically superior bulls that were in demand.

More recent times have seen a resurgence in interest in deep inseminations. A study in Estonia reported by Kurykin *et al.* (2003), for example, showed that with deep intracornual insemination of PG-synchronized heifers at a fixed time (80–82 h) after the second of two PG injections, a sperm dose of 2 million sperm was as efficient as a dose of 40 million. It should be noted that intracornual AI requires greater care on the part of the inseminator because of the risk of perforating the uterine wall due to the tonicity of the uterus at oestrus and the danger of rupturing the ovulatory follicle while palpating the ovaries *per rectum* to determine the probable site of ovulation.

AI by laparoscopy

Although there is ample evidence showing that a single laparoscopic insemination in sheep with frozen–thawed semen can result in acceptable conception rates, its use in commercial practice is limited by its expense and even more so by animal welfare considerations. Among the alternative insemination procedures studied is that of transcervical AI, which involves the passage of an instrument through the cervix and into the uterus; unfortunately, this technique, quite apart from not being possible with all sheep, has serious welfare implications because of the manipulations of ewe and cervix involved.

Fixed-time AI for sheep and goats

For any thought of applying AI commercially under small farm and flock conditions, the need for oestrus detection must be eliminated and all members of the selected group inseminated at a predetermined hour. This calls for precise control of oestrus and ovulation, using oestrus control technology, not only among cyclic ewes in the autumn breeding season but also in sheep during other seasons of the year when they would be in anoestrus. In the 1970s, methods were developed which enabled the application of a new fixed-time approach to sheep AI (Fig. 2.7). France was the main country at the forefront of such developments; by 1991, 740,000 ewes (representing 10% of all breeding ewes) were inseminated in that country, nearly all after applying oestrus control technology (see Chapter 5). The French sheep AI industry has been based

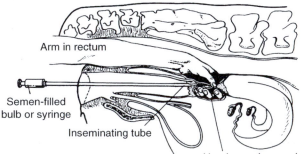

Fig. 2.6. The rectovaginal method of inseminating the cow (after Hammond *et al.*, 1983).

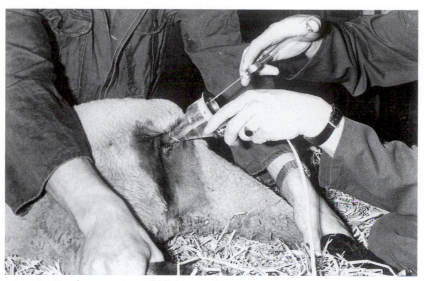

Fig. 2.7. Inseminating the ewe using cold-light speculum.

on distributing fresh semen, used on the day of collection, from more than 20 centres around the country; demand for AI in milk- and mutton-sheep in France is generally concentrated in the spring/early summer period.

Trials conducted at the sheep AI centre of the Roquefort Confederation were detailed by French workers in a mid-1990s report. They reported reductions in sperm doses from 500 million to 380 million sperm in the 1970s and continuing efforts in the 1990s with 30,000 ewes in which a further reduction in sperm dose was made. In Scotland, workers dealt with data recently obtained for 2072 Suffolk ewes in 25 flocks spread throughout the UK that were partici- pating in a national Sire Reference Scheme. In this programme, sheep were inseminated at 50–56 h after withdrawal of sponges (45 mg fluorogestone acetate (FGA)) and PMSG treatment (400 iu); although concep- tion rates to AI averaged 60–70% in most flocks, their results showed that a small but significant number of ewes were not actually in oestrus at the time of insemination, apparently due to a persistent corpus luteum.

In France, a widely employed oestrus control treatment for goats uses an FGA vaginal sponge left in place for 11 days and injections of $PGF_{2\alpha}$ and PMSG 2 days before sponge removal. Conception rate in goats after oestrus control and fixed-time AI may be adversely affected by the fact that, in some herds, 20% of females may be pseudo- pregnant at the start of the synchronization treatment. It also appears that repeated oestrus control treatments (progestogen– PMSG) may result in an immune reaction, resulting in delayed oestrus. Studies by workers in France showed that the negative effect of repeated PMSG treatment on subse- quent fertility in alpine goats was due to a humoral immune response involving the major histocompatibility complex.

Transcervical AI in sheep

In sheep, numerous studies have been reported on a variety of insemination pro- cedures. The tortuous nature of the cervix in the ewe prevents routine transcervical intrauterine AI; laparoscopy, laparotomy or special equipment and procedures are used typically to inseminate sheep by the intra- uterine route, but most sheep producers have neither the skills nor the equipment required for this procedure. As mentioned earlier, several studies in sheep have exam- ined the possibility of depositing semen in the uterus by transcervical insemination;

all too often, results have been disastrous. According to one report, numerous trials in sheep have been carried out in New Zealand using the Canadian transcervical technique, but with only 30% conceiving; the same article mentions a high incidence of cervical damage and stress as a result of this insemination procedure. In the USA, one recent study examined the possibility that the hormone oxytocin could be used to induce uterine tetany and cervical dilation prior to AI; workers reported a pregnancy rate of 38% for laparoscopic AI and a 0% value for transcervical insemination. Other studies in the same laboratory showed a 93% fertilization rate after laparoscopic AI and 28% after transcervical insemination, with no evidence that oxytocin assisted the latter process. In Scotland, workers in the Scottish Agricultural College used oxytocin to dilate the cervix prior to AI but found

that this did not permit complete cervical penetration and had an adverse effect on conception rate.

AI in pigs

Steps in the insemination procedure in pigs are shown in Fig. 2.8. Stimulation of the sow during AI is quite different from that she would receive during a natural mating. It is known that myometrial contractility is an essential component in the fertilization process because it is a mechanism by which boar sperm are transported to the site of fertilization. For such reasons, several research groups have sought to stimulate certain physiological events (e.g. uterine contractility) associated with breeding, using hormonal supplements to the pig semen dose; such products have included oestrogen, oxytocin, prostaglandin and

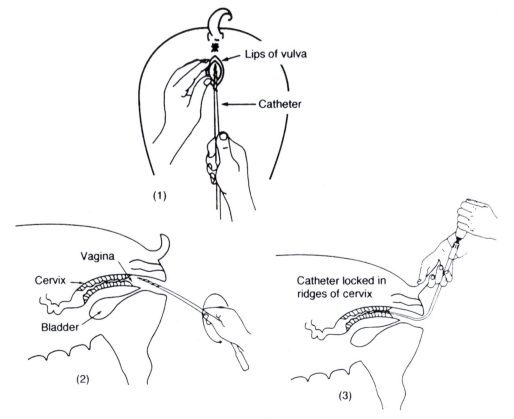

Fig. 2.8. Stages in the insemination of the sow (after Glossop, 1991).

relaxin analogues. Where $PGF_{2\alpha}$ has been added to a semen dose at the time of AI, there has been evidence of significant improvements in fertility.

Hormonal additions to diluents

There is evidence that the addition of PG to diluted boar semen may increase the reproductive performance of pigs bred by AI. Although the benefits of adding PG have not been apparent on all farms, there have been reports of increases in conception rates and in total number of piglets born alive, generally ranging from 1 to 20% and from 0.1 to 1 pig per litter, respectively. Cheng et al. (2001) found that diluted boar semen does not inactivate PG within 72 h of supplementation; for that reason, PG can be added to diluted boar semen at the time of processing rather than being added to semen just prior to insemination. A trial by Horvat and Bilkei (2003) in a large indoor Hungarian production unit with repeat breeder sows also suggested that $PGF_{2\alpha}$ could improve fertility when added to AI semen compared with normal insemination. The mode of action of PG as a semen additive is not well understood (Maes et al., 2003). Apart from PG as a semen additive, it is also known that administration of $PGF_{2\alpha}$ at the time of AI may increase both farrowing rate and litter size in pigs.

AI in horses

An AI dose for mares consisting of 500 million progressively motile sperm is considered 'standard' by most clinicians. The outcome of AI in mares inseminated between 6 h before and 6 h after ovulation was recently reported by workers in Italy, using frozen–thawed semen; pregnancy rates 14 days after AI were 38.0% in the pre-ovulation group and 38.3% in the post-ovulation group. In carrying out AI, the operator, using a sterile plastic sleeve, guides the inseminating catheter into the uterus by way of an index finger inserted in the cervix (Fig. 2.9). The mare's cervix is shorter and simpler in structure than that of the cow; the cervical canal is open throughout the oestrous cycle but closed during pregnancy.

AI in deer

Various insemination techniques have been studied for use in farmed deer. In the USA, working with three species of farmed deer (fallow, red and white-tailed deer), Willard et al. (2002) examined the site of semen deposition (vaginal, cervical or uterine) and

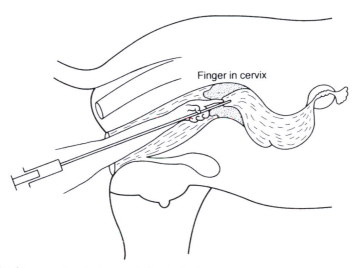

Fig. 2.9. Artificial insemination in the mare (after Foote, 1974).

the time taken for transvaginal AI using a standard speculum-guided insemination 'gun'.

Low-dose inseminations

Although the effects of low-dose inseminations on fertility, as a preliminary to using sexed sperm, have been frequently studied in heifers, information for lactating dairy cows is more limited. In Finland, workers evaluated pregnancy results after AI with a dose of 2 million frozen–thawed sperm compared with 15 million; for most bulls, the 2 million sperm dose was too low and the average pregnancy rate dropped by 15% (Andersson *et al.*, 2003). The same workers noted a slight trend for better results when the 2 million sperm dose was deposited into the uterine horn ipsilateral to the site of impending ovulation.

In the current methods available for AI in pigs, thousands of million sperm are used per insemination in a large volume of liquid (70–100 ml), which is deposited through the cervix into the uterus at insemination. Approximately 90% of the sperm inseminated cannot be recovered from the uterus within an hour or two of insemination. Only about 10,000 sperm reach the uterotubal junction and about 1000 reach the sperm reservoir in the caudal isthmus, in which the sperm cells can survive without reduction in their fertilizing ability until the time of ovulation. It is believed that sufficient sperm to ensure subsequent fertilization are established in the isthmus reservoir within 1 h of mating. The rapid loss of sperm is caused mainly by the back-flow of semen in the first few minutes after insemination and by intensive uterine phagocytosis by polymorphonuclear leucocytes that invade the uterus within 2 h of insemination.

Taking such factors into account, procedures in horses and in pigs have been examined that involve insemination of small numbers of sperm close to the uterotubal junction to establish a sperm population in the isthmus reservoir that is sufficient to ensure optimal fertilization. A paper by Morris *et al.* (2000) dealt with the hysteroscopic insemination of small numbers of sperm at the uterotubal junction of preovulatory mares, which achieved fertilization as effectively as with natural service (3–15 thousand million ejaculated sperm) and with the accepted minimum sperm dose (500 million) used for conventional uterine-body insemination in mares. According to these authors, the simplicity of the technique offered a practical means of exploiting new breeding programmes that require very small numbers of sperm in horse breeding. In Spain, Martinez *et al.* (2001) demonstrated that endoscopic non-surgical deep intrauterine inseminations can be performed quickly in sows and that normal farrowing rates and litter sizes can be expected using a small number (100-fold reduction) of porcine sperm. The authors concluded that non-surgical deep uterine insemination could have a high economic impact on the use of fresh and frozen semen in the AI industry in pigs and could help in making sexed semen a commercial possibility in this species.

2.2.5. Do-it-yourself insemination (DIY-AI)

There has been a large increase in the sale of frozen semen directly to the dairy and beef farmer over the past 25 years, with herd owners assuming responsibility for semen purchase, handling insemination of the cow and record-keeping. In the USA, as long ago as 1980, it was estimated that more than half of the cows bred by AI in that country were inseminated by herd-owners or stockpersons. In the UK, it is possible for a farmer to be licensed to carry out inseminations in his herd after attending an approved training course. As the costs involved in setting up DIY-AI can be substantial, those farmers holding licences tend to have larger than average herds. Conception rates achieved by DIY-AI are likely to be a matter of numbers and experience. In Ireland, it is recommended that farmers and stockpersons using DIY-AI should undertake a refresher course at least every second year. However, despite the perceived or potential advantages of DIY-AI, comparisons show that farmers may not obtain higher conception rates than those achieved by professional inseminators in matched herds.

With many dairy farmers around the world introducing DIY-AI into their herds, efforts have been made to evaluate the fertility and financial impact of initiating a DIY-AI programme in comparison with keeping a bull. Factors that influence the successful outcome of AI include the semen used, time of year, proficiency of the operator, age of cows and farm effects. Some workers have found little or no difference between owner inseminators (DIY-AI) and technicians employed by cattle breeding centres; others have shown a 5% increase in conception rates over the first 3 years after initiating DIY-AI, presumably due to the acquisition of greater skill during this period. In Ireland, a move towards increased use of DIY-AI in larger herds was initially associated with a significant fall in calving rate; farmers are now encouraged to monitor their performance using NRR% and enrol annually in a refresher course. There are certainly those who believe that one reason for suboptimal reproductive performance in commercial dairy herds lies in the high proportion of semen which is now sold directly to producers and the fact that professional inseminators have been replaced by less clinically qualified personnel; as mentioned elsewhere in this text, current figures indicate that the pregnancy rate after a single AI service is rarely higher than 50%, which is far removed from the 60% or higher commonly recorded in the 1960s.

DIY-AI in sheep

In Sweden, AI in sheep is generally conducted on a DIY basis, often using ram semen frozen and packed in German mini-tubes (0.25 ml) or in French mini-straws, which are longer and have a smaller diameter compared with the mini-tubes. Since the temperature in the sheep barn in Sweden is often below 0°C during the breeding season, this can present problems in semen handling. A recent study by Swedish workers indicated that using a somewhat lower

temperature over a longer period (50°C, 9 s), rather than 70°C for 5 s, could facilitate the use of frozen–thawed semen under farm conditions in Sweden.

2.2.6. Measuring effectiveness of AI

It is only to be expected that much information on fertility levels in cattle is to be found in the records of cattle AI centres, although this generally takes the form of non-return rates (NRRs) rather than actual calving rates. The 30–60 days NRR% is frequently used; this represents the proportion of cows that failed to return to service by 30 days after the end of the month in which they were inseminated. The NRR overestimates the actual proportion of cows that are pregnant and takes no account of cows that are bred to bulls rather than submitted for reinsemination. However, as a statistic widely employed throughout the cattle AI industry, it can permit valid comparisons of bull and cow fertility, inseminator efficiency and other factors influencing the efficiency of this breeding method. In horses, the validation of NRR as a parameter for stallion fertility was attempted in one study using data from 6361 matings/inseminations; it was concluded that the NRR at 28 days was a valuable parameter for early assessment of fertility of stallions.

2.3. Semen-sexing Technology

The scientific literature contains an abundance of accounts claiming varying degrees of success in separating the two sex-determining types of sperm in humans and domestic animals; convincing evidence of success was less easy to find until recent years. Although differences in physical, biochemical and immunological properties of X- and Y-chromosome-bearing sperm have been suggested by many workers, the only certain difference appears to be those contributed by the sex chromosomes themselves. Of historical interest, it may be noted that semen-sexing by flow-cytometry was actively researched in the early 1980s at the National Institute of Medical Research in London and a semen-sexing patent was granted to that institution in 1986, covering Europe and Canada (Fig. 2.10).

In the USA, the sexing technology devised by Johnson at the US Department of Agriculture (USDA) Beltsville research centre was patented in 1989, the technology subsequently being licensed by the USDA for use in mammals in two categories: humans and animals. In humans, the Genetics and IVF Institute in Fairfax, Virginia, was granted a licence to use the sexing patent in 1994 and workers have already used the technology in assisted human reproduction. The animal licence was granted to Animal Biotechnology Cambridge (ABC) Ltd (later to Mastercalf, a subsidiary of ABC) in 1992. A second licence was granted to the Colorado State University Foundation. Mastercalf was purchased by XY Inc. of Fort Collins, Colorado and that company is, in association with various partners, such as

the Cogent Cattle Breeding Company in the UK, currently attempting to commercialize the sexing method in various countries.

2.3.1. Factors influencing the sex ratio

Sex is determined in farm mammals by the sex chromosome content of the sperm produced; females are produced by gametes containing an X-chromosome and males by sperm carrying the Y-chromosome. The sex ratio in a litter of piglets at birth (the so-called secondary sex ratio) is determined by the primary sex ratio in the litter (the sex ratio immediately after fertilization) and eventual selective prenatal mortality. In several farm and other mammals, variation has occasionally been reported in the sex ratio of offspring, which is apparently influenced by genetic and environmental factors. The way in which chromosomes segregate at meiosis in the testes ensures

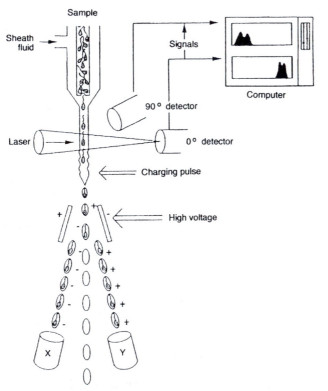

Fig. 2.10. Diagrammatic representation of flow cytometer.

that sperm carrying X- and Y-chromosomes are produced in equal numbers. There are, however, several interesting aspects to the actual ratio of males to females that are born under some conditions; it may not always be a matter of equal numbers of each sex. It now seems probable that, in species living in socially structured herds or flocks, females in good body condition or of high social rank produce more male than female offspring; increased dietary energy appears to be the factor that skews the sex ratio in favour of males. In deer, the suggestion is that male fetuses may be more vulnerable than females to the mother's nutritional stress (arising from high population densities or other causes of food deprivation) because of faster male growth rates *in utero*; male fetuses require more nutrients and so may well be more adversely affected by food restrictions.

Others who have speculated on the cause of sex ratio skewing suggest that one sex may signal its presence to the mother more strongly; alternatively, the uterine tract environment may favour embryos of one sex more than the other (Roberts *et al.*, 2002). It is known that expanded female cattle blastocysts produce about twice as much interferon (IFN) as male blastocysts. It is also evident from various studies that male embryos have a greater ability to survive in a glucose-rich medium; the uterine glucose environment may be greater in well-fed cows, thereby providing male embryos with a survival advantage. The higher production of IFN by female blastocysts may be a factor in their survival in a glucose-rich uterine environment. Data presented by Roberts *et al.* (2003) emphasize the high sensitivity of embryos, particularly females, to glucose; one possible explanation is that female embryos are compromised because of the presence of two transcriptionally active X-chromosomes, which causes imbalance in glucose metabolism.

2.3.2. Sorting technology

The semen-sorting technology currently applied commercially by Cogent in the UK is based on work at the American Beltsville Agricultural Research Center. It has been recognized for some time that the X-chromosome is larger and carries more DNA than the smaller Y-chromosome; in domestic livestock the DNA difference between X- and Y-bearing sperm varies from 3.5% to 4.2%. The first calves born from semen sexed by the Beltsville technique were produced by IVF. Clearly, far fewer sperm are required for fertilization when IVF is employed, although the techniques require modification to take account of the reduced motility and viability of sorted sperm (Zhang *et al.*, 2003). According to Galli *et al.* (2003c), the use of sexed semen for IVF would increase if an efficient intracytoplasmic sperm injection (ICSI) procedure could be employed.

Although there is a genuine interest among many cattle farmers in semen sexing, not all researchers and commercial concerns have expressed enthusiasm about sexing bull sperm by flow-cytometry; there are those taking the view that the high cost and lower pregnancy rates associated with flow-cytometry sorting make the approach impracticable for widespread use. For such reasons, some commercial concerns have supported work which seeks to detect sex-specific differences in sperm surface antigenicity in cattle. One approach by a Canadian biotechnology company was based on the assumption that bull and boar sperm have sex-specific proteins on their surface and that these can be separated using appropriate antibodies. Under this scenario, addition of male antibody would permit X-chromosome sperm to be filtered out, without causing cell membrane damage, enabling the sperm to be used fresh (pigs) or frozen (bull) in the normal way. Although some Canadian researchers expressed optimism about developing a viable immunological sperm-sexing procedure, elsewhere researchers have shown reservations on the possibility of identifying membrane proteins specific to X- or Y-bearing sperm. In Brazil, Matta *et al.* (2001) reported testing a monoclonal antibody against a male-specific protein for sexing semen, claiming that there was a cytolytic effect on male gametes and no effect on female cells; they reported almost 80% of female embryos after IVF. It is probably wise never to say that the seemingly impossible cannot happen.

2.3.3. Advantages of sex control

Cattle

For cattle producers the commercial availability of sexed semen may well be a factor improving animal welfare. If dairy farmers were able to use heifer-producing semen, this would be one way of ensuring that their maiden heifers produce smaller calves, likely to be born much more easily than bull calves. The same farmers would be able to avoid the birth of dairy bull calves, which otherwise might well be destined for early slaughter. For selective breeding purposes, the availability of sexed bull semen will enable young dairy sires to be proved much more efficiently, using their sperm to produce a preponderance of heifer calves.

One cattle production possibility that could be greatly facilitated by the use of sexed semen would be the single-sex bred heifer system of beef production; in this the birth of heifer calves could be the means of maintaining the system on a continuous basis, each female being slaughtered at a young age after producing a calf to replace herself. When a dam is slaughtered shortly after her first calf is weaned, there is a marked increase in biological efficiency because the dam herself assumes the role of slaughter offspring and most of the conventional maternal overhead cost of producing a calf disappears by becoming part of productive growth. As noted by Seidel (2003), the main challenge of this production system is to get all the key life-cycle events – pregnancy, calving, weaning and 3 months of fattening after weaning – fitted into a period not exceeding 28–30 months; in the USA, carcasses are severely discounted after 28 months of age. In Ireland, available evidence on meat quality has indicated that there is essentially no difference in quality between once-calved and maiden heifers; however, convincing the meat trade may be a different story. A summary of possible advantages of sex control in cattle is detailed in Table 2.11.

Horses

There could be interest among certain sectors of the horse breeding industry in the prospect of being able to produce foals of the desired sex. Gender plays an important part in determining the success of horses in several areas of their sporting activities. As observed by Allen and Antczak (2000), geldings are favoured for 3-day events and mares preferred for polo playing, to take but two examples. Using sperm sorting, foals of a predetermined sex could be produced either by the embryo transfer approach or by low-sperm-dose AI. In Colorado, Buchanan et al. (2000) compared pregnancy rates in mares inseminated with reduced sperm numbers in the tip of the uterine horn (horn ipsilateral to the preovulatory follicle); breeding with a 25 million

Table 2.11. Possible advantages of sex control in cattle production.

Dairy cattle
  More heifer progeny from genetically valuable cows as herd replacements for milk production
  Heifer calves from maiden heifers for easier births
  More bull progeny for beef production, especially from 'cull cows'
  Ensuring birth of bulls as potential sires from best cow × sire matings
  Avoiding freemartins in multiple births

Beef cattle
  More bull calves for beef rearing
  Ensuring bull progeny as potential sires from best cow × sire matings
  Ensuring birth of heifer replacements from genetically good cows
  Avoiding birth of freemartins in multiple births

progressively motile sperm dose resulted in a pregnancy rate of 57%.

Pigs

There could be useful commercial advantages in the use of sexed semen in the pig world. Preselecting the sex of piglets could have a significant impact on pig production; the application of sexing technology to genetic programmes could accelerate genetic progress. The production of all-boar litters in pig production units could mean that half the pigs would automatically become 15% more efficient, there would be no need for 'split-sex' feeding, nutritional requirements could be met more accurately and the end-product could be made more uniform. The obvious need is for an insemination protocol that establishes pregnancies using low sperm concentrations; researchers have been able to use surgical intratubal insemination to deposit porcine sperm directly into the oviduct but this is not practicable on the farm. What is required is a procedure that provides for non-surgical deep intrauterine insemination with a low sperm dose. In Spain, Vazquez *et al.* (2003) achieved farrowing rates of 25 and 77% using 70 million sorted and non-sorted sperm, respectively, and 32 and 81% for 140 million sorted and unsorted sperm, respectively, in post-weaning sows hormonally treated for oestrus control.

Sheep

Trials reported by David Cran in Scotland demonstrated, for the first time, that pregnancy in sheep could be established and lambs of predetermined sex obtained by the deposition of a low dose of sorted semen (flow-cytometer technique) close to the uterotubal junction of the ewe.

2.3.4. Problems and prospects in semen sexing

Although current sperm-sexing technology is expensive, which severely limits the extent of commercial applications, those working in the field expect that much simpler equipment, specifically designed for sperm sorting, will become available within a few years and that this will permit more extensive use of the technique on the farm. In the meantime, it is known that the ability of sperm to withstand sorting procedures may be strongly influenced by the individual bull (Zhang *et al.*, 2003). For those engaged in sperm sorting, it can be important to identify those males whose sperm most readily tolerate the sexing procedure. Using conventional field trial methods, literally hundreds of inseminations are required to obtain reliable fertility data. However, some have shown that a heterospermic method may be used to test the *in vivo* fertility of bull sperm rapidly (Flint *et al.*, 2003); a test requiring the insemination of about 60 females was able to provide information, based on the genotype of embryos, that was able to identify bulls of relatively high or low fertility.

2.4. Future Developments in AI Technology

Despite the long history of AI and its successful application to cattle breeding programmes, there are still areas of semen technology in which improvements can be made. Rejection of ejaculates, due to low motility, before or after freezing, is still common in AI centres; some bulls that apparently present a normal sperm picture and have an acceptable NRR of 60–70% may still yield ejaculates of which more than 50% are unsuitable for use.

2.4.1. Sperm survival at ambient temperature

Fresh semen diluted *in vitro* remains viable for a limited period of time; much research has been conducted over the years with bull sperm to understand the various factors which may influence sperm survival in a non-frozen state. Despite the formulation of numerous diluents, the fertile lifespan of sperm is maintained for only 3–5 days when stored in a liquid environment at high dilutions and at ambient temperatures. In contrast to their limited lifespan *in vitro*, sperm remain viable for several weeks when stored in the cauda section of the epididymis; here they are quiescent, presumably to conserve energy required for the fertilization process. Clearly, conditions in the cauda epididymidis favour the survival of sperm; this is a part of the epididymis characterized by low pH and hyperosmotic pressure, with sperm stored at high density, at a warm temperature and under low oxygen tension. Bearing such facts in mind, some researchers have attempted to mimic these epididymal conditions *in vitro*. In Belgium, De Pauw *et al.* (2003a) demonstrated that the simple action of coating bull sperm for less than 5 min with egg-yolk diluent during collection had a marked effect upon several sperm characteristics after 4 days of storage in a simple salt solution at pH 6 and 300 mOsm/kg; such results may be encouraging for future improvements in the preservation and fertilizing ability of bull sperm. In a further paper, De Pauw *et al.* (2003b) showed that their sperm-coating technique could preserve bull sperm characteristics and the penetrating capacity of fresh bull sperm stored in egg-yolk-containing diluent for up to 6 days.

2.4.2. Sperm encapsulation

Microencapsulation of bovine sperm may have a useful part to play in future developments in cattle AI, especially in beef cattle. Microcapsules with a semi-permeable membrane, designed to release sperm over an extended period of time, may be a means of ensuring that they are in the cow's oviduct at the optimal time for fertilization to be achieved. In other words, microencapsulation may provide the means of reducing the importance of the timing of insemination.

Various researchers have given due consideration to modifying factors such as sperm dose and sperm packaging so that fertilization might be more readily achieved after a fixed-time AI schedule. In this category is research on the encapsulation of sperm, which is aimed at providing sperm in the animal's reproductive tract over a much greater time span than is possible with conventional frozen–thawed semen. Ways and means of increasing the lifespan of sperm in the reproductive tract of the cow are of practical interest. Sperm encapsulation technology has the potential to increase the lifespan of sperm at body temperature and permit the progressive release of viable sperm over several days. It may be possible to increase sperm lifespan during micro-encapsulation by adding agents that stabilize membranes, inhibit peroxidation and decrease calcium uptake. For many years, the way in which chicken sperm survived for days in the hen's oviduct remained a mystery. Clearly, avian sperm are capable of remaining viable and functional for several days in the sperm storage tubules in the oviduct; sperm encapsulation technology may eventually be able to do much the same for farm mammals.

2.4.3. Cryopreservation and freeze-drying of sperm

Although millions of sperm are normally used to inseminate the different farm-animal species, in all cases, only a minute fraction of these sperm reach the site of fertilization. There are those who believe that the differences among species in the ability of their sperm to survive freeze–thawing is related to their tolerance of osmotic stress (Guthrie *et al.*, 2002). These workers see it as critically important that the osmotic behaviour of sperm be determined and that cryopreservation protocols are adjusted to make it possible for sperm with appropriate motility and survival ability to be inseminated. It is likely that future research will continue to be concerned with semen storage (Vishwanath, 2003; see Table 2.12).

Although it has been known for many years that mouse sperm do not freeze well, it now appears that an alternative approach to long-term storage in this species may lie in freeze-drying. The production of mouse

Table 2.12.   For and against fresh and frozen
semen storage.

Liquid-stored semen	Frozen–thawed semen
Advantages	
Low sperm numbers	Long-term storage
High sire utilization	Flexibility of use
Inexpensive storage	
Ease of use in the field	
Disadvantages	
Limited shelf-life	High sperm numbers
	Expensive to store

pups from freeze-dried sperm has recently
been reported by Japanese researchers; they
used low temperature and pressures to
remove water from the sperm and stored
them at 4°C for periods of up to 3 months. In
reconstituting the gametes, it was simply a
matter of adding water, removing the sperm
heads and injecting them into mouse
oocytes; of 57 oocytes injected, 54 survived
and 49 began development *in vitro*. The
transfer of 46 embryos to the uterus of recipient mice resulted in the birth of 14 normal
young. Freeze-dried mouse sperm are
motionless and are not fertile in the conventional sense; however, ICSI can enable them
to achieve fertilization. In farm animals, a
study by Keskintepe *et al.* (2002) was the
first to report the production of apparently
normal cattle blastocysts after the injection
of frozen–thawed bull sperm that were
selected, freeze-dried and stored at 4°C until
use. Other work in the USA, this time by Lee
et al. (2003a) working with pigs, has demonstrated that cytologically dead boar sperm,
which had been freeze-dried, were capable
of fertilizing oocytes by way of ICSI and of
delivering exogenous genetic material for
the production of transgenic pig embryos.

Success with freeze-drying led to the
exploration of desiccation as a preservation
protocol for mouse sperm storage. Desiccation offers the possibility of entire handling
and subsequent storage of sperm at ambient
temperatures, significantly reducing the
complexity of operations associated with
low-temperature preservation. In the USA,
Bhowmick *et al.* (2003) explored the feasibility of convective drying as a method of
desiccation. Such convective drying, using
an inert gas (nitrogen), may offer a simpler
and less expensive alternative to freeze-
drying. The workers desiccated mouse
sperm to < 5% moisture at ambient temperature; after overnight storage at 4°C, they used
desiccated sperm to produce live fetuses via
ICSI.

2.4.4. *In vitro* spermatogenesis

In the past decade, there has been considerable progress in the development of techniques that may eventually result in the
ability to completely recapitulate spermatogenesis *in vitro*. Sperm production involves
complex endocrine and autocrine/paracrine
regulation of germ cell proliferation and
differentiation. Such studies will enable the
molecular mechanisms regulating the process of spermatogenesis to be determined.
In practical terms, *in vitro* spermatogenesis
may provide the basis for treating certain
forms of male infertility in humans and of
genetically modifying the male germ line
in farm animals. A review by Parks *et al.*
(2003) provides an overview of male germ
cell and testis development, the process of
spermatogenesis *in vivo* and how an understanding of the *in vivo* process may help in
achieving sperm production *in vitro*.

3

Embryo Transfer

The first embryo transfer (ET) dates back to Walter Heape more than a century ago in England. Many years were to pass before this novel form of reproductive technology was to reach the farm; when it did, it was predominantly in the breeding of cattle. For the last 30 years of the 20th century, ET came to play an increasingly important role in the genetic improvement of dairy cattle in many countries. The technology has been used in many aspects of reproductive research since the 1950s.

3.1. Advantages of Embryo Transfer

3.1.1. Cattle and buffaloes

Significant commercial expansion of cattle ET really began with the introduction of non-surgical flushing in the early 1970s and grew further with the introduction of simple non-surgical transfer techniques towards the end of that decade. Although the surgical transfer of embryos, in suitably experienced hands, probably remains the method of achieving the highest pregnancy rates (70%), thought of any extensive use of the technology is only feasible using a non-surgical transfer technique essentially similar to that used in routine artificial insemination (AI) (Fig. 3.1). It is also important for animal welfare considerations to use non-surgical rather than surgical procedures.

ET in cattle is an example of a reproductive technology introduced into commercial practice to meet certain specific problems in cattle breeding that arose in the early 1970s. The need arose to expand small populations of valuable cows, mostly breeds originating in continental Europe (e.g. Limousin, Simmental, Maine Anjou, Salers, Gelbvich) that had been imported into North America or the UK. In due course, demand for such exotic cattle decreased and cattle ET technology turned increasingly towards dairy cattle breeding improvement programmes, mainly in Holstein cattle. ET technology gave considerable impetus to the drive for more efficient dairy cattle; it is now possible for many genetically superior bulls to be produced by way of these methods and to select dams more intensively for breeding herd replacements.

International trade in cattle embryos developed when effective methods of cryopreservation arrived on the scene in the early 1970s. However, before such trade could be established with confidence, it was necessary to have regulations in place to minimize the risks of disease transmission associated with the long-distance movement of farm-animal embryos. Although regulations were in place for live animals and for semen, they had to be appropriately modified to deal with embryos. Towards that end, the International Embryo Transfer Society

Fig. 3.1. Non-surgical embryo transfer in the cow. The development of non-surgical transfer methods in the mid-1970s was a major milestone in progress towards extensive application of embryo transfer technology in cattle. Those with skill and experience in artificial insemination could train to become competent in embryo transfer without difficulty.

(IETS) did much in formulating uniform standards for the identification of embryos; this ensured that embryos could be traced accurately as well as showing that they had been processed according to the appropriate health certifying standards (see Thibier and Stringfellow, 2003). Current practice in ET in domestic animals involves many procedures, such as the recovery of embryos, handling, evaluation and storage of embryos, management of recipients and transfer of embryos.

The success and economics of a commercial cattle ET programme is dependent on several factors: (i) skill and experience of the ET operator; (ii) selection and management of recipient animals, which must be healthy, cyclic and reproductively normal; (iii) close synchrony of oestrus between donor and recipient; (iv) quality of embryos transferred; and (v) methods used in embryo handling and transfer on the farm. The success of cattle ET is usually measured in terms of the pregnancy rate in recipient animals. In general terms, using single embryos, pregnancy rates achievable after non-surgical transfer are broadly in line

with expectations for the same category of recipient animal when bred normally (i.e. 50–70% pregnant to first service). With frozen–thawed embryos, pregnancy rates are usually some 10% below those found with fresh embryos.

The development of ET technology in cattle started in the early 1970s, largely the result of the pioneering efforts of Tim Rowson in Cambridge. In the same way that frozen semen revolutionized the AI industry, embryo freezing allowed the global commercialization of cattle with élite genetic qualities (Table 3.1). Currently, cattle ET technology is used in numerous countries around the world, an estimated half-million embryos being transferred annually. The technology relies heavily on the stimulation of selected donor cattle to produce large numbers of embryos after the induction of multiple ovulations by gonadotrophin treatment (superovulation). Europe is the world's second largest user of embryo technologies after North America, responsible for some 20% of the world's transfers.

In general, the use of ET is restricted to nucleus herds and some of the largest

Table 3.1.   Pioneers in the cryopreservation of gametes and embryos.

Author(s)	Date	Subject
Spallanzani	1776	Freezing of sperm in snow and subsequent thawing and retrieval of motile sperm
Polge and associates	1949	Freezing of avian and mammalian sperm using glycerol as cryoprotectant
Whittingham	1971	Cryopreservation of mouse embryos and birth of live young
Wilmut and Rowson	1973	Cryopreservation of calf embryos and birth of live young
Willadsen and associates	1976	Development of effective techniques for freezing sheep and cattle embryos

commercial herds; it is not routinely used in most of the relatively small-scale European herds. In terms of the genetic gains made possible by ET in cattle, these are primarily the result of increased selection intensity in females and a reduction in the generation interval. When young cattle are used as donors, however, phenotypic measures of performance, such as lactation records, are unlikely to be available; for such reasons, selection tools capable of identifying superior cattle in the absence of phenotypic records can be particularly useful.

Towards that end, efforts to characterize the genomes of cattle and other farm livestock have increased greatly; such characterization should enable appropriate molecular tools to be developed for the improvement of animal performance. Recent advances in molecular biology can provide new tools for improving cattle genetics. Gene mapping of cattle has entered the stage where it is possible to think of genetic markers being used to select animals with desirable genetics; marker-assisted selection (MAS) is something that can be applied at the early embryo stage, thereby allowing the transfer of desirable quality embryos to recipient animals.

Do-it-yourself ET

The high cost of cattle ET has been one reason why the technique has not been exploited as readily as AI by the commercial dairy farmer. There is no reason, after appropriate training, why ET techniques cannot be handled by the suitably trained farmer or stockperson. The benefits of this would be twofold: lower costs and increased flexibility in applying the technology. Provided the farmer has already mastered the AI technique and is thoroughly experienced in its use, do-it-yourself (DIY)-ET, like DIY-AI, may have a place on the farm. Although the transfer technique requires skill, with appropriate 'hands-on' experience, there is no reason why satisfactory pregnancy rates with non-surgical transfers cannot be achieved. In the USA, one country where DIY-ET is in operation, the proponents of this approach report costs per pregnancy being halved.

Buffaloes

Various authors have noted the need to identify, multiply and distribute buffaloes that are of superior genetic merit; this certainly involves the use of various assisted reproduction technologies, including ET. The birth of the world's first buffalo calf after ET was in the USA in 1983; the many studies by Martin Drost and colleagues in Florida also provided information used in the preparation of a valuable training manual dealing with the selection of donors and recipients, superovulation, non-surgical embryo recovery, flushing and holding media and embryo handling and evaluation, as well as ET itself. Although the induction of multiple ovulation follows the same lines as in cattle, the general experience in buffaloes has been disappointing, typically showing a very poor yield of transferable quality embryos. It is known that the ovaries of the buffalo contain smaller numbers of antral follicles than those of cattle; this is likely to be one factor explaining the reduced response to superovulation. However, it is by no means the only factor;

a report by Baruselli *et al.* (2000) has shown results strongly indicating that low embryo recoveries reported in buffaloes may be explained by failure of oocytes to gain entrance to the oviduct (Table 3.2); others suggest that high oestradiol levels from multiple follicles may result in low recovery rates due to reverse peristalsis in the oviduct.

Despite the general anatomy of the reproductive organs and the reproductive physiology of buffaloes and cattle being very similar, there has apparently been no report of natural or artificial hybridization betweeen the two genera; this may not be too surprising, given the marked disparity in chromosome numbers between the two (48–50, buffalo; 60, domestic cattle).

3.1.2. Sheep and goats

The reasons for interest in sheep ET are largely the same as those for cattle. The technique can and has been used to expand the population of particular breeds or strains of sheep in demand; a further consideration is in importing and exporting sheep in the form of frozen embryos rather than animals on the hoof. For many years, the recovery and transfer methods developed by Tim Rowson in the mid-1950s continued to be used as standard practice (Fig. 3.2); indeed, the methods developed at Cambridge for sheep were to lead on to successful ET in cattle. In expert hands and used for research, the Cambridge procedures proved to be valuable; for commercial practice, however, such surgical interventions were often to be regarded as unacceptable on

welfare grounds. A committee appointed by the Norwegian Ministry of Agriculture in 1994, for example, decreed that, for ethical reasons, further trials on ET in pigs, sheep and goats should not be permitted in that country until suitable non-surgical techniques were developed.

Use has been made of ET technology to expand goat populations of scarce and expensive animals. A recent illustrated French article briefly described the technique of ET in goats; data were provided on the results of ET carried out at a testing station, a breeding station and a farm in France. In that country, in the late 1980s, for example, the technology was used in expanding populations of Angora goats imported from New Zealand, Texas and Australia. Demand for high-quality pure-bred Angora and Cashmere goats in Europe and Australasia has resulted in extensive use of ET technology. It would not be unreasonable to expect, after one flushing of embryos and the remating of the doe shortly afterwards for a normal pregnancy, that the rate of population expansion could be six to seven times faster than under normal circumstances.

The advances made in sheep and goat embryo production and ET technology have been described in several publications during the past decade (Cognie *et al.*, 2003). Oestrus can now be readily controlled on a year-round basis in goats using progestogens, and superovulation can be induced with suitable follicle-stimulating hormone (FSH) products; embryos are usually collected 6–8 days after breeding by laparotomy or laparoscopy. The development rate of transferred embryos reported by various authors is approximately 50%, whether embryos are fresh or frozen; ethylene glycol is the recommended cryoprotectant. ET can be the means of avoiding the transmission of disease from infected donors to their offspring. It is now well recognized that ET in small ruminants has the potential to facilitate the safe movement of germplasm around the world, provided appropriate disease prevention measures are taken in the management and handling of embryos and animals. The cryopreservation of valuable breeding material, in the shape of

Table 3.2. Superovulation and embryo/oocyte recovery in the buffalo (from Baruselli *et al.*, 2000).

Day of slaughter	No. follicles at oestrus	No. corpora lutea	No. embryos and oocytes
3.5 (*n* = 4)	16.7 ± 4.9	8.5 ± 4.0	3.5 ± 2.6
4.5 (*n* = 4)	16.0 ± 6.3	10.2 ± 3.8	3.5 ± 2.6
5.5 (*n* = 4)	19.0 ± 6.9	8.8 ± 4.6	2.7 ± 2.8
Total	17.2 ± 5.7	9.2 ± 3.8	3.2 ± 2.6

Fig. 3.2.   Texel lamb and its Suffolk-cross foster mother.

embryos, can also be used in the event of infectious diseases, such as the devastating outbreak of foot-and-mouth disease in the UK in recent years.

3.1.3. Pigs and horses

In pigs, historically, there have been two main difficulties limiting the commercial application of ET. On the one hand, the long-term cryopreservation of pig embryos has not been possible; on the other, there has been the need to use a surgical transfer procedure. However, in recent years, encouraging progress has been made in the vitrification of pig embryos and in the development of non-surgical transfer techniques. In Germany, Wallenhorst and Holtz (2002) compared conventional surgical transfer of pig embryos with a semi-endoscopic approach in attempts to develop an improved ET technique, recording an 88% pregnancy rate with the conventional technique and a 47% rate with the semi-endoscopic approach. The outcome of other efforts is described by Cuello *et al.* (2002), who dealt with the first pregnancies achieved by non-surgical transfer of vitrified/thawed unhatched pig blastocysts.

Effective surgical procedures for the collection and transfer of pig embryos were established in the mid-1960s; since that time, ET technology has been used by many research workers. Although the first recorded successful transfer in pigs was reported by Kvasnicki more than 50 years ago in the former Soviet Union, much of the research and development work in this species over the years has taken place in Europe, North America and Australasia. An important part of the technique's early use was in disease control; the technique permitted greater flexibility in the introduction of new genetic material into 'closed' herds (Table 3.3). Although there are some reports of pig embryos being shipped around the world in the fresh state, the export–import side of ET in pigs has been seriously restricted by the fact that embryos could not be frozen in the same way as those of cattle and sheep.

Undoubtedly, the commercial application of ET in pigs has been limited by the requirement for tedious and expensive surgical interventions in the recovery and transfer of embryos. In this, the story in pigs is quite different from that in the cow, where effective non-surgical recovery and transfer procedures were rapidly developed. Apart from that, welfare concerns in exposing the pig to surgical procedures must be given

Table 3.3. Reasons for using embryo transfer in pigs (from Martin, 1984).

Group no.	Reason	No. of recipients (%)	No. of herds
1	Establish new herds from herds with pseudorabies	133 (43)	4
2	Make additions to specific-pathogen-free (SPF) herds	67 (26)	13
3	Obtain boars for closed commercial herd	61 (23)	11
4	Obtain more offspring from superior gilts and sows	19 (7)	8
		280	36

due consideration; all the more reason, therefore, for concentrating efforts on the development of acceptable non-surgical methods. Although considerable interest has been expressed by the pig industry in this possibility, a low success rate was achieved in some of the early research efforts. However, workers in Spain have devised a new flexible catheter which can be inserted deep into a uterine horn; using such an instrument, they were able to record a pregnancy rate of 70.8% and an average of 6.9 ± 0.7 piglets (Martinez *et al.*, 2003). In other work, non-surgical deep intrauterine transfer of vitrified/warmed embryos resulted in a 62.5% pregnancy rate and an average litter size of 6.25 ± 0.4. Such results are encouraging and provide reasonable grounds for believing that an effective and practical technique for commercial ET in pigs is possible.

Disease outbreaks in pig breeding herds can cause much damage to the farmer and the cost of eradication can be considerable. For such reasons, farmers would welcome the possibility of infected gilts or sows producing healthy piglets by way of ET, using zona-intact embryos. Although such embryos are highly sensitive to chilling and cryopreservation, it has been possible to preserve them by vitrification; several research groups have now published information on successful techniques. In Japan, for example, Misumi *et al.* (2003) demonstrated that zona-intact porcine embryos at the compact morula to early blastocyst stages can be preserved using a simple vitrification method.

Embryo handling and storage techniques for use in pigs enable porcine embryos to be transported far and wide at a reduced cost and with minimal animal welfare problems and disease risks compared with the transport of live animals. As pig embryo collection and transfer techniques become increasingly effective, it becomes all the more important to have optimal handling and storage techniques to maintain embryo quality until the time of transfer. At the present time, long-term preservation techniques, such as cryopreservation and vitrification, are promising but there is also interest in storing pig embryos for much shorter periods (24 h or so) to enable them to be sent long distances within the same country to waiting recipients. Although, for animal breeding purposes, there is not the same need in pigs for extra offspring from particular females, there is the need to transport valuable porcine germplasm from one country to another without the risk of disease transmission; this is reason enough for developing ET technologies in pig breeding and production.

Horses

Information on a wide range of factors influencing the successful outcome of the ET technology in horses is now available, largely the result of valuable contributions by workers such as Allen in Cambridge, Hinrichs and Squires in the USA, Dell'Aquila in Italy and Palmer in France. In considering possible commercial benefits of ET to the horse industry, four areas where the technology may be applicable have been identified: (i) in old valuable mares no longer able to give birth; (ii) in young mares (i.e. 2-year-olds) yet to reach full growth and development; (iii) in sports horses, to avoid the need to interrupt their training, racing or showing in competitions; and (iv) in chronic miscarriers. Clearly, in all this, it would also be useful to identify those horses that are genetically superior in their various performance traits; in Ireland, for example, some attention has been paid in the recent decade to the genetic evaluation of show-jumping animals.

In horses, the first foals from ET were born as a result of work in Japan in 1973 and in Cambridge in 1975. In the Irish Republic, the first ET foal was born in 1984, using non-surgical recovery and transfer procedures (Fig. 3.3); efforts were subsequently made in sports horses to apply ET commercially in that country. Elsewhere, in the 1980s, commercial ET programmes were described by workers in the USA and in France. By the end of the 20th century, it was estimated by North American workers that about 1500 foals were being produced by ET in North America, with Argentina and Brazil being clear leaders in the application of the technology in South America. Countries such as Australia are also showing increasing acceptance of the technique.

In France, the first foal produced as a result of ET was born in 1986 and the number of such foals born annually has increased continuously since then. A recent article by a French author lists the equine ET centres in France and discusses the performance of foals produced by ET. According to this author, further expansion of ET in sports

Fig. 3.3.   Embryo transfer foal and surrogate mother.

horses is likely to depend on the successful freezing of horse embryos and the development of effective superovulation techniques for mares. According to other reports, success with horse ET in France has been moderate due to the use of older and subfertile mares as donors and the inefficiency of superovulation techniques; results achieved annually since 1995 show about 45% of recovered embryos used for transfer and of these 40–70% resulted in a pregnancy. Those describing developments in northern Europe noted that the first ET foal in Scandinavia was born in Finland in 1984 and in Sweden in 1986; at that time, no ET had been carried out in Norway and there has been limited ET activity in Denmark.

Although application of ET in horses has increased steadily during the past 20 years, several factors unique to the horse have limited its uptake in this species. There is, for example, no commercially available hormonal preparation for inducing multiple ovulations in the way that is commonplace in cattle; most embryos are recovered from single-ovulating mares. However, factors affecting pregnancy rates after transfer are much the same as in the cow, and include method of transfer, synchrony of donor and recipient, embryo quality and management of the recipient. As observed by North American authors, one major improvement in equine ET technology in recent years in the USA has been the ability to store embryos at 5°C and to be able to ship them to a centralized station for transfer into waiting recipient mares. Embryos can be collected by practitioners on the farm, cooled to 5°C in a passive cooling unit and shipped to an ET station without a major decrease in fertility. Unlike cattle, equine embryos have proved difficult to freeze; it has been possible to freeze only small day-6 embryos with any great success.

Registration of ET foals

In many ways, the economic incentive for ET in horses is likely to differ markedly from that in cattle. The horse industry traditionally tends to be much more conservative in its approach to breeding. It has been remarked that the speed of winning horses has progressed little in the past century (Long *et al.*, 2003); compare this with improvements in milk yield of the dairy cow. In the USA and in virtually all other countries, the two major racing breeds, Thoroughbreds and Standardbreds, do not accept ET foals in their registries. However, in 2002, the largest registry in the USA, the American Quarter Horse Association, did approve the unlimited registration of foals from a mare during a given year using ET. According to Squires *et al.* (2003), this is likely to more than double the number of transfers performed each year in that country.

Avoiding genetic defects

One application of ET in horses could be in avoiding serious genetic defects by the DNA testing of embryonic material. A biopsy prior to transfer would allow for the determination of whether or not the embryo was a carrier of a defective gene; discarding carrier embryos would be financially beneficial and the percentage of healthy offspring from valuable mares could be increased.

Research applications

Many studies have been reported in horses in which interspecific and extraspecific pregnancies have been established by way of ET. In England, Cambridge workers have studied some of the developmental, endocrinological and immunological problems associated with the establishment of the xenogenic fetus to term in mares carrying interspecific or transferred extraspecific foals. Other research areas in which ET has been used include the effect of maternal size on fetal and postnatal development in the horse. In England, in common with the classical findings of Arthur Walton and John Hammond in the late 1930s, who used AI to reciprocally cross Shire horses and Shetland ponies, results recently presented by Cambridge workers have shown that, in the mare, maternal size affects fetal growth, presumably by limiting the area of uterine endometrium available for attachment of the diffuse epitheliochorial placenta.

Studies using ET technology in the horse have also examined postnatal cardiovascular function after manipulating fetal growth; in England, data reported by Giussani *et al.* (2003) showed that fetal growth acceleration as well as fetal growth restriction, resulting from between-breed ET in the horse, can lead to altered postnatal cardiovascular function. Such findings suggest that deviations in the pattern and rate of fetal growth both above and below the normal trajectory may influence events in the later life of the animal.

3.1.4. Deer and camelids

The rapid growth in deer farming in countries such as New Zealand and the growing interest in the conservation of various deer species have stimulated a number of research groups to develop ET and associated procedures that can be used with these animals. The protocols developed for ET in deer are usually based on those used in sheep and cattle. A description of the methods used, covering superovulation, embryo collection, recipient management, freezing embryos and other useful information was provided in articles in the early 1990s; much of what was said then holds true today.

New World camelids

It is believed that establishing practical and effective ET methods for the llama could be of considerable benefit to the American llama industry; it could, for example, facilitate the importation of genetically valuable embryos from South America to improve herds in the USA.

3.2. Growth and Development of Embryo Transfer Technology

3.2.1. Historical

Tim Rowson and associates in Cambridge, working with sheep in the mid-1950s, showed the usefulness of ET as a research tool. The technique provided the means for testing in sheep the relative importance of genetic and environmental factors for the developing sheep embryo. Although the birth of the first calf by ET occurred in the USA in the early months of 1950 (see Betteridge, 2000), it was to be a further two decades before Rowson's work in Cambridge led to commercial application of ET technology.

3.2.2. Superovulation techniques

The earliest descriptions of superovulation date back to Smith and Engle in 1927, who used anterior pituitary preparations to induce a fourfold increase in the ovulation rates of rats and mice. A few years later, Cole and Hart in the USA demonstrated that the blood serum of pregnant mares would induce multiple ovulations in rats, establishing the basis for what was to become the most widely used gonadotrophin in the treatment of farm animals. Pregnant mare's serum gonadotrophin (PMSG) is a glycoprotein found in the blood of the mare between days 40 and 130 of gestation and is unique among gonadotrophins in possessing both FSH and luteinizing hormone (LH) biological properties within the one molecule. It is now known that PMSG is secreted by specialized trophoblastic cells that invade the mare's endometrium between days 3 and 40 of gestation; for such reasons, the term equine chorionic gonadotrophin (eCG) rather than PMSG is preferred by many. The name notwithstanding, early research on superovulation in cattle and other farm animals invariably involved the use of PMSG. In the first decade of human *in vitro* fertilization (IVF), on the other hand, superovulation was usually by way of human menopausal gonadotrophin (hMG), countless thousands of healthy babies being born as testimony to the safety and efficacy of this urinary gonadotrophin.

Although, initially, relatively undefined preparations such as PMSG were used for superovulation in cattle and other farm ruminants, these were subsequently replaced by purified pituitary extracts from pigs, horses and sheep. One practical consequence of using pituitary preparations, because of their much shorter half-life, was the need to administer them by multiple injections rather than a single administration.

Such variability in superovulatory response in cattle, which is reflected in buffaloes, sheep and goats, is known to be related to differences in the gonadotrophin preparation, total dose of FSH administered and duration and timing of treatment and the use of additional hormones in the superovulation regimen. There is evidence that pretreatment with recombinant bovine growth hormone (recombinant bovine somatotrophin (rBST)) or increased dietary intakes, which induce an increase in the population of small follicles, can significantly improve the response to standard superovulatory protocols; it is evident that ovarian status can be manipulated in various ways to improve superovulatory response.

Cattle

The treatment of farm animals, especially cattle, to induce additional ovulations (i.e. superovulation) has been the object of much research during the past half-century and has been a major consideration in the development of commercially acceptable ET technology. Cattle have featured prominently in superovulation research in view of their primary importance in the production of meat and milk in most countries around the world. The induction of multiple ovulations is the first step towards enabling the genetically superior cow to produce a greater number of calves. Although a genetically superior bull may sire a million calves by AI in 3 years, a cow in the same period normally gives birth to no more than three calves.

However, despite extensive research in such areas as follicular dynamics and the use of gonadotrophins in combination with gonadotrophin-releasing hormone (GnRH), prostaglandins (PGs), progestogens, oestrogens and rBST, there are few indications of significant improvements in the overall efficacy of cattle superovulation treatments during the past two decades (Mapletoft *et al.*, 2002). One report from the mid-1980s, dealing with more than 2000 superovulated beef cows, recorded an average of 11.5 oocytes/embryos recovered, of which 6.2 embryos were viable; however, 24% of collections failed to yield any viable embryo, 64% produced fewer than average and 30% yielded 70% of the embryos. Figures recorded in 2002 after commercial superovulation treatments in 25 European countries would seem to bear this out (Table 3.4).

Physical manipulation of the follicular wave is a relatively new procedure used in the cattle ET industry and few reports have appeared in the scientific literature to justify its use. In cattle, sheep and goats, the growth of large antral follicles occurs in several distinct waves during the oestrous cycle (see Chapter 5), each wave being characterized by the simultaneous emergence of a cohort of follicles and the establishment of one (cow) or more (sheep and goats) dominant follicles that continue to develop to ovulatory size, while apparently suppressing the growth of other follicles in the cohort. This is in contrast to events in the pig, where growth of ovulatory-size follicles is limited to the follicular phase of the cycle. The presence of large dominant follicles at the time of gonadotrophin administration adversely affects superovulatory response.

In the USA, Shaw and Good (2000) examined a database of 2031 recoveries over a 3-year period, including cows which had undergone dominant follicle ablation prior to FSH stimulation; the evidence suggested that, while dominant follicle ablation may increase ovulation rate among donors, factors other than follicular dominance are important determinants of FSH responsiveness and embryo quality. In cattle, there is ample evidence showing that differences among cows in the number of growing follicles in their ovaries probably account for part of the variability observed in the number of ovulations and the embryo yields in response to gonadotrophin treatments.

Some problems in superovulating cattle are not due to treatment but rather to the nature of the cattle themselves. Beef cattle, being less accustomed to handling, may be much less docile than dairy cattle. Taking an extreme example, it may be noted that, after about 10 years of ET in Spanish fighting cows, ET practitioners still could not obtain more than 1.6 embryos per flush, markedly below the normal yield. The wildness of the animals made handling very difficult; apart from feeding time, the cows never saw a human; in dealing with such fractious animals, using PMSG rather than FSH would have an advantage in only needing to inject donor animals once. Conventional superovulation protocols developed for domestic cattle may be less efficient or even not applicable to non-domestic ungulate species or indigenous breeds of cattle, due either to their particular reproductive physiology or to the stress of handling. Age is a consideration influencing the superovulatory response in cattle. In Brazil, Oliveira *et al.* (2002) concluded that Nelore donor cows over 13 years of age do not respond satisfactorily to exogenous hormone stimulation; on that basis, they recommended their exclusion from ET programmes.

Buffaloes

It is now well recognized that a major limitation to the use of ET as a means of genetic improvement in the buffalo is the extremely low response to superovulatory treatment. Many studies show an average of one or two transferable embryos per cow for river buffalo and one embryo for swamp buffalo; it is also noted that often no more than one-third of ovulations are accounted for as embryos. In India, reviewing progress over a 5-year period, workers in that country recorded an embryo yield averaging 2.6 per donor, of which only 1.3 were viable; transfer of embryos resulted in a pregnancy

Table 3.4. Cattle embryo production by superovulation (data collected from 25 countries in Europe in 2002; from Lonergan, 2003).

Number of flushed donor cows	18,294
Number of transferable embryos	102,996
Mean number per flushed donor	5.63

rate of 17%. Such results are markedly below those found with cattle. The belief is that this is due to the lower number of follicles and the higher rate of follicular atresia in comparison with cattle.

Studies of buffalo ovaries have reported 12,636 primordial follicles in cyclic buffalo heifers, a much lower figure than that reported in cattle. On the other hand, ovarian follicular dynamics appear to be similar to those in cattle; although the two-wave cycle is the most common in the buffalo, both three-wave and one-wave cycles are observed. Unlike much of the evidence for cattle, Indian studies concluded that the superovulation response of buffaloes is not affected by the presence of a dominant follicle at the initiation of FSH treatment. Studies reported by Carvalho *et al.* (2002) in buffaloes demonstrated that a GnRH agonist bio-implant containing the agonist deslorelin prevented ovulation after stimulation of ovarian follicular growth with FSH; ovulation was induced by injection of exogenous LH. It is believed that treatment with the GnRH agonist induced down-regulation of the anterior pituitary, leading to an absence of pulsatile secretion of LH and FSH. One advantage of the GnRH agonist– LH protocol, in comparison with conventional superovulation treatments, lies in the fact that ovulation can be accurately controlled; when ovulation does occur, it occurs over a relatively short period, which may permit breeding by a single fixed-time insemination.

Although superovulation is a key step in producing embryos from the buffalo, it is important that the donor animal returns to cyclical breeding activity as soon as possible after embryo recovery. An early return is important if the donor is to be superovulated on several occasions over a relatively short time period; it is also important if the donor is to become pregnant soon after embryo collection. In India, Misra and Pant (2003) showed that, in the superovulated buffalo, the mean interval to oestrus after PG treatment was considerably prolonged (by about 12 days) compared with the cyclic buffalo, in which oestrus occurred after about 3–4 days; the delay was believed to be attributable to an elevated concentration of progesterone produced by multiple corpora lutea and its negative feedback effect on LH secretion. As in superovulated cattle, the return-to-oestrus interval lengthened significantly with increased numbers of corpora lutea.

Pigs

For research purposes and for developing ET technology in pigs, prepubertal gilts have often been regarded as an inexpensive source of embryos. These animals also have the advantage that the superovulation treatment is not complicated by the presence of an oestrous cycle, which means that gonadotrophin treatment can be applied at any time. Doses of 250–2000 iu PMSG, in combination with human chorionic gonadotrophin (hCG) injected 2 days later, have been used to superovulate (Table 3.5).

Horses

It became evident at an early date that superovulation techniques regarded as satisfactory in the cow were of little avail in the mare. A review in the mid-1990s concluded that ovulation rates for mares following superovulation treatment are much lower than for cattle. This difference has been attributed to the relatively limited area available in the ovulation fossa for ovulation to occur in mares compared with cattle, combined with the large size of the equine preovulatory follicle. Owing to the unique morphology of the mare ovary (cortex within the medulla), the mare's preovulatory follicle must migrate through dense connective tissue and rupture at a specialized site, the ovulation fossa. It is believed

Table 3.5. Superovulatory responses in prepubertal gilts (from Baker and Coggins, 1968).

PMSG	hCG	Mean no. of ovulations
250	500	7.2
500	500	12.5
1000	500	19.6
2000	500	45.8

that activity of the hormone relaxin may be one of the mechanisms facilitating follicle migration and ovulation. It is probable that the number of ovulations in mares may be limited physiologically by the size of the follicular cohort that can be stimulated by gonadotrophins.

There are two methods which may be used for superovulation in the mare: the administration of FSH preparations or suppression of the inhibin feedback by anti-inhibin immunization One recent paper noted that superovulation techniques in mares based on FSH had been used for some 15 years; during this period, the superovulatory response increased but the number of viable embryos produced per cycle did not. Immunization against inhibin can be active (vaccination against inhibin) or passive (injection of anti-inhibin antibodies); as well as its use in superovulation, the inhibin-inhibition technique should contribute to a greater understanding of the mechanisms regulating follicular growth and ovulation in the mare. Active immunization follows the lines of classical protocols, with a response becoming evident after some time; passive immunization is conducted with one injection of antibodies per cycle, with immediate effects on ovulation rate being evident.

Sheep

In sheep, French workers have focused efforts in the past decade on synchronizing follicular wave emergence by treatment with exogenous gonadotrophin (Cognie et al., 2003); this approach involves the use of a GnRH agonist or antagonist combined with a progestogen treatment to suppress endogenous gonadotrophin and follicular development, followed by exogenous gonadotrophins administered over 4 days. Pretreatment over a 10-day period with an antagonist/progestogen suppressed large follicles, doubled the number of small follicles and improved response to FSH by 50%; a uniform cohort of follicles was recruited by FSH treatment, synchronization of oestrus occurred 20–24 h after removal of the progestogen sponge and an LH preparation injected intravenously 32–36 h after

sponge removal resulted in ovulation 20–28 h later. Breeding the donor ewes was by way of a timed insemination 48–50 h after the end of progestogen treatment.

Knowledge that part of the variability in superovulatory treatment in sheep is associated with the presence or absence of a large growing follicle at the onset of gonadotrophin treatment has led some workers to develop what is known as the 'day 0 protocol'. In this, treatment with FSH initiated soon after ovulation (day 0) was shown to increase follicle recruitment, ovulation rate, embryo quality and the number of embryos recovered when compared with a treatment regime applied 3 days after ovulation, when an active growing follicle is present in the ovary.

Goats

Two major limiting factors were identified at an early stage in ET programmes in goats: variability of response to superovulatory treatments and the premature regression of corpora lutea. Conventional protocols for superovulation techniques used commercially in goats have generally consisted of a long priming (11–18 days) period with progesterone/progestogen-releasing devices and the administration of gonadotrophins prior to or around the time of progesterone/progestogen withdrawal. Studies by Menchaca et al. (2002) in Uruguay explored a similar approach to that employed in sheep. They showed that the presence of a dominant follicle at the time of initiating gonadotrophin treatment had an adverse effect on superovulatory response. The study supported the advantage of the 'day 0 protocol', with treatment starting soon after ovulation, when the first follicular wave is emerging and no dominant follicle is present.

Workers have examined alternatives to the conventional gonadotrophin approach to superovulation. Some suggest that the neutralization of inhibin bioactivity may have potential. Studies reported by Medan et al. (2003) sought to determine the effect of passive immunization against inhibin (48 h before induced luteolysis) on FSH secretion and ovulation rate in goats; they were able to show that the immuno-neutralization of endogenous inhibin in cyclic goats led to increased secretion of FSH and multiple ovulations (4.2 vs. 1.8 in controls).

Superovulated goats frequently suffer from premature luteal regression, which results in progesterone concentrations returning to basal levels as early as 3–6 days after the onset of oestrus; the lack of pro-gestational support in goats with premature luteal regression leads to the loss of most embryos before collection on days 6 or 7 after the onset of oestrus. It appears that premature luteal regression is much more common when superovulation is induced with PMSG than with FSH; for such reasons, PMSG has been largely abandoned, despite the advantage offered by a one-injection rou-tine. Even with FSH-based superovulation protocols, a substantial proportion of does may show premature luteal regression.

It is believed that premature luteal regression in the superovulated goat may be due to the activation of the luteolytic release of $PGF_{2\alpha}$ by high levels of oestradiol; such high levels are thought to come from follicles which continue to be recruited and stimulated in the early luteal phase, due to the long half-life of PMSG. In an effort to combat this problem, GnRH or LH has been administered after the onset of oestrus to eliminate large unovulated follicles. In Mexico, workers have used such means to maintain adequate progesterone con-centrations in all the animals until day 6, when embryo collection is normally scheduled.

Progestogen–gonadotrophin treatment is effective in the induction of super-ovulation in the doe both during the breed-ing season and during anoestrus. A typical protocol employed in the breeding season has been one in which fluorogestone acetate (FGA) sponges are used over an 11-day period (Fig. 3.4), with FSH being adminis-tered over 3–4 days, starting 2 days before progestogen withdrawal and around the time at which prostaglandin is adminis-tered. French workers have also shown that superovulation in the goat could be induced repeatedly by employing gonadotrophins which either had no potential (caprine FSH) or a low potential (ovine FSH) to induce the production of anti-gonadotrophin antibodies.

Studies in Spain by Gonzalez-Bulnes *et al.* (2003) evaluated superovulatory response in goats as affected by follicular status (number of follicles capable of responding) at the time of initiating FSH treatment, which was given twice daily from 60 h before to 24 h after removal of a 16-day intravaginal treatment; results indicated that ultrasonography could be used as a practical criterion for the selection of donor goats, which could avoid treatment of poor-responding animals.

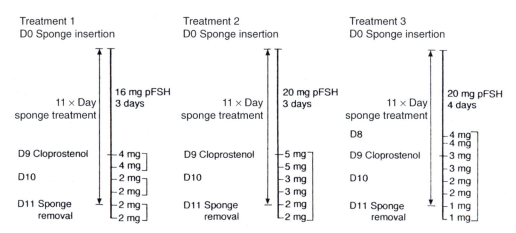

Fig. 3.4. Treatment for superovulation in the goat. D0–D11, day 0 to day 11; pFSH, porcine FSH (from Baril *et al.*, 1988).

Deer

Superovulation procedures for deer generally involve synchronization of the oestrous cycle with exogenous progestogen and stimulation of follicle development with exogenous FSH preparations. A typical protocol for red deer involves a 12-day controlled internal drug release (CIDR) treatment, with FSH administered for 4 days (days 8–12) prior to removal of the intravaginal device; the schedule for fallow deer is similar except for a longer intravaginal progesterone period (14 days) to take account of the longer oestrous cycle length in this species.

New World camelids

In New World camelids, South American workers have reported on the superovulation of adult llamas in Chile, using FSH treatment at 12 h intervals over 5 days; embryos were recovered by non-surgical uterine flushing 7 days after mating.

3.2.3. Embryo recovery and evaluation

In cattle, non-surgical embryo recovery procedures combined with non-surgical transfer techniques became available in the late 1970s and have been widely used commercially since that time. The donor animal is flushed by non-surgical procedures a week after breeding and an average of four to six transferable embryos are recovered. Where necessary, this could lead to an average of 50 freezable embryos per donor per year, resulting in the birth of about 30 calves after transfer of embryos to recipients. Repeated collection of embryos is used in multiple ovulation and embryo transfer (MOET) breeding programmes, where the aim is to speed up the genetic improvement of beef and dairy cattle.

Timing of embryo recovery

The timing of embryo recovery is determined in the different farm animals by the rate of embryo development and by the location of embryos at various times after ovulation in the female reproductive tract. Studies reported by Sartori *et al.* (2003) in the USA showed that shallow uterine flushing in the cow (catheter tip inserted just beyond the uterine bifurcation) gave improved embryo recovery results compared with deep uterine flushing (close to tip of horn). Although the chronological events of *in vivo* embryo development are well-known for cattle, information for the water buffalo is much more limited. Several studies indicate that the rate of embryonic development in the buffalo is 24–36 h earlier than in cattle.

A prerequisite for successful non-surgical embryo recovery is knowledge of the time the embryo descends into the uterus. Workers dealing with this species have concluded that most buffalo embryos reach the uterus by 108 h post-oestrus; timing embryo recovery around 150 h post-oestrus (day 6 of the cycle) is suggested as optimal for recovering embryos at the compact morula and blastocyst stages. A further consideration in the buffalo is the possibility that some proportion of oocytes fail to enter the oviduct in the superovulated animal.

Horses

As recorded in the literature, the horse embryo first divides about 20–24 h after ovulation and reaches the uterus either as a morula or an early blastocyst between 144 and 169 h after ovulation. The fact that the equine embryo spends much longer in the oviduct than pig embryos (48 h) and farm ruminant embryos (72 h) has been commented upon by various authorities; it is suggested that one possible advantage of the equine embryo leaving it so late to enter the uterus is that it allows additional time for inflammation induced at the time of mating to be resolved. For practitioners with embryo recovery in mind, such tubal characteristics may not be welcome, particularly those interested in embryo cryo-preservation (tubal embryos being smaller and more viable). Most embryos collected from donor mares are from spontaneous single-ovulating animals and are recovered

7 or 8 days after ovulation (day 0 = day of ovulation). Semen quality can influence embryo recovery indirectly; mares inseminated with fresh semen are more likely to produce an embryo than those inseminated with either cooled or frozen–thawed semen.

Surgical recovery of embryos from the mare's tract is performed using procedures based on those previously developed in cattle. The straight and readily distensible cervix of the mare, together with its simple T-shaped uterus, suspended in the abdominal cavity beneath the rectum, makes nonsurgical flushing and ET procedures easier than in cattle. A paper by Squires *et al.* (2003) describes current recovery procedures, which the authors note have remained essentially unchanged for the past two decades. As described, a large-bore, Foley-type catheter is inserted through the mare's cervix, and the cuff inflated with 80 ml of air; the catheter is withdrawn against the internal os of the cervix, and 1–2 l of medium is infused into the mare's uterus. The fluid is then drained out of the uterus through a filter cup; the procedure is repeated at least three times, for a total of 3–6 l medium per embryo-recovery attempt (Fig. 3.5).

Sheep and goats

Laparoscopy is regarded by some as an effective and minimally invasive procedure for use in the recovery of embryos in sheep and goats. In the early days of embryo recovery in sheep and goats, it was customary to use surgical intervention (laparotomy); this method, however, could result in adhesions limiting the repeated use of donor animals. In Germany, workers have developed an effective laparoscopic technique for flushing embryos from the goat oviduct, recovering an average of 77% of pronuclear-stage goat embryos; the lower incidence of postoperative adhesions after using the procedure appeared to justify its use.

Camelids

In camelids, various reports deal with the time of entry of the embryo into the uterus. A report by Palasz *et al.* (2000) dealt with the time of embryo arrival in the uterus of the llama superovulated with PMSG; their observations suggested that llama embryos arrived in the uterus 6.5 days post-mating, primarily at the hatched blastocyst stage.

Fig. 3.5. Large volumes of flushing fluid in the mare.

Fate of unfertilized oocytes

In the mare, it is difficult to be as certain as with other farm mammals about the effectiveness of non-surgical recovery attempts because of the marked tendency for unfertilized oocytes to be retained in the Fallopian tubes of the mare rather than passing into the uterus; embryos can only be recovered from the mare's uterus if fertilization has taken place. In instances of single-ovulating mares where no embryo is obtained, it may not be clear whether the problem lies with fertilization or with flushing itself. It should be noted that the early equine embryo is unique among farm animals in being able to control its passage through the oviduct, apparently, by release of PGE_2, which induces relaxation of the tube's smooth muscle. It is believed that the horse embryo needs to develop beyond the two- to four-cell stage in the oviduct before its humoral signal is sufficiently strong to facilitate its transport to the uterus. In the UK, Cambridge workers recently demonstrated the efficacy of a simple and practical method of speeding up the passage of embryos through the oviduct by the local application of PGE_2; this enabled them to recover morulae from the uterus by non-surgical procedures.

Embryo evaluation

As part of the developing cattle ET scene in the 1970s, it was realized that, to achieve optimal pregnancy rates in recipient animals, it was essential to have meaningful information on the chronological and morphological development of the bovine embryo. Since 1931, when Hartman and associates first described a cattle embryo, many reports have appeared on methods of evaluating the normality and viability of embryos. Many methods centred round morphological features of the embryo, such as uniformity of cell size, shape of embryo and its colour and overall dimensions; various embryo classification schemes have been developed based on such features. Embryo classification schemes should be based on easily recognizable morphological features and should be backed with firm evidence on the pregnancy rates to be expected with each of the grades. It is likely that techniques for assessing embryos *in vitro*, using objective non-invasive measures, will become increasingly important (Table 3.6).

Most evaluation procedures in horses take account of morphological features, such as pronuclear formation, uniformity of cell size, shape of embryo and its colour and

Table 3.6. Criteria used in evaluating cattle embryos (from Lonergan, 1992).

Developmental stage	Identifying features
Morula	Individual blastomeres are difficult to discern from one another. The cellular mass of the embryos occupies most of the perivitelline space
Compact morula	Individual blastomeres have coalesced, forming a compact mass. The embryo mass occupies 60–70% of the perivitelline space
Early blastocyst	This is an embryo that has formed a fluid-filled cavity or blastocoel and has the general appearance of a signet ring. The embryo occupies 70–80% of the perivitelline space. Visual differentiation between trophoblast and inner cell mass may be possible at this stage of development
Blastocyst or midblastocyst	Pronounced differentiation of the outer trophoblast layer and the darker, more compact inner cell mass is evident. The blastocoel is highly prominent, with the embryo occupying most of the perivitelline space
Expanded blastocyst	Overall diameter of the embryo dramatically increases (1.2–1.5×), with a concurrent thinning of the zona pellucida to approximately one-third of its original thickness
Hatched blastocyst	Embryos can be undergoing the process of hatching or may have completely shed the zona pellucida
Hatched expanded blastocyst	A re-expanded embryo with a large blastocoel and round, very fragile appearance or, in later stages, an elongated shape

overall dimensions. In pigs, survival of embryos after non-surgical transfer tends to be related to the average diameter of the blastocysts transferred (Hazeleger et al., 2000a); since diameter is related to the number of cells, a simple assessable parameter such as diameter of a blastocyst provides valuable information about the quality of an embryo and its likelihood of survival after transfer.

3.2.4. *In vitro* culture and cryopreservation of embryos

Storing the fresh embryo

As part of normal cattle ET operations, there is an obvious need for the short-term storage of embryos recovered from donor cattle until such time as they are either transferred to a waiting recipient or undergo cryopreservation. The general view is that the viability of the bovine embryo, usually at the blastocyst stage, begins to decline after 12 h storage at normal room temperature (20–30°C) in an appropriate medium. The choice of media for embryo collection and temporary storage has ranged from complex tissue culture media (e.g. TCM-199) to much simpler formulations such as Dulbecco's phosphate-buffered medium with bovine serum (D-PBS). For those operating under farm conditions, the requirement may simply be for PBS containing antibiotics and 2% fetal calf serum (FCS) for embryo collection and PBS with 10–20% FCS for temporary storage prior to transfer. In a recent survey in the USA on media used for recovering and storing cattle embryos temporarily, the two most widely used alternatives to D-PBS were Em-Care and ViGro Plus, both commercially available.

Rabbit oviduct as an incubator

Early studies in the storage of horse embryos included the use of the rabbit, which has been employed in Cambridge and elsewhere as an *in vivo* incubator for sheep and cattle embryos. Cambridge

workers have described the development and viability of horse embryos after long-distance transport in the rabbit oviduct. In Cambridge in the summer of 1975, Welsh ponies were flushed and their embryos placed in the oviducts of rabbits, which were then driven to Poland by car; the embryos remained in the rabbit for 48 h before transfer to recipients, two of which, in due course, gave birth to healthy foals.

The conservation of equine embryos is regarded as technically more difficult than that of other farm mammal embryos. Equine embryos have been cultured in a variety of systems with variable results. According to Squires et al. (2003), one of the major changes in horse ET technology in the recent decade is the ability to store embryos at 5°C; cooling and storing embryos at that temperature enables embryos to be shipped to a centralized station for transfer into recipient mares. This is where horse breeders and ET practitioners do not have their own recipients and, to meet the need, there are several large recipient stations that handle shipped embryos. Embryos are collected on the farm and transported to recipient stations in a passive cooling device (e.g. Equitainer: Hamilton Thorne Biosciences, Beverly, Massachusetts) at 5°C either by commercial airline or by overnight carrier, working to a time schedule of 12–30 h between collection and transfer.

In Colorado, studies have shown that horse embryos < 600 μm in diameter can be cooled for 22 h and then cryopreserved without this reducing their viability after transfer to recipient animals (Maclellan et al., 2003). In France, Moussa et al. (2003) used Emcare holding solution (EHS) and ViGro holding plus (VHP) and concluded that these media offered a satisfactory alternative to Ham's F-10 for 24 h cooled storage of equine embryos; it was also found that larger embryos may have a better viability after 24 h of cooled storage than smaller embryos.

Freezing embryos

The freezing of mammalian embryos was first shown to be possible in 1971, when

David Whittingham and colleagues in London obtained live mice pups after the transfer of frozen–thawed embryos that had been frozen using either glycerol or dimethyl sulphoxide (DMSO). Some milestones in the freezing of mammalian embryos are detailed in Table 3.7. It is believed that sperm and embryos are capable of remaining viable at a temperature of −196°C (liquid nitrogen) for perhaps 1000 years or more, the only source of damage at such a temperature being direct ionization from background radiation. For normal purposes, however, there is little need to think in terms of storage for other than a few months or years. Many studies have shown that the usual length of cryopreservation of cattle embryos in liquid nitrogen does not affect their viability after thawing.

Cambridge workers were at the forefront in developing effective techniques for the freezing of cattle embryos, the work of Chris Polge, Ian Wilmut and Steen Willadsen being particularly valuable. It was found that slow freezing (0.3°C/min) of cattle embryos to low subzero temperatures (−80°C) required slow thawing; slow freezing of embryos to relatively high subzero temperatures (−25°C to −35°C), on the other hand, required rapid thawing (360°C/min). Such findings, initially described by Willadsen in the freezing of sheep embryos, subsequently formed the basis of the cryopreservation technique that was to be widely adopted in commercial practice; since the

late 1970s, the method has been the standard technique in freezing the embryos of many species, including human embryos.

The first calf born after transfer of a frozen embryo in the early 1970s at Cambridge (Frosty II) did much to stimulate such research. In the course of the next three decades, countless thousands of cattle embryos were frozen and thawed for transfer in countries around the world. In cattle, as in other domestic species and humans, the study of cryobiology as it relates to embryo preservation is one of the most intensively researched area of embryo biotechnology. Numerous protocols of cryopreservation were to be proposed (conventional slow freezing, ultra-rapid freezing and vitrification), each methodology with its advantages and disadvantages.

Disease control is an important consideration in the export and import of embryos and recent years have seen moves to minimize or abandon altogther the use of animal proteins (serum or bovine serum albumin (BSA)) in the medium employed in bovine embryo freezing. A study reported by George *et al.* (2002) substituted plant protein (wheat peptones) for BSA, the workers finding that the substitution did not affect blastocyst survival and quality.

Ethylene glycol as the cryoprotectant

During recent years, ethylene glycol has been effectively employed as a cryoprotectant for bovine embryo preservation. The molecular weight of this agent (62.1) is lower than that of glycerol (92.1), propylene glycol (76.1) and DMSO (78.1) and it seems possible that its beneficial effect is partly due to its high permeability; the fact that it permeates the embryo rapidly also eliminates the need for the stepwise dilution of the cryoprotectant at the time of thawing. In a survey of the cattle ET industry in North America, workers reported the growing popularity of ethylene glycol and direct embryo transfers, recording that, in 1997, 55.4% of frozen–thawed embryos transferred in the USA and 87.6% of those in Canada were frozen using the agent. In France, workers have recently reported

Table 3.7. First young born after transfer of frozen–thawed embryos.

Year	Species	Researcher(s)
1971	Mouse	Whittingham *et al.*
1973	Cow	Wilmut and Rowson
1974	Rabbit	Bank and Maurer
1974	Sheep	Willadsen
1975	Rat	Whittingham
1976	Goat	Bilton and Moore
1982	Horse	Yamamoto *et al.*
1984	Human	Zeilmaker *et al.*
1985	Hamster	Ridha and Dukelow
1988	Cat	Dresser *et al.*
1989	Pig	Hayashi *et al.*
1989	Rhesus monkey	Wolf *et al.*

studies with 2134 transfers in which ethyl-ene glycol was compared with a glycerol–sucrose combination; they recorded an improved success rate with ethylene glycol (55.4% vs. 47.2%). Further information on the use of ethylene glycol in the freezing of cattle embryos in the USA is provided by Hasler (2002).

Vitrification

For successful commercial application, a simple process that permits direct embryo transfer and gives high pregnancy rates is what is required. The most significant steps in the cryopreservation of cattle embryos in recent years include the ability to freeze and transfer embryos in straws without dilution and the development of the open pulled straw (OPS) method for efficient vit-rification of embryos and oocytes. In vitrifi-cation, ice-crystal formation is prevented by using high concentrations of cryoprotec-tants and high cooling and warming rates. Although vitrification as a method of cryo-preserving embryos appeared on the scene in the mid-1980s as an alternative to the traditional slow freezing of cattle embryos, its suggested advantages (simplicity, cost, speed) had little impact on commercial ET operations and its application remained largely confined to research studies.

Vitrification is an ultra-rapid cooling technique based on direct contact between the vitrification solution, containing cryo-protectant agents, and liquid nitrogen. The protocols for vitrification are simple, allow-ing cells and tissue to be placed directly into the cryoprotectants and then plunged directly into liquid nitrogen. It may be noted that the literature of cryopreservation tech-nology makes a distinction between 'thaw-ing' as applied to embryos and oocytes preserved by conventional freezing, and 'warming', which is the term used in bring-ing embryos back to ambient temperature after vitrification. In human-assisted repro-duction, vitrification protocols are starting to be employed, with several births reported using protocols that have been successfully applied to bovine oocytes and embryos (Liebermann et al., 2002). Milestones in the

application of vitrification to various mam-malian species are detailed in Table 3.8. Vitrification has the attraction of avoiding the need for expensive equipment, as required in a conventional cryopreservation programme.

In physical terms, vitrification is a pro-cess of solidification in which crystalline ice does not separate and there is no concentra-tion of solutes, as in conventional freezing; there is an abrupt increase in the viscosity of the holding medium, producing a glasslike solid. Very high cooling rates are employed but initial exposure to the vitrifying solution is at refrigerator temperature and very brief to avoid adverse effects from cryoprotectant toxicity. Warming rate is also rapid to avoid crystal formation as the temperature returns to normal. Using the standard French mini-straw as an embryo container, vitrification enabled a maximum cooling rate of about 2000°C/min. Vajta's OPS method, on the other hand, permits much higher cooling and warming rates (> 20,000°C/min); the method involves loading the cattle embryos into a straw previously heat-pulled to half the diameter and thickness of the wall.

In the freezing of embryos, the methods originally used by Whittingham and others with mice were found to be entirely unsuc-cessful with pigs; the three approaches taken towards the cryopreservation of the pig embryo are shown in Fig. 3.6. It early became evident that pig morulae are extremely sensitive to cooling below 15°C and it even-tually became clear that this sensitivity to cooling and freezing was the result of their high lipid content; it also became clear that pig embryos at the expanded and hatched

Table 3.8. Milestones in vitrification of embryos.

Year	Species	Researcher(s)
1985	Mouse	Rall and Fahy
1986	Cow	Massip et al.
1986	Hamster	Critser et al.
1988	Rat	Kono et al.
1989	Rabbit	Smorag et al.
1990	Sheep/goat	Scieve et al.
1994	Horse	Hochi et al.
1998	Pig	Kobayashi et al.

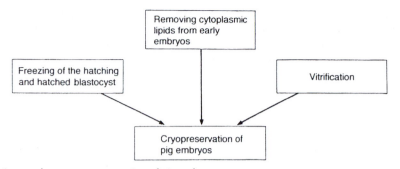

Fig. 3.6. Approaches to cryopreservation of pig embryos.

blastocyst stages have a higher tolerance to cooling than early blastocysts or morulae.

International trade in embryos

Freezing cattle embryos is a valuable means of avoiding the disease hazards normally associated with conventional methods of moving live cattle from one continent to another; 30 years of international trade with frozen embryos has apparently not resulted in the transmission of a single infectious disease agent. It is clear, from demonstrations in the UK and other countries, that there is much public concern about the welfare of farm animals during transportation, whether this is in the course of their export from the country or during lengthy journeys within the country. In the case of breeding animals, clearly such problems need not arise if the cattle are transported in the form of embryos, which can be done at a much reduced cost.

The techniques employed to ensure that frozen embryos are free of pathogens include the use of specific pathogen-free donors, the washing and trypsin treatment of embryos, or a combination of these methods. Health certification procedures for international trade generally require embryos to be examined under the microscope over all surfaces to ensure that there is no material adhering to the zona pellucida (ZP); cattle embryos are apparently less 'sticky' than those of sheep and pigs, which eases the problem of removing materials and cells adhering to ZP. All the indications are that cattle embryos, when appropriately treated according to the standards agreed by the IETS, should be free from bacterial and viral infections; in the preparation and pre-freeze treatment of embryos, it is essential that the integrity of the ZP be preserved.

Horse embryos

The freezing of equine embryos is not yet accepted practice in the horse ET industry because of the limited success with the technique. Embryos < 250 μm in diameter can be cryopreserved successfully if glycerol is used as the cryoprotectant; success with embryos > 250 μm in diameter is much less likely. Cooling is controlled so that most of the water leaves the cells before intracellular ice forms and glycerol is removed after thawing without undue osmotic swelling of cells. A recent review by French workers describes the structure of horse embryos, the principles of freezing, methods of studying the effects of freezing, the effect of cryoprotectants on embryos and features of embryos and their freezing specific to horses. The authors make it clear that equine embryos are much more sensitive to low-temperature preservation than cattle embryos; to date, the greatest success in freezing has been with small, day-6 embryos.

Pig embryos

The vitrification of pig embryos by Vajta's ultra-rapid cooling method (OPS) was tested by Berthelot *et al.* (2000) in France; the method allowed cryopreservation of blastocysts but it was concluded that the method would have to be improved before success was possible with morulae. A further report by Berthelot *et al.* (2003) reported more than 200 piglets born using the OPS procedure; the authors note that the method is reliable, efficient, simple and easy to implement. In the USA, Dobrinsky *et al.* (2000) documented microfilament (MF) alterations during pig embryo vitrification using an MF inhibitor to stabilize MF during cryopreservation; they demonstrated that cytochalasin-B did not improve morula/early blastocyst viability but did significantly improve the survival and development of expanded and hatched blastocysts. It was concluded that the pig embryo cytoskeleton can be affected by vitrification and that MF depolymerization prior to vitrification improves blastocyst developmental competence after cryopreservation; the vitrified embryos were capable of producing live, healthy piglets after transfer.

Camelid embryos

In Argentina, Palasz *et al.* (2000) suggested that ethylene glycol may be the preferred cryoprotectant for the cryopreservation of llama embryos. Other work in the same country evaluated the viability of vitrified llama embryos after transfer to recipient females (Aller *et al.*, 2002); their results demonstrated the effectiveness of the vitrification technique for the cryopreservation of llama hatched blastocysts.

3.2.5. Donor–recipient synchrony

Over the years, much evidence has accumulated on the importance of synchrony between donor and recipient in terms of their cycle stage. Exact synchrony should be the aim, but recipients out of phase by ± 1 day are generally regarded as acceptable, although some reduction in pregnancy rate is to be expected; cattle that are out of synchrony by as much as 2 days would not normally be used because of the reduced pregnancy rates. Some workers have looked at ways of making synchrony as exact as possible. In Arkansas, for example, the use

of an electronic oestrus detection system to continuously monitor cattle permitted more precise timing of ET and resulted in improved pregnancy rates; Rorie *et al.* (2002) reported data suggesting that continuous monitoring of embryo donors and recipients and selection of recipients with synchrony of ±12 h could improve pregnancy rates (Table 3.9).

Buffaloes

Many studies have dealt with factors affecting the pregnancy rate following non-surgical ET in the buffalo. Indian workers have recorded the highest pregnancy rate (42%) when donors and recipients were closely synchronized; according to these researchers, pregnancy rate was compromised when recipients were in oestrus either 12 h before (14%) or 12 h after (19%) oestrus in donors; asynchrony beyond 12 h on either side resulted in conception failure. Such findings may indicate the need for synchrony to be even closer in buffaloes than in cattle to achieve pregnancies.

Horses

In the mare, as in the cow, recipient management is a critical factor influencing the establishment of pregnancy after ET. The highest pregnancy rates are achieved with recipients that ovulate either the day before the donor or up to 3 days after the

Table 3.9. Embryo–recipient synchrony and pregnancy rates (from Rorie *et al.*, 2002).

Oestrus synchrony category (h)	Number of embryo transfers	Pregnancy rate (Mean ± SEM)
−12 to −24	37	51.4 ± 8.2
0 to −12	67	58.2 ± 6.1
0	9	66.7 ± 16.6
0 to +12	78	61.5 ± 5.6
+12 to +24	37	48.6 ± 8.2
0 to ±12	126	62.7 ± 4.4[b]
±12 to ±24	102	50.0 ± 4.9[a]

Numbers within columns with different superscripts differ significantly ($P = 0.054$). SEM, standard error of the mean.

donor mare. According to Squires *et al.* (2003), the most important points in selecting a recipient are uterine tone and cervical tone prior to transfer; mares showing excellent tone have higher pregnancy rates than recipients with marginal uterine and cervical tone. Most recipients are intact mares, but alternatives include ovariectomized, progesterone-treated mares and non-cycling, progesterone-treated mares early in the breeding season.

Sheep and pigs

In sheep, it is well accepted that the main variables affecting ET are embryo quality and the degree of synchrony between donors and recipients (Alabart *et al.*, 2003). In a study reported by Hazeleger *et al.* (2000b), pig embryo survival was determined after non-surgical transfer to recipients with a variable synchrony of ovulation. They concluded that recipients should ovulate between 24 h before and 12 h after donors; transfers to recipients ovulating 18–36 h after the donors led to very low pregnancy rates.

3.2.6. Non-surgical embryo transfer procedures

Cattle

Work in Ireland in the mid-1970s and elsewhere showed that it is possible to establish pregnancies by a non-surgical procedure involving the use of the standard Cassou inseminating instrument. The embryo is loaded, held in a small volume of medium (e.g. phosphate-buffered saline (PBS) supplemented with 15% bovine serum), into a plastic straw (usually 0.25 ml capacity). At transfer, the straw is inserted into the inseminating instrument ('gun') in the usual way and the same procedure followed as for AI, the main difference being that the embryo is deposited around the mid-horn position (ipsilateral horn); before carrying out the transfer, the recipient animal is given an epidural anaesthetic and tranquillizer. During the past quarter-century, several variants of the standard transfer instrument have been marketed, with appropriate modifications to ensure that the embryo is deposited safely in the uterus. Factors that have been shown to influence the success of non-surgical ET in cattle are shown in Fig. 3.7.

Horses

Pregnancy rates in mares after non-surgical ET are usually lower than for surgical transfer and are highly dependent upon operator experience and expertise to pass the sterile transfer pipette through the closed dioestrous cervix. Cambridge workers have reported a new transfer method which obviates the need for such operator skill; they reported an 85% pregnancy rate, a marked improvement on rates previously achieved (50–55%) using the conventional transfer method (Wilsher and Allen, 2003). It was concluded that the new technique is manipulatively uncomplicated, simple to perform and remarkably successful.

3.2.7. Selection and management of recipients

Maiden heifers are usually preferred as recipients in conventional cattle ET operations; quite apart from being free of problems arising from previous pregnancies, such animals are likely to cost less and be easier to acquire than cows. However, in terms of ease of transfer, the parous cow has a definite advantage; various studies have shown that 10% or so of heifer recipients may be difficult, if not impossible, to use for cervical transfers. Certain categories of maiden heifers as recipients may also pose very real welfare problems. In the UK, for example, the use of beef-type heifers as recipients for embryos from large dairy breeds (e.g. Holstein–Friesian) and double-muscled beef breeds has occasionally resulted in a proportion requiring surgery to deliver the fetus. Clearly, it is undesirable for unsuitable embryos to be transferred to recipients; legislation has been enacted in

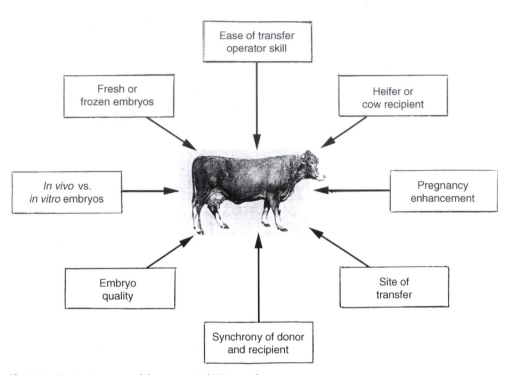

Fig. 3.7. Factors in successful non-surgical ET in cattle.

several countries to prohibit transfer of embryos likely to produce large calves.

The importance of minimizing stress in recipient animals is rightly emphasized in various reports. Any routine treatment (e.g. antiparasitic) should take place at least 3 weeks prior to transfer; changes in the feeding regimen should be prohibited for 3–4 weeks before and after transfer. Recipients should be located where they can be easily and quietly handled on the day of transfer. Several authors stress the need for ET teams to examine their procedures to reduce stress and improve the welfare of all animals involved in their activities.

Use of NSAIDs

Ibuprofen is a non-steroidal anti-inflammatory drug (NSAID) which has a number of beneficial actions in addition to its analgesic and antipyretic effects. There is evidence that substances that inhibit cyclo-oxygenase enzyme isoforms may improve IVF outcome in humans; treatment may take the form of a daily dose of 100 mg

aspirin. It is also known that ibuprofen improves pregnancy rates in recipient cattle after ET and may be a useful, effective and safe adjunct to assisted reproduction in cattle (Elli *et al.*, 2001).

3.3. Practical Applications of Embryo Transfer

3.3.1. Embryo transfer and breeding improvement

Although MOET is the name of a famous champagne, in farming circles it usually refers to multiple ovulation and embryo transfer, a method of increasing reproductive potential. MOET can be used to increase female selection intensity and the rate of genetic improvement. The method can also be used to breed replacements from younger females, thereby reducing the generation interval and speeding up genetic improvement by that means. The initial thinking on MOET was that it could produce substantial increases in the rate of genetic improvement in any species in which the natural reproductive rate is low and that, if high rates of ET could be achieved, the rate of genetic improvement could even be doubled. It was recognized that serious limitations to the effectiveness of MOET programmes in cattle could be the low average and high variability of embryo numbers. However, as shown in Chapter 4, it is now feasible to think in terms of overcoming such limitations by the non-surgical harvesting of oocytes (ovum pick-up (OPU)), with subsequent *in vitro* maturation and IVF, as the means of yielding large numbers of transferable embryos.

3.3.2. Embryo transfer and gender preselection

The first success with gender preselection in cattle was a calf born in 1975 in Canada after an embryo was sexed by chromosomal analysis. Many thousands of sexed calves have been born since that time, many of them by way of embryo sexing. In the Czech Republic, Lopatarova *et al.* (2003) reported that sex determination in bisected cattle embryos was an excellent method of increasing the numbers of sex-desired calves from a limited number of embryos; sexing was achieved by detection of male-specific, Y-chromosomal DNA sequences, using PCR.

Obtaining embryos of the desired sex for use in cattle ET could be one way of improving the biological and economic efficiency of beef and dairy cattle production systems. Until the present time, as mentioned above, the way to achieve this particular goal was to apply PCR technology to a biopsy sample taken from an embryo (Table 3.10). The problem with this approach is that frozen embryos after biopsy may have low pregnancy rates and half the available embryos may have to be discarded because they are not of the required sex; all this can be costly and time-consuming. Now, however, there is the possibility that embryo sex can be predetermined using frozen sperm sorted by

Table 3.10. Embryo sexing (by PCR) in cattle (based on data from Bousquet *et al.*, 2003).

	In vivo	In vitro
Number	879	3032
Males	429 (48.8%)	1696 (55.9%)
Females	407 (46.3%)	1168 (38.5%)
Unknown	43 (4.9%)	168 (5.6%)
Transferred fresh	326	1004
Pregnancies (day 30)	186 (57%)	532 (52%)
Pregnancies (day 60)	(55.2%)	(48.9%)

flow-cytometry to inseminate donor animals in an ET programme. In Argentina, Panarace *et al.* (2003) have demonstrated that sexed semen can be successfully used for this purpose in both cows and heifers. Details of semen sexing are in Chapter 2.

3.3.3. Genetic preservation of endangered breeds

There is considerable interest in programmes for the preservation of livestock breeds that may be in danger of extinction, in both developed and developing countries. Of the 3831 breeds or breed varieties of donkeys, buffaloes, cattle, goats, horses, pigs and sheep believed to exist or to have existed in the past century, 618 (16%) have apparently become extinct. A few years ago, the Food and Agriculture Organization (FAO)'s Global Data Bank for Domestic Livestock, which carried 2047 entries, showed 221 cattle breeds to be at risk, most of these (60%) in the developed countries; intensification of agriculture in the Western world has led to a greater reliance on a small number of breeds, with consequent neglect of the remainder. In developing countries, threats to genetic diversity usually take the form of increased use of AI and indiscriminate crossbreeding of indigenous breeds. The intensification of farming in these countries can mean that indigenous breeds are in danger of being pushed to extinction because native farmers, aiming at greater productivity, employ exotic breeds such as Holsteins and Friesians.

According to the FAO, it is essential to use the largest number of animal species for breeding in order to preserve genetic resources. A thousand species have been lost during the last century and today it is estimated that one-third of breeding animals are threatened with extinction. There are those who regard it as an obligation to humankind to preserve biological diversity, and this applies to farm animals as well as to many wild species; apart from scientific considerations, there are also likely to be cultural and ecological reasons. It is believed that the domestication and use of animals by humans dates back some 10,000–12,000 years. Although a relatively minute fraction of the total number of animal species was domesticated by humans (Table 3.11), within these species many hundreds, if not thousands, of distinct breeds were established over the years by humans in countries around the world. As observed by Hodges (2002), this vast array of breeds is a human heritage worthy of conservation. However, although humans were responsible for building up numerous breeds of farm animals over the centuries, in the past 50 years they have also been

Table 3.11. The limited extent of animal domestication.

Insects	Bee; silkworm
Birds	Chicken; turkey; duck; goose; swan; guinea-fowl; peafowl; pigeon; ostrich
Small mammals of limited utility	Dog; cat; rabbit
Important mammals	Horse; donkey
	Arabian and Bactrian camels
	Llamas and alpacas
	European and zebu cattle; buffalo, yak; gayal; banteng; sheep and goat
	Deer
	Pig

responsible for the destruction of many of these same breeds.

The disappearance of many breeds has usually been in the name of progress, driven by intensification of food production methods, which has favoured the most productive breeds (Fig. 3.8). In Ireland and the UK, the brown and white of the Dairy Shorthorn has been totally replaced by the black and white of the Holstein–Friesian.

Fig. 3.8. Norfolk Horn sheep – Cambridge 1954.

The predominant use in the developed countries of breeds with high economic potential has seen Large White and Landrace pigs sweep the board in Europe and these two breeds plus the Duroc and Hampshire in North America. In poultry, it has been White Leghorn-derived hybrids for egg production and White Plymouth Rock plus White Cornish-derived hybrids for meat production. Fortunately, there have been a far-sighted few who saw the coming threat of animal monoculture and have advocated appropriate animal conservation measures.

ET technology is now regarded as a vital tool for genetic preservation of endangered species and breeds; it enables the establishment of embryo banks and embryos to be transferred into populations with decreased biodiversity. In native breeds of cattle, ET can be used to preserve genetic lines with good maternal characteristics, fertility, acclimatization to extreme climatic or nutritional conditions and natural resistance to disease. Selection of breeding animals with the most desirable traits and the dissemination of their genes throughout the population are the keys to increasing the productivity and profitability of dairy cattle.

Interspecific embryo transfer

The transfer of embryos of a given species to the uterus of a different species is performed in animal conservation programmes where there is a lack of a suitable number of recipients. It is believed that interspecific ET could prove to be a key technique in the conservation of endangered species. It is evident that compatability between different species is only partial in most cases. It is, for example, well established that the oocytes of some species can be matured and fertilized in the oviduct of a different species and that the resulting embryos can develop normally once transferred to the reproductive tract of a female of their own species. It is clear that farm-animal embryos are normally tolerant and tolerated in the oviduct of a different species (e.g. the rabbit) until they reach the blastocyst stage.

However, most interspecific ET is unsuccessful and, in those few instances where pregnancies are established, most fetuses do not survive to term. In the genus *Ovis*, there has been success in the transfer to domestic sheep (*Ovis aries*) recipients of embryos from moufflon (*Ovis musimon*) and Armenian red sheep (*Ovis orientalis*). The first report of successful interspecific pregnancies in the genus *Capra* was published in 1999, showing that it was possible to achieve pregnancies after transfer of ibex embryos into domestic goats; it was also shown that live ibex kids can be produced when a domestic goat embryo shares the uterus with an ibex embryo.

3.4. Future Developments

3.4.1. Embryo transfer as a research tool

It should be noted that active research programmes, involving ET in horses, are currently being conducted in several countries, particularly the UK, the USA, France and Italy, providing a valuable flow of new information. Special mention should be made of Cambridge studies in which ET was used to produce horse foals from jenny donkey recipients, horse and donkey foals from F1 hybrid mules and Przewalski's horse and Grant's zebra foals from domestic horse mares; such studies have illuminated the various developmental and immunological factors that permit the intrauterine survival and growth of the equine fetus. On a more practical level, the Cambridge studies have demonstrated the tolerance of the horse uterus and the maternal immune system to the development of embryos from other members of the genus *Equus*. Research with embryos has already shown maternal age to be an important cause of embryo mortality in mares; embryo survival has been shown to be significantly higher using embryos from normal than from subfertile mares, pointing to defective oocytes being ovulated by aged, subfertile mares.

4

In Vitro Embryo Production

4.1. Advantages of *In Vitro*-produced Embryos

There are many reasons for interest in embryos that can be produced in the laboratory, rather than recovered from the living animal. Progress in cellular and molecular embryology in farm animals has been difficult in the past due to the limited availability of suitable experimental material at an acceptable cost. Although oocytes and embryos can come from superovulated donor animals, this is likely to be expensive and not always free from animal welfare concerns. For such reasons, *in vitro* pro-

duction (IVP) techniques, particularly those based on ovaries recovered after the donor's demise, have received much attention in the past 10–15 years (Galli and Lazzari, 2003). Europe has been at the forefront of applying such technologies (Fig. 4.1).

4.1.1. General considerations

In the commercial cattle embryo transfer (ET) industry, *in vitro* embryo production is now an alternative to conventional means of obtaining embryos for transfer, using immature oocytes collected from the

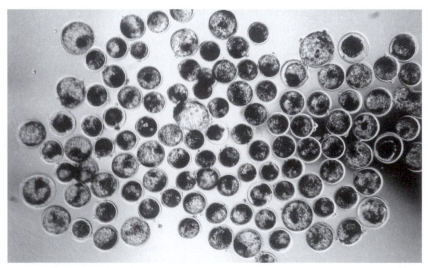

Fig. 4.1. *In vitro*-derived cattle embryos *en masse*.

©I.R. Gordon 2004. *Reproductive Technologies in Farm Animals* (I.R. Gordon)

ovaries by ovum pick-up (OPU) of donor cattle of differing ages and physiological states. Reliable methods are now available for the maturation and fertilization of bovine oocytes *in vitro*; culture methods, although still imperfect, enable embryos to be grown to the stage at which they are suitable for transfer or cryopreservation. *In vitro* embryo production involves three steps, which have been developed to the greatest extent in cattle: oocyte *in vitro* maturation (IVM), *in vitro* fertilization (IVF) and embryo culture (IVC (*in vitro* culture)).

During the recent decade, IVP of bovine embryos has become a routine research tool in many laboratories; on the farm, the method is employed in commercial cattle breeding programmes in several countries. Statistics gathered for 25 European countries in the year 2003 showed more than 8000 *in vitro*-produced (IVP) embryos for commercial use, oocytes being collected either by aspiration from abattoir ovaries or by OPU from live donors (Table 4.1). Although commercial uptake of such embryo-based biotechnology in cattle remains limited in the context of overall cattle ET activity, on the research front the ability of laboratories to generate large numbers of embryos is likely to have a considerable impact on the accumulation of biological knowledge on factors influencing early embryonic development in this species. In due course, marked improvements in embryo production technology can be expected, which will help to improve the efficiency of cattle production. OPU is a very flexible procedure which does not interfere with the productive and reproductive career of donors, yet allow the production of more embryos than conventional super-ovulation. None the less, IVP of cattle embryos remains technically demanding and requires specific laboratory expertise and equipment to ensure the production of high-quality embryos.

Human IVF

Human IVF has led to the birth of an estimated 1 million and more babies world-wide since the day in 1978 when Louise Brown was born in England (Henig, 2003). At that time, media attention verged on the hysterical, with reports suggesting that test-tube babies were 'the biggest threat since the atom bomb'. Although such predictions turned out to be quite wrong, there are those, some with the benefit of hindsight, who have introduced a strong note of caution about applying human-assisted reproduction techniques without a solid body of research evidence in support. It is now recognized that the unregulated nature of human IVF failed to put appropriate emphasis on follow-up studies on the children born after IVF.

There are those who have commented that techniques with such an impact on human welfare should have been under government-sponsored regulation from the start so that appropriate follow-up information would be guaranteed. In the USA, for example, the National Institute of Child Health and Human Development never funded human IVF research in any form. There are those, such as Winston (2003) in the UK, who draw attention to worrying aspects of more recent advances, such as intracytoplasmic sperm injection (ICSI). In this context, there is an obvious need for those engaged in the animal field to share their experiences with those in the human field.

Table 4.1. Cattle embryo transfer activity in Europe in 2002 (from Lonergan, 2003).

	Number of embryos transferred
From *in vivo*-produced embryos	90,371
From *in vitro*-produced embryos	8,167
Total number of embryos transferred	102,176
Proportion of IVF embryos transferred	8.29%
Proportion of frozen embryos transferred	54.6%

4.1.2. Cattle and buffaloes

In vitro embryo production procedures are applied worldwide with different objectives for various livestock species and exotic, wild and endangered animals. In breeding improvement programmes, there are several ways in which *in vitro* embryos may be useful; they may be produced at earlier ages and in much greater numbers than is possible with conventional ET technology. Genetic gain depends on the intensity and accuracy of genetic selection and the generation interval; the possibility exists in cattle that the generation interval could be shortened by at least a year using IVP embryos from juvenile or prepubertal heifers. The ability to repeat OPU–IVP procedures once or even twice a week means that the number of calves that can be obtained per unit of time can be four to five times as large as with conventional ET, even after taking account of the moderate rate of IVP blastocyst production under current culture conditions.

Looking to commerce, the technology may be used to produce embryos that can be applied in beef production. According to New Zealand workers, for example, the benefit of using IVP to produce beef calves in dairy herds would be most marked when the beef calf replaced a dairy calf of a type (e.g. Jersey) unsuitable for beef production. In fact, the first company to attempt the commercialization of *in vitro* embryo production was set up in Ireland in 1987; the aim of Ovamass was the production of high-quality, low-cost beef embryos for twinning, using technology developed and patented by workers in University College, Dublin (Fig. 4.2). At that time in Ireland, there was a scarcity of calves for beef rearing and it was thought opportune to apply the technology. Unfortunately, the problems encountered in the early days of applying the technology were to prove a bridge too far; by the time technical difficulties were overcome, the demand for beef calves had gone.

Buffaloes

In vitro embryo production in buffaloes, for both research and commerce, progressed much more slowly than in cattle. The birth of the first IVP calf was reported in 1991 in India in the laboratory of Madan and associates. The smaller follicle population in the ovaries and lower rate of IVM made the technology much less successful in buffaloes. Chinese workers, for example, showed that buffalo ovaries contained an average of five oocytes per ovary compared with 14.3 for cattle. The general consensus is that much more research is required before buffalo IVP embryos can be produced efficiently for commercial use. The need is to develop methods that will enable greater numbers of oocytes to be obtained from animals. Follicle aspiration by OPU at regular intervals and, perhaps at some point in the future, culture techniques enabling the growth of pre-antral follicles may have particular relevance for this species. In the meantime, Italian workers have shown the OPU technique to be much superior to superovulation in terms of monthly yield of transferable quality embryos (2 vs. 0.6). The buffalo is an important livestock resource, occupying a critical niche in many ecologically disadvantaged agricultural systems, providing milk, meat and draught power. Because of its high milk production, the Mediterranean buffalo is in high demand around the world, as well as important for the local economy; this calls for special efforts to rapidly multiply genetically superior males and females. The application of OPU technology, together with multistep embryo production *in vitro*, could help in meeting this need (Table 4.2).

4.1.3. Sheep and goats

The use of embryo production in sheep, particularly embryos derived from juvenile donors, has considerable room for improvement, with further research required to improve oocyte cytoplasmic maturation, as well as the cryopreservation of embryos. Reviewing progress in advanced reproductive technologies in sheep, Cognie *et al.* (2003) note that the optimization of IVP procedures is a major challenge to those involved in the production and propagation of transgenic and cloned animals.

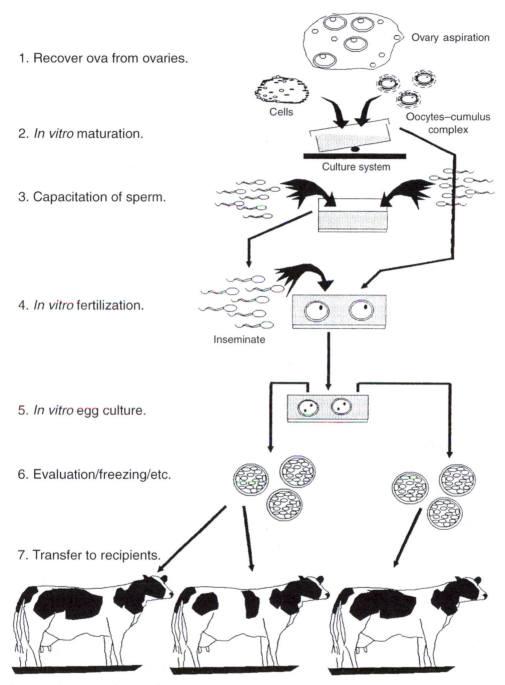

1. Recover ova from ovaries.

2. *In vitro* maturation.

3. Capacitation of sperm.

4. *In vitro* fertilization.

5. *In vitro* egg culture.

6. Evaluation/freezing/etc.

7. Transfer to recipients.

Fig. 4.2. Ovamass embryo production system.

According to Koeman *et al.* (2003), IVP of goat embryos is a rapidly advancing field; the production of valuable transgenic goats, capable of producing substances of pharmaceutical value in their milk, has encouraged the development of techniques that can support the propagation of these animals. The use of prepubertal goats would

Table 4.2.   Repeated oocyte recovery in buffaloes (from Zicarelli *et al.*, 1996).

Times of OPU	No. of buffaloes	No. of follicles		No. of oocytes		Recovery rate %
		n	Mean ± SD	*n*	Mean ± SD	
First	8	50	6.3 ± 6.0	43	5.4 ± 6.0	83.2
Second	8	66	8.3 ± 2.8	54	6.8 ± 3.0	82.5
Third	7	52	7.4 ± 1.9	42	6.0 ± 2.3	81.3
Fourth	8	36	4.5 ± 1.2	27	3.4 ± 2.1	68.8
Fifth	8	52	6.5 ± 3.5	46	5.8 ± 3.6	87.9

SD, standard deviation.

be regarded as particularly helpful; collection of oocytes at an earlier age would reduce the generation interval and accelerate the propagation of genetically valuable animals. For such reasons, much research is currently devoted to factors involved in determining the developmental competence of oocytes from prepubertal and adult goats. In goats, unlike cattle, a limiting factor in many countries to obtaining goat oocytes from abattoir sources is the relative scarcity of commercial enterprises capable of providing consistent supplies of suitable ovaries. In Spain, for example, there are commercial abattoirs where goats are slaughtered regularly for meat, but these animals are generally only 2 months old; although oocytes from such animals are capable of undergoing maturation, they do not have the same capacity for fertilization and normal development as those from older animals.

Some workers suggest that the recovery of oocytes by non-invasive transvaginal ultrasound-guided follicle aspiration (OPU) for *in vitro* embryo production may prove to be a viable alternative to conventional ET in the goat. In contrast to does subjected to repeated laparoscopic oocyte recoveries, the problem of adhesions is much less likely to arise.

4.1.4. Pigs and horses

IVP of porcine embryos is of considerable interest, not only because of its economic significance to production agriculture, but also to its great potential for application in biomedical research. Many believe that techniques such as ET, which permit simultaneous control and manipulation of the genetic quality and health status of selected animals, will become increasingly important for internationally based pig-breeding companies. The development of effective embryo production methods in the pig is likely to be important in facilitating the application of this technology in several areas; for commercial application, the combination of IVP embryos with effective non-surgical transfer techniques would have considerable potential. Instrumentation for non-surgical ET has recently been developed and considerable progress has been made in the cryopreservation of porcine embryos by vitrification. However, despite much research, the successful production of IVP embryos lags some way behind that of cattle, the problem of polyspermy remaining a major issue.

In the normal mating of pigs, in which the incidence of polyspermy is lower than 5%, it is believed that there are two major mechanisms responsible for keeping this problem under control: (i) the number of capacitated sperm reaching the site of fertilization in the pig's reproductive tract is less than 100 cells per oocyte; and (ii) the exocytotic reaction of oocytes to prevent further sperm penetration. In laboratory embryo production, on the other hand, abnormally high numbers of competent sperm reach the oocyte surface almost simultaneously and mechanisms designed to prevent penetration by more than one spermatozoon prove less than adequate.

Pigs have become increasingly important in biomedical research and there is considerable interest in the potential use of transgenic pigs in such research. Most attempts to produce transgenic pigs use *in vitro*-matured (IVM) oocytes and early embryos; for that reason, there is a need to produce large numbers of developmentally competent oocytes. In the past two decades, as described by Day (2000), the technology has progressed from research on the formation of male and female pronuclei in pig oocytes matured *in vitro* to the production of litters of live piglets after transfer of advanced cleavage-stage IVP embryos to recipients. During the most recent decade, blastocyst development rates approaching those in cattle and sheep have been achieved.

None the less, pig embryo production presents problems, such as incomplete cytoplasmic maturation, leading to the development of blastocysts with reduced cell numbers and compromised viability. Evidence presented by Kidson *et al.* (2003) indicated that the presence of porcine oviductal epithelial cells during IVM enhanced the quality of cytoplasmic maturation of the pig oocyte and the subsequent blastocyst cell proliferation; although the yield of blastocysts was reduced in the presence of oviductal cells, those blastocysts that did form were of superior quality. It is possible that the oviductal cells provided beneficial factors, such as growth factors or scavenger ions, which have a role in the functioning of the ooplasm. In Spain, Selles *et al.* (2003) used a porcine IVF system to assess the quality of frozen–thawed boar semen prior to its commercial use, to verify the quality of stored semen and to evaluate new sperm freezing procedures; it was concluded that this method of evaluating semen was preferable to examining semen by the more conventional route.

Horses

One advantage of IVP technology in horses would be as an adjunct to conventional ET activities, where it could be employed to obtain foals from subfertile mares, competition mares and mares that otherwise might have difficulty in achieving or maintaining a pregnancy of their own. As in cattle, there might be value in providing a genetic salvage service in instances where the death of a valuable mare occurs without warning. It is also possible that the production of horse embryos by *in vitro* techniques could find a useful place in the study of many aspects of reproduction in this species, as well as enabling progress to be made in areas of commercial interest. The availability of live horse embryos less than 6 days of age, a requirement for certain lines of research (e.g. embryo cryopreservation), is currently limited unless there is recourse to costly surgical intervention; the laboratory production of horse embryos could be a useful solution to such problems.

Compared with cattle, however, progress in producing IVP equine embryos has been much slower. Cabianca *et al.* (2003) note that techniques are still in a preliminary phase and far from being ready for field application. This is partly due to difficulties in obtaining abattoir ovaries as a source of oocytes and in recovering oocytes by OPU in the live animal. In many countries, the slaughter of horses is found in only a limited number of places and the number of mares slaughtered is often quite low. A further difficulty, for reason of distance, was the fact that oocytes could not be aspirated until 18–24 h after ovary collection. In contrast to the cow, follicle aspiration yields a low recovery rate in the mare. In Denmark, workers found the oocyte recovery rate to be higher for aspirations at intervals of 23 or more days than for aspirations at 6-day intervals (35.8 vs. 18.4%). Difficulties in achieving an acceptable fertilization rate in the mare turned the attention of researchers to ICSI, a procedure used with great technical success in human assisted reproduction; in horses, reports from several countries show sperm injection to be much more effective than conventional IVF. Those IVF- and ICSI-derived foals that have been born have usually followed the transfer of early embryos into the Fallopian tubes of surrogate mothers.

4.1.5. Deer

The development of *in vitro* embryo production systems for red deer (*Cervus elaphus*) has been reported by New Zealand workers (Berg *et al.*, 2002); they recorded the birth of five normal calves after transfer of IVP embryos into synchronized recipients. This was evidence that developmental competence of oocytes recovered from slaughterhouse ovaries could be achieved by the techniques described. Maturation of red deer oocytes followed a similar time course to that reported for cattle and sheep oocytes (i.e. maturation took 24 h). The development of effective IVP techniques in deer could be valuable in the *in vitro* hybridization of different deer species which do not readily hybridize naturally (e.g. Pére David's and red deer).

4.2. Growth and Development of Technology

4.2.1. Historical

As a result of increasing knowledge of essential factors influencing the maturation of mammalian oocytes and sperm capacitation, a method of maturing and fertilizing cattle oocytes *in vitro* was developed in Dublin in the mid-1980s. By the autumn of 1987, this work had led to the birth of 18 calves to 13 surrogate cattle and a considerable body of evidence supporting the view that embryos could now be produced capable of establishing normal pregnancy rates in cattle (Fig. 4.3). Initially, an *in vivo* culture period in the sheep oviduct was employed to enable embryos to reach the blastocyst stage for transfer, but, late in 1987, the first calves were born from a totally IVP system, the culture of embryos being by way of the oviductal cell co-culture method previously developed in Cambridge. The world's first commercial company for cattle embryo production (Ovamass Ltd) was established in Dublin and during 1988 transfers of laboratory-produced embryos to more than 1000 recipient cattle were carried out in 58 herds in the Irish Republic and Northern Ireland.

The indications at the time suggested that the *in vitro* embryos were capable of establishing pregnancy rates and tolerating freezing much the same as *in vivo* embryos produced by superovulation procedures. In the 1988 field trials, the Ovamass embryos, after freezing, yielded results little different from those commonly experienced at that

Fig. 4.3. IVF calves and their surrogate mothers.

time with *in vivo*-derived embryos (pregnancy rates of 50.3% for 803 'fresh' transfers vs. 43.1% for 308 frozen embryo transfers). With the abandonment, under commercial pressure, of the sheep oviduct as the early embryo culture system and the development of somatic-cell co-culture systems came evidence that the embryo's ability to survive had seriously deteriorated; pregnancy rates in many cases with frozen embryos were halved, making commercial application uneconomic and impractical.

In 1989, Ovamass was assimilated into Animal Biotechnology Cambridge (ABC), the Irish-based laboratory concentrating on refining embryo production technology and ABC in Cambridge devoting its attention to embryo sexing and cloning technology. Despite the fact that IVP technology was greatly strengthened by the research and field trials conducted in Ireland in the early 1990s, when it came to applying the technology in commercial practice, insurmountable financial difficulties arose. ABC relocated its research staff from Ireland to Cambridge and concentrated attention on the semen-sexing side of its research and development activities. In this latter regard, the first calves born from semen sexed by the Beltsville sorting technique were the result of ABC's efforts in the UK. Elsewhere in the world, *in vitro* embryo production was commercialized in countries such as Japan and Italy as a means of producing premium beef calves from dairy herds (Galli and Lazzari, 2003).

4.2.2. Abattoir materials

Cattle ovaries

Although superovulation remains the chosen method of producing high-quality bovine embryos for most commercial ET purposes, cost may well make this prohibitive for many research programmes. Abattoir materials, on the other hand, form an inexpensive and readily available source of oocytes for research in embryo production and for use in cloning and the production of transgenic animals. In the production of cattle embryos from slaughterhouse ovaries,

it was necessary to develop methods that permit the recovery of several good-quality oocytes per ovary; the number actually recovered will vary with different collection procedures. In Dublin, oocytes were collected from heifer cattle slaughtered for beef in local abattoirs and their ovaries brought to the laboratory within an hour or so of slaughter. A comprehensive account of cattle oocyte sources and the methods used in recovering them from abattoir ovaries is provided in articles and books; it is important that appropriate time limits and temperatures are observed in the collection of oocytes (Fig. 4.4).

Equine ovaries

In horses, recent studies have shown that the equine ovary may be stored after excision from a slaughtered mare for at least 6–8 h at temperatures of 37–27°C during transport to the laboratory. Where oocytes from slaughtered animals are used in embryo production, the time taken between the abattoir and the laboratory and IVM culture may be lengthy and can, like the temperature during transport, adversely affect oocyte developmental competence.

Sheep ovaries

In sheep, storage times of 4, 8 and 24 h after slaughter were found to have a negative effect on oocyte competence but temperature during storage had its own effect, allowing a higher maturation rate at 22°C than at 5 or 37°C. Temperature effects were also evident in studies in which oocytes from ovaries held at 20°C during transport showed greater competence than those transported at 30 and 37°C.

4.2.3. Ovum pick-up (OPU) technique

An alternative to oocyte collection from slaughterhouse ovaries is OPU from live donor females (Fig. 4.5). Workers have developed OPU devices that are practical and economical for routine use in oocyte retrieval, without causing negative effects

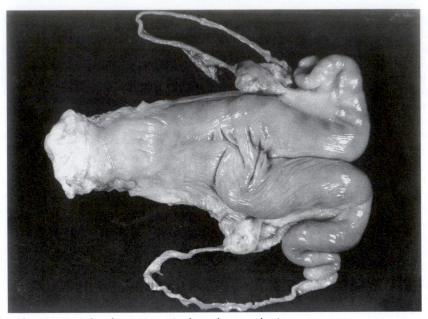

Fig. 4.4. Abattoir material as the starting point for embryo production.

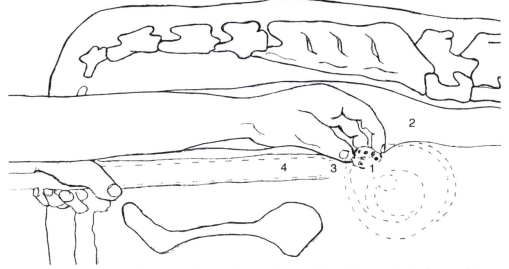

Fig. 4.5. Ovum pick-up in the cow. Schematic diagram of the position of the ovary (1) in the cow during transvaginal puncturing of follicles. With one hand in the rectum (2) the ovary is manipulated against the anterior part of the vagina, just beside the cervix (3). With the other hand, the transducer (4) is advanced into the vagina and as the needle is pushed forward it penetrates the vaginal wall and punctures the ovary (from Kruip *et al.*, 1991).

on ovarian structure and subsequent ovarian function. Although donor cows used for OPU vary considerably in their yield of oocytes, there is evidence that a certain oocyte number may be characteristic of each donor animal and determined by its genetic constitution (Lopes *et al.*, 2003).

In using OPU, a scanner with an intravaginal sector probe and a guided needle is employed; the needle is connected to a test-tube and to a vacuum pump to aspirate the follicular fluid and the oocyte contained within it. Using an ultrasound scanner with good resolution, it is possible to envisage ovarian follicles down to 2–3 mm in size; the recovery procedure can be carried out on the farm with the donor sedated and confined in a crush. It is possible to collect oocytes not only from cyclic heifers and cows but even from animals in the early months of pregnancy, from those in the early post-partum period and from prepubertal heifers.

After collection, oocytes are maintained in media such as Dulbecco's phosphate-buffered saline (PBS) or tissue culture medium (TCM)-199 (HEPES-buffered); it is clear that oocytes are extremely sensitive to temperature shock, making it important to monitor temperature carefully during the collection procedure. Among the methods available to improve cattle embryo production efficiency is the follicle-stimulating hormone (FSH) pre-stimulation of donors prior to OPU (Merton *et al.*, 2003). In the Czech Republic, Cech *et al.* (2003) also demonstrated that pre-stimulation of dairy cows with FSH increased the number of oocytes recovered by OPU in the first trimester of pregnancy.

Equipment, methods and practical applications of OPU are described by Christie *et al.* (2002), who also drew attention to the benefits and associated problems of this new technology. Although OPU/IVF can produce more embryos over a given period of time than conventional superovulation and embryo recovery, the *in vitro*-derived embryos have proved to be significantly less viable than their *in vivo* counterparts; there is also a higher incidence of abortion, perinatal mortality and fetal and placental abnormalities and a tendency to increased birth weights among IVP embryos. None the less, a large majority of IVF pregnancies do result in normal healthy calves and the technique provides a valuable new means of producing additional offspring from donors unable to produce calves by any other means (Table 4.3).

OPU in lactating cattle

A study by Walters *et al.* (2002) suggested that conditions associated with early lactation in Holstein cows had a negative effect on oocyte quality; in Ireland, however, Rizos *et al.* (2003b) showed that, in the early post-partum period, lactating Holstein–Friesian cows submitted to OPU, while having fewer follicles for aspiration than nulliparous heifers, yielded oocytes with similar developmental competence.

Seasonal effects on oocyte quality

In Israel, Roth *et al.* (2002) used pre-stimulation treatments with FSH and growth hormone (GH) in attempts to improve oocyte quality in the autumn period. This was on the basis of evidence that conception rate in dairy cows during autumn remains low even after summer temperatures decline, due to a residual effect of heat stress on oocyte quality.

OPU and cow well-being

In France, Chastant-Maillard *et al.* (2003) reported on the consequences of trans-vaginal follicular oocyte recovery (OPU) for the well-being of cows; the evaluation relied on the physiological measurement of

Table 4.3. Calves after OPU and superovulation (adapted from Bousquet *et al.*, 2003).

	In vitro embryos		Conventional embryos
	Adult cows	Prepubertal heifers	Superovulated donors
Frequency of collections	4/60 days	4/60 days	1/60 days
Collections/year	24	24	6
No. transferable embryos/year	96	48	24
Pregnancy rate	53%	53%	60%
Number of calves	50	25	14

stress, milk production criteria, immune status and a histological examination of ovaries. The authors concluded that repeated OPU by transvaginal follicular puncture had no adverse effect on the welfare of the animals. Unlike the application of conventional cattle ET, which can interfere markedly with the resumption of normal reproduction in the donor animal, the OPU method may appeal because it can be applied at frequent intervals without adverse effects on reproduction; as mentioned above, the cow may even be pregnant at the time OPU is carried out.

Buffaloes

The numbers of oocytes recovered from buffalo ovaries are usually smaller than those found in cattle; collecting oocytes from abattoir ovaries by aspiration, numbers per ovary often average no more than about one. Techniques used for oocyte recovery include slicing, follicle puncture and aspiration, with the highest recovery rate achieved by slicing. In Pakistan, for example, workers found slicing better than aspiration for oocyte recovery (3.85 vs. 1.76 per ovary). As in cattle studies, there are reports with buffaloes showing that cleavage rates and blastocyst yields are significantly higher with cumulus-intact oocytes than with cumulus-free oocytes.

Goats

In dealing with valuable donor goats of superior genetics, transgenic founder animals or goats of a rare blood-line, some workers have recommended a transvaginal ultrasound-guided oocyte recovery procedure. Although the number of oocytes harvested per female is likely to be lower than with a laparoscopic approach, a non-invasive procedure enables post-collection adhesions to be held to a minimum.

Deer

In deer, two oocyte recovery approaches have been used in New Zealand: recovering oocytes from abattoir ovaries and collecting oocytes from live animals. An ample supply of oocytes can be recovered from venison abattoirs and farmed red deer can readily be handled to meet the needs of OPU techniques. Transvaginal ultrasound-guided OPU is regarded as a relatively non-invasive, reliable and safe technique for red deer (Berg and Asher, 2003); Berg and Asher report OPU to be a good technique for recovering oocytes but scarcity of suitable follicles (3–7 mm diameter) limits its usefulness at the present time. The same authors note the potential to overcome this problem by using exogenous hormones and follicular management. The OPU technique also has the potential to permit collection of oocytes from early pregnant animals, which could help in increasing the number of embryos produced from genetically élite hinds.

4.2.4. Evaluation and maturation of the oocyte

Selection criteria

The methods of selection of the cumulus–oocyte complexes (COCs) are usually based on parameters such as the morphology of the cumulus, the combined morphology of the cumulus and of the ooplasm, the size of the follicle and the oocyte and the level of follicular atresia. COCs with a compact and complete cumulus mass and a uniform appearance seem to present a higher developmental ability (Fig. 4.6). Many reports have proposed classification schemes based on the compactness and number of layers of cumulus cells surrounding the oocyte and on the appearance of the oocyte itself. Oocytes with the highest developmental competence are expected to possess an even, smooth, finely granulated cytoplasm, surrounded by fewer than three compact layers of cumulus cells. There are those suggesting the need to revise the criteria employed for selecting oocytes. Data reported by Vassena et al. (2003) suggest that oocytes collected during the static or regressing phases of the follicular wave are preferable to those

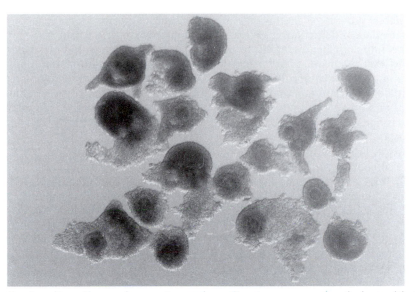

Fig. 4.6. Good-quality cattle cumulus–oocyte complexes. Oocytes are assessed on the basis of their morphology and that of the surrounding cumulus cells. The oocytes shown here have at least five layers of cumulus cells, which are tightly packed.

collected from follicles in the growing phase; it appears that the effects of early follicular atresia are beneficial to oocyte competence, although the reasons for this phenomenon are still unclear. Results reported by Alm and Torner (2003) showed that the staining of bovine COCs with brilliant cresyl blue before IVM could be used to increase the number of developmentally competent oocytes and to act as a marker of oocyte quality for techniques such as cloning. Brilliant cresyl blue stain determines the intracellular activity of glucose-6-phosphate dehydrogenase (G6PD), which is known to play a critical role in cell growth.

Gene expression in oocytes

A study reported by Dalbies-Tran and Mermillod (2003) compared gene expression pattern in bovine oocytes before and after IVM, using heterologous complementary DNA (cDNA) array hybridization; they were able to generate a transcription profile of bovine oocytes and a list of candidate genes for differential regulation during maturation. They showed that about 300 identified genes were expressed in their bovine model. Most messenger RNAs appeared

stable during IVM but 70 transcripts underwent a significant differential regulation. Some of these transcripts may be used as markers of oocyte quality for the improvement of oocyte culture conditions.

Factors involved in maturation

It is evident, from many reports, that granulosa cells play a critical role in the growth and development of the oocyte during its intrafollicular life. Intercellular communication between cumulus cells and the oocyte occurs via paracrine factors and through gap junctions. Cumulus cells facilitate the transfer of nutrients and factors essential for oocyte development, such as metabolites, amino acids, signal transduction molecules and other factors. Cumulus cells are known to play an important role in the regulation of cytoplasmic and nuclear maturation of the oocyte. In more recent times, the importance of the two-way communication axis between oocytes and granulosa cells has been recognized; oocytes secrete paracrine factors that regulate a broad range of granulosa cell functions by modulating fundamental control elements. It is apparent that oocytes not only promote growth of granulosa cells but also regulate differentiation processes within the follicle. Cumulus cells have a phenotype that is distinct from that of mural granulosa cells (which line the wall of the follicle) and it appears that the maintenance of this phenotype is dependent on oocyte secretions.

Meiosis of follicle-enclosed oocytes is maintained in prophase of the first meiotic division and oocytes do not spontaneously resume meiosis during oocyte growth and follicle development. It is believed that arrest of the meiotic process is probably secured by the presence of follicular purines (e.g. hypoxanthine), which maintain high levels of cyclic adenosine monophosphate (cAMP) in the oocyte. It is only in response to the preovulatory luteinizing hormone (LH) surge that oocytes in preovulatory follicles overcome the meiosis-arresting effect of follicular factors and resume meiosis, proceeding to metaphase of the second meiotic division. The oocyte itself is known to lack gonadotrophin receptors, and LH action is mediated through the attached cumulus cells.

Bovine oocytes in the cow's ovaries are arrested at the diplotene stage of the first meiotic prophase, the germinal vesicle (GV) stage; this is the stage where intense transcription of the decondensed genome enables accumulation of RNA for the period of oocyte maturation and early embryonic development. The resumption of meiosis is characterized by germinal vesicle breakdown (GVBD), chromosome condensation and spindle formation. The oocyte then proceeds towards metaphase, anaphase, telophase I and, without any chromosome decondensation, reaches metaphase II; meiosis is arrested again at this stage and this second block is normally released by events at fertilization. Some of the notable events in the development of knowledge about oocyte maturation are noted in Table 4.4.

Maturation system

The maturation system employed in many laboratories involves the use of TCM-199, supplemented with 10% fetal calf serum (FCS) and gonadotrophins (FSH, LH), in 5% carbon dioxide in air at 38.5°C. After incubation for 24 h, the bovine oocyte is mature, extrudes the first polar body and is ready for fertilization. Under conditions in which donor cattle dealt with by OPU are located far from the processing laboratory, maturation may be initiated during transport from the collection site to the laboratory, using portable battery-powered incubators. Under optimal IVM conditions, more than 90% of oocytes can be expected to reach metaphase II; just prior to fertilization, cumulus cells are partially removed to leave fewer cell layers surrounding the oocyte.

A wide range of compounds have been examined as supplements to the maturation medium, with many of them showing positive effects. In Brazil, for example, Silva *et al.* (2003) demonstrated that embryo production could be improved by adding penicillamine to the medium. Also in Brazil, Martins *et al.* (2003) investigated the

Table 4.4. Towards an understanding of oocyte maturation.

Year	Event	Researcher(s)
1935	Observation of spontaneous resumption of meiosis in rabbit oocytes	Pincus and Enzmann
1939	Observation of resumption of meoisis in human follicular oocytes	Pincus and Saunders
1955	Detailed study of maturation of rabbit oocytes	Chang
1962–1965	Nuclear maturation *in vitro* achieved in several mammalian species	Edwards
1968	Chronology of nuclear maturation in cattle oocytes	Sreenan
1977	Appreciation of importance of cytoplasmic and nuclear maturation	Thibault *et al.*
1978	Less than 1% of ruminant oocytes attained developmental competence after artificial maturation	Moor and Warnes
1984	Crucial support of follicle cells in maturation of sheep oocytes	Staigmiller and Moor

role of catalase on maturation and reported that it could enhance embryo production; it was suggested that the mechanism may be by reducing or inhibiting peroxidation damage of oocytes and/or embryos. In the USA, Goncalves *et al.* (2003) reported that bovine oocytes treated with osteopontin (an acidic glycoprotein detected in bovine oviductal fluid) increased rates of cleavage and embryonic development *in vitro*, suggesting a facilitatory role for this protein in early embryo development. Studies reported by Kuzmina *et al.* (2003) indicated that GH and granulosa cells could have a beneficial effect on the oocytes of prepubertal cattle when used in maturation medium; they showed similar blastocyst yields from calves (1–2 months old), heifers and cows.

Effect of LH

It is clear from numerous studies that LH produced by the pituitary gland is essential for oocyte meiotic maturation and ovulation in the cow and the other farm mammals. The action of LH on its target cells is mediated by way of binding to specific receptors on the cell membrane. Although there are no LH receptors (LHRs) on the surface of oocytes, they are to be found in the follicle somatic cells. It is thought that the *in vivo* LH surge stimulates the preovulatory follicle in which LHRs have been expressed in cumulus cells, leading to the resumption of meiosis in the oocyte. In recent years, some researchers have attempted to hold oocytes at the GV stage before IVM, on the understanding that oocytes may require time to acquire developmental competence during meiotic arrest. Work reported by Shimada *et al.* (2003) in Japan suggests that the induction of LHR expression in cumulus cells while holding oocytes at the GV stage may be important for producing IVM oocytes with high developmental competence. Studies reported by Kolle *et al.* (2003) have demonstrated that GH induces cumulus expansion during IVM of oocytes by stimulating cell proliferation and inhibiting apoptosis, showing that GH is capable of exerting direct and indirect effects on oocytes.

Growth factors

A whole range of growth factors have been examined to determine their effect on oocyte maturation. In China, Luo *et al.* (2002) presented results indicating that vascular endothelial growth factor (VEGF) had a beneficial effect, as evident in a significant increase in the maturation rate (90.5 vs. 78.2%). According to Xu *et al.* (2003), there is need to examine the optimal epidermal growth factor (EGF) and FSH combination to employ in the maturation of cattle oocytes.

Buffaloes

Embryo production efficiency in buffaloes is lower than that in cattle, although the reason for this difference is not clear. According to Neglia *et al.* (2003), buffalo oocytes may be more sensitive to environmental stress than cattle oocytes; these Italian workers showed that the source of oocytes was a factor significantly affecting post-fertilization embryo development. They noted that oocytes recovered from abattoir ovaries may not be immediately processed, as in the case of those recovered by OPU; this may impose a stress that subsequently reveals itself in embryo development. According to some studies, the optimal duration of IVM for bubaline oocytes is 24 h.

Goats

Although caprine oocytes matured and fertilized *in vitro* have resulted in the birth of live young, the process is regarded as very inefficient, with less than a third of the embryos resulting from IVM and IVF developing to the morula stage *in vitro*; the indications are that only a small proportion of caprine oocytes selected for IVM complete cytoplasmic maturation, which confers the ability to support early embryonic development. Various authors have stressed the need to develop chemically defined IVM systems. In the USA, Bormann *et al.* (2003) demonstrated the importance of vitamins during IVM for the subsequent development of goat embryos; they showed that addition of minimum essential medium (MEM) vitamins to synthetic oviductal fluid (SOF) maturation medium significantly increased blastocyst development and viability. As with other farm animals, workers have found the developmental competence of oocytes recovered from young prepubertal goats to be much inferior to that found with oocytes recovered from adult animals (Table 4.5).

Sheep

Numerous supplements to maturation, fertilization and culture media, including serum, gonadotrophins, oestradiol, amino acids and growth factors, have been tested in attempts to improve the production of sheep embryos. Several studies have demonstrated that the addition of gonadotrophins, oestradiol and EGF to the maturation medium results in improved conditions for maturation and fertilization and a relatively high rate of blastocyst formation. In the USA, Grazul-Bilska *et al.* (2003) were among those who found that EGF was a useful additive to maturation media to increase *in vitro* embryo production in FSH-treated sheep; they reported that EGF supplementation of the medium doubled the formation of blastocysts from 13% to 27%. It was evident, from research many years ago in sheep in Ireland, that survival of the embryos of mature ewes was greater than that of ewe lamb embryos; it is now clear that oocytes from prepubertal ewes generally lack the competence found in those of mature ewes.

Pigs

It is also known that oocytes from sows have a higher maturation potency than oocytes from prepubertal gilts. It is also clear that IVM of sow oocytes in the presence of FSH not only enhances nuclear maturation but also improves the developmental competence of oocytes, as shown by an increased yield of blastocysts. In Holland, Schoevers *et al.* (2003) showed that the presence of FSH during the first half of the entire culture period enhanced metaphase II formation and induced cumulus-cell expansion. In the USA, Abeydeera *et al.* (2000) showed that the addition of EGF to a protein-free maturation medium improved the developmental competence of pig oocytes; there was evidence that part of the beneficial effect of EGF was due to its stimulatory effect on intracellular glutathione (GSH) synthesis.

Horses

It is evident, from many studies, that IVM of equine oocytes needs to be refined; it is thought that a greater understanding of the factors that maintain the horse oocyte in meiotic arrest and regulate the resumption of nuclear maturation may permit the development of protocols that will improve the quality of cytoplasmic maturation. The literature shows the oocyte maturation rate varying among laboratories and usually lower than that reported for other farm animals; typically, 40–70% of equine oocytes reach metaphase II in contrast to the outcome in the cow, where more than 90% of oocytes can be expected to progress to metaphase II. There are, however, reports from Italy suggesting that certain selection criteria for equine oocytes may be useful in improving embryo production in this species (Dell'Aquila *et al.*, 2003).

Table 4.5. Competence of prepubertal and adult oocytes (cleavage and development of prepubertal and adult goat oocytes after IVM and culture to the blastocyst stage 6–8 days post-insemination) (from Cognie *et al.*, 2003).

Age	Oocytes	Cleavage (%)	Blastocyst yield/cleaved eggs (%)		
			6 dpi	7 dpi	8 dpi
2 months old	249	51[a]	0	10	19[a]
Adult	51	78[b]	20	50	65[b]

a vs. b: $P < 0.001$.
dpi, days post-insemination.

Among farm animals, horses usually have a considerably longer reproductive lifespan than cattle, sheep and pigs; for that reason, age of the mare may be an important consideration in determining oocyte quality. Reproductive efficiency certainly declines with advancing age in the mare; studies reported by some workers recently found evidence of morphological anomalies in oocytes from old mares (> 19 years old) not evident in oocytes from young mares (3–10 years old). A further clear difference between horses and other farm mammals lies in the number of ovarian follicles found in the ovaries. It is estimated that the equine ovary contains a low number of primordial follicles: about 36,000 vs. 120,000 in the cow and 160,000 in the ewe. It is also clear that the hormonal environment of the mare's preovulatory follicle differs from that in other farm animals during the resumption and completion of meiosis. The short, massive, preovulatory LH surge evident in ruminants is not seen in the mare; instead, there is a slow increase lasting several days, with maximum LH values observed a day after ovulation.

Deer

In oocyte maturation in deer, methods are usually based on those developed and used in cattle, with oocytes either recovered from ovaries recovered at abattoirs or collected by the laparoscopic aspiration of follicles in hormone-treated females; a typical medium would be TCM-199, supplemented with FCS, gonadotrophins, oestradiol and additional granulosa cells.

Prepubertal oocytes

As with goats, there is much evidence showing that developmental competence *in vitro* is lower in prepubertal cattle and sheep oocytes than in those in adults (Table 4.6). Efforts in Australia to overcome this problem used hormonal induction of oestrus and ovulation in donor animals. A report by Morton *et al.* (2003) used a 10-day norgestomet implant treatment in combination with pregnant mare's serum gonadotrophin (PMSG) and FSH before collecting oocytes; blastocyst formation after IVM/IVF/IVC was significantly increased after such stimulation (0% in controls vs. 41% in stimulated ewe lambs).

Prematuration of oocytes

It is evident, from studies reported in the past two decades, that conventional oocyte maturation protocols may not expose the oocyte to certain factors present in normal preovulatory follicle development; for such reasons, IVM oocytes may show lower competetence than those matured *in vivo*. Although as yet poorly understood, it is well-known that the ooplasm stores mRNA and proteins to provide maternal control during the first cleavages of embryonic development before the embryonic genome is activated. Studies reported by Duque *et al.* (2002), for example, have shown that the presence of retinoic acid during a 24 h

Table 4.6. Developmental competence of adult and calf oocytes.

Year	Donor treatment and age		Blastocyst %	Researcher(s)
1993	None	Calf	16	Palma *et al.*
	None	Adult		
1995	FSH	4 months	9	Revel *et al.*
	None	Adult	21	
1995	FSH	< 8 months	0	Looney *et al.*
	None	Adult	33	
1996	FSH	4 months	6	Damian *et al.*
	None	Adult	33	
1997	None	3–5 months	12	Mermillod *et al.*
	None	Adult	32	

prematuration period improved cytoplasmic granular migration, embryonic development, cryopreservation tolerance, total cell numbers and bovine embryo quality. Although a number of selective inhibitors of cyclin-dependent kinases have been shown to reversibly inhibit meiotic resumption in cattle oocytes for 24 h, much more needs to be determined about their effect on subsequent events in the life of the oocyte. In Ireland, Lonergan *et al.* (2003c) described the morphological changes occurring in bovine oocytes following exposure to either butyrolactone-1 (BL-1) or roscovitine for 24 h at concentrations known to be consistent with normal development; although some modifications at the ultrastructural level were induced in the oocytes, it was not clear whether this would improve embryo viability. A report by Edwards *et al.* (2003) evaluated maturation and fertilization of bovine COCs after culture in roscovitine for 24 or 48 h; they showed that the oocytes could be maintained at the GV stage for 48 h after removal from follicles without altering maturation or fertilization characteristics.

Pig and goat oocytes

Studies in pigs have also used agents that may be used in prematuration treatments in that species. GVBD in pig oocytes is regulated by the activation of maturation-promoting factor (MPF). In Taiwan, Ju *et al.* (2003) used roscovitine to maintain pig oocytes in the GV stage; their study showed that roscovitine-treated oocytes resumed meiosis after removal of the inhibitor. In goats, studies reported by Tan *et al.* (2003) found that both BL-1 and roscovitine inhibited meiotic resumption in a dose-dependent manner and that the effect was reversible, without compromising the oocytes' subsequent maturation competence.

4.2.5. Sperm preparation and *in vitro* fertilization (IVF)

Certain preliminaries must occur before sperm are in a position to effect fertilization. In embryo production, one of the first steps is the selection of sperm for use in IVF. A common practice is to select frozen–thawed sperm on the basis of a Percoll separation method; in Hungary, Somfai *et al.* (2002) showed that sperm with a higher viability and acrosome integrity could be obtained by Percoll separation than by a 'swim-up' method. IVF is a complex procedure involving oocyte maturation, sperm separation and sperm capacitation. Sperm capacitation is the biochemical modification sperm must undergo within the female tract before the cell can bind to the zona pellucida and undergo the acrosome reaction (AR). Capacitation is possible *in vitro* in the absence of reproductive-tract fluids and several compounds are known to induce *in vitro* capacitation (Table 4.7); the most common of these is the glycosaminoglycan (GAG) heparin. Although there are anecdotal reports from commercial clinics suggesting that heparin may not be needed for capacitation of frozen–thawed bull sperm prepared for IVF by centrifugation through Percoll, this is not supported by Mendes *et al.* (2003); they showed that heparin improved cleavage rates and embryo production *in vitro*, even when sperm were centrifuged through Percoll. The same workers found the commercial preparation Puresperm to

Table 4.7. Artificial capacitation of bovine sperm *in vitro*.

Year	Method	Researcher(s)
1982	High-ionic-strength (HIS) medium	Brackett *et al.*
1983	Bovine follicular fluid	Fukui *et al.*
1984	Standard-ionic-strength medium	Iritani *et al.*
1984	Heparin	Parrish *et al.*
1985	Elevated pH	Cheng
1985	Ionophore A23187	Hanada
1986	Liposomes	Graham *et al.*
1988	Percoll gradient/ hypotaurine	Utsumi *et al.*
1988	Caffeine	Niwa *et al.*
1989	TEST yolk	Ijaz and Hunter
1989	Oviductal cell monolayer	Guyader *et al.*

be a useful alternative to Percoll when separating cryopreserved bull sperm for IVF.

Sperm maturation in the live animal

Mammalian sperm must undergo epididymal maturation, capacitation and the AR to be able to fertilize the oocyte. Studying epididymal sperm maturation in pigs, Burkin and Miller (2000) concluded that porcine sperm develop zona pellucida binding sites on the acrosomal ridge while they reside in the corpus region of the epididymis, thereby gaining the ability to fertilize oocytes. Studies reported by workers in Japan indicate that an increasing percentage of goat sperm acquire the potential to undergo the AR and fuse with the oocyte plasma membrane during transit through the caput epididymidis. During capacitation, several biochemical modifications occur in the sperm's surface membrane; such changes are essential in permitting sperm–oocyte binding and the AR. During the AR, hydrolytic enzymes are released by exocytosis to enable the sperm to penetrate the zonal pellucida.

Assessing capacitation status of sperm

For many years, methods available for assessing capacitation of sperm were cumbersome and time consuming and did not always reflect the true nature of physiological events occurring in the cell. Now it is possible to evaluate the acrosomal status of sperm using FITC-PSA and the chlortetracycline fluorescence assay. These methods are rapid, reproducible and reliable and detect an increase or decrease in physiological ARs; the method has been used to evaluate the capacitation status of pig, stallion, bull, buffalo, goat, ram and human sperm.

Male effects on sperm quality

Workers involved in laboratory embryo production have attempted to assess different sperm quality parameters to predict *in vitro* fertility of bulls (Fig. 4.7). In Ireland, studies reported by Ward *et al.* (2003) have shown that the sire used in IVF can have a profound effect on the proportion of oocytes developing to the blastocyst stage and that this effect can be related to the time of first cleavage after insemination for the individual sire; the kinetics of these early cleavage divisions can be used to discriminate between bulls of high and low field fertility. Various studies have identified paternally linked differences in the fertilization process between bulls of high and low fertility that become evident during the first cell cycle.

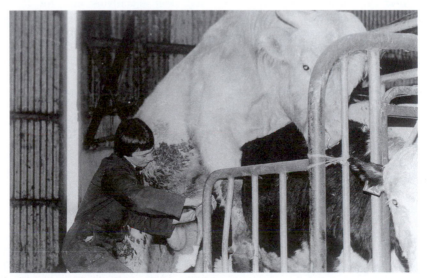

Fig. 4.7.   The bull as a factor affecting embryo production.

It has also been shown that a beneficial paternal effect from sperm recovered from bulls of high *in vitro* fertility is evident during the first G1 phase after fertilization; this beneficial effect is shown by an earlier onset and longer duration of the first DNA replication in both male and female pronuclei, which can result in higher rates of blastocyst formation. Studies reported by Comizzoli *et al.* (2003) examined the interaction between male and female pronuclei and compared glycolysis and pentose phosphate pathway (PPP) activities in bovine oocytes fertilized with sperm from bulls of high or low fertility; they showed that male pronucleus formation is necessary for the onset of the S phase in the female pronucleus and that the component promoting an early S phase in both pronuclei is metabolic and linked to an up-regulation of the PPP during the male pronucleus formation. The authors suggest that it would be of interest to examine whether differences in the level of PPP after fertilization are also to be found in humans, in which marked differences exist in the *in vitro* fertility of sperm from different donors.

Frozen semen

In cattle, frozen semen is generally employed for IVF and a Percoll-based separation method employed to isolate a motile sperm fraction after thawing; although there are other approaches, the Percoll gradient technique is well suited for commercial use. The two media usually employed for IVF are a tyrode-albumin-lactate-pyruvate (TALP)-based medium and an SOF-based medium, with heparin as the capacitating agent. The sperm dose used will depend on the bull, among other factors, and is determined by carrying out IVF tests with differing sperm concentrations. IVF is usually achieved by co-incubating sperm and oocytes for 18–20 h; the oocytes are denuded of cumulus cells before transfer to the IVC system used for development to the blastocyst stage.

Stallion sperm

A major obstacle to progress in equine IVF is that neither an optimal capacitation stimulus nor a suitable medium for supporting viability of stallion sperm during culture *in vitro* has been established. Although numerous capacitation treatments have been evaluated, the only treatment that has resulted in reasonable fertilization rates in equine IVF is calcium ionophore A23187.

Boar sperm

A study reported by Matas *et al.* (2003) examined the effect of different sperm preparation methods in pigs on the AR, oocyte penetration time and early embryo development; their findings indicated that treatment with Percoll (layered in a Percoll gradient) yielded the best results in their pig embryo production system.

Deer sperm

In red deer, New Zealand workers have shown that sperm can be capacitated *in vitro* by supplementing the fertilization medium with heparin, as in cattle; the same authors also found that epididymal sperm could be successfully employed in IVF. Elsewhere, however, in both New Zealand and France, it has been shown that a 20% oestrous sheep serum supplement was effective for the *in vitro* capacitation of red deer ejaculates and for IVF of red deer oocytes (Comizzoli *et al.*, 2000). The fertilizing ability of different deer semen samples, including epididymal sperm, in various supplemented IVF media, was examined by Comizzoli *et al.* (2001), using zona-free bovine oocytes to decondense deer sperm chromatin; the authors suggest that this form of heterologous IVF may be a useful tool in permitting a clearer understanding of unsuccessful homologous IVF in deer.

Using ICSI as the fertilization approach

For various reasons, there may be a need to explore novel means of achieving fertilization; some of the possible approaches are illustrated in Fig. 4.8. As a result of poor IVF results, often with only 15–30% of

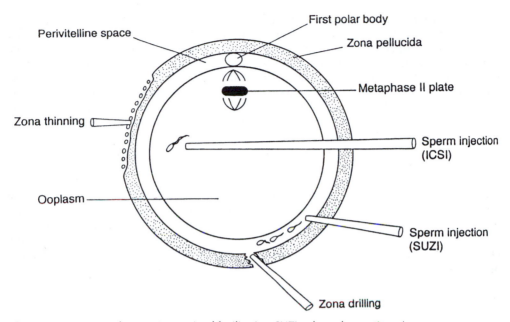

Fig. 4.8. Four approaches to micro-assisted fertilization. SUZI, subzonal sperm insertion.

oocytes being fertilized, ICSI has been employed in dealing with horse oocytes. Various workers have shown that ICSI is capable of increasing fertilization rates compared with IVF, and several foals have been born from IVM horse oocytes fertilized by sperm injection. In Italy, studies have demonstrated that a blastocyst rate of about 30% can be achieved after ICSI of IVM equine oocytes. According to Squires *et al.* (2003), the ICSI technique provides the possibility of obtaining pregnancies from stallions that may have low sperm numbers or poor semen quality. Although ICSI is the method of choice for fertilizing horse oocytes *in vitro*, the reasons for this are not well understood and embryo development rates have remained low.

For such reasons, a study undertaken by Tremoleda *et al.* (2003) in The Netherlands sought to characterize the nuclear and cellular events occurring in horse oocytes during fertilization after sperm injection; it was concluded that, until conventional IVF becomes more reliable, ICSI probably remains the best way to produce equine embryos, to perform fundamental research into the cellular and molecular events of fertilization, to investigate infertility and to understand the cellular basis of early pregnancy failure in horses. ISCI has proved to be a valuable addition to the technology of assisted reproduction, especially in humans and, to a more limited extent, in horses. The technique also provides an opportunity for research into cell-cycle control and the mechanisms involved in sperm-induced oocyte activation. The same sperm injection technique also has relevance as a sperm vector system for transgenic animal production (see Chapter 13) and with freeze-dried sperm for which the maintenance of motility is not required (see Chapter 2).

ICSI in cattle

Species differences in the response to ICSI are now well recognized. In humans, mice, hamsters and rabbits, sperm injection alone is apparently sufficient to activate the oocytes for further embryonic development. In cattle some maintain that additional activation after ICSI is necessary and in pigs there is evidence that activation of oocytes may be beneficial. In comparison with ICSI in humans, the outcome of sperm injection in cattle has been disappointingly poor. In efforts to overcome this, attempts have generally been made to activate the oocytes in conjunction with sperm injection. It has been thought that exogenous oocyte activation treatment is a prerequisite for sperm-head decondensation, male pronuclear formation, cleavage and embryonic development. However, the same activation treatment is likely to induce parthenogenesis and abnormal forms of fertilization after ICSI.

Studies in Japan have now demonstrated that promising rates of cleavage, embryonic development and embryo normality can be achieved in cattle ICSI without exogenous oocyte activation treatment (Wei and Fukui, 2002); the same laboratory reported the birth of several normal young after the use of the ICSI technique. In Italy, Galli et al. (2003c) are also among those showing that ICSI blastocysts produced without activation can establish pregnancies that go to term. Although it does not necessarily follow that exogenous activation is not required in cattle, the routine would be simpler if it could follow the lines now well established in human assisted reproduction.

ICSI in goats

Workers in Canada have reported on the efficacy of ICSI in goats. Results from work in the laboratories of Nexia Biotechnologies Inc. suggest that the use of tail-cut goat sperm and a piezo micropipette-driving system is an efficient approach to the production of viable embryos and young in that species (Wang et al., 2003).

ICSI in pigs

In the USA, Lee et al. (2003b) provided evidence that IVM porcine oocytes could be activated by the injection of frozen–thawed sperm in the absence of additional activation stimuli. However, activation treatment significantly improved fertilization and blastocyst development of sperm-injected oocytes; in fact, activation treatment more than doubled blastocyst production. Such

evidence may suggest that the amount of oocyte-activating factor(s) released by the sperm was only sufficient to partially initiate the physiological cascade associated with fertilization. Studies in Japan by Nakai et al. (2003) sought evidence that IVM–ICSI oocytes have the ability to develop to normal viable piglets; results of the group indicated that normal development in IVM pig oocytes was possible, even in the absence of a sperm centrosome during fertilization, by injection of a sperm head. The authors suggest that the successful birth of piglets, as demonstrated in their study, paved the way for the utilization of boar genetic resources, including not only non-motile sperm before and after freezing but also sperm preserved by methods other than freezing; one possibility is a freeze-drying method in combination with sperm injection. Such a procedure could enable sperm to be preserved on a long-term basis at room temperature.

4.2.6. Other fertilization approaches

Gamete intra-Fallopian tube transfer (GIFT) has been used in horses to deal with stallions having low sperm numbers or in situations in which frozen semen is in limited supply or sexed semen is being employed. In the GIFT technique, oocytes and sperm are transferred to the mare's oviduct. Results from studies by Squires et al. (2003), who inseminated fresh semen into the oviduct, showed an 82% pregnancy rate compared with an 8% rate with the use of frozen–thawed sperm; the authors note that further work is required to determine why pregnancy rates are depressed when using cooled or frozen semen compared with fresh semen for GIFT transfer. When techniques such as GIFT are used in the horse, the culture and subsequent transfer of donor oocytes may pose additional risks for disease transmission; although the zona pellucida provides pathogen protection, the cumulus cells surrounding the oocyte, which are inevitably transferred with the gamete, could harbour intracellular viral pathogens.

4.2.7. *In vitro* embryo culture

Traditionally, the commercial development of bovine embryos derived from IVM, IVF and IVC has employed co-culture of somatic cell lines, such as buffalo-rat liver (BRL) cells or Vero cells in complex TCM supplemented with serum for embryo culture. Although such culture systems support high rates of blastocyst development, the commercial application of these techniques has been associated with problems in the resultant pregnancies. Currently, the IVC of the cattle zygote and early embryo is one of the most active research areas. In the mid-1980s, it was first shown that oviductal cells could be employed in co-culture systems and subsequently other cell types, such as BRL or Vero cells; however, such co-culture systems introduced a further biological variant into an already complex problem.

For such reasons, in the last decade, efforts have focused on the use of semi-defined media in the absence of feeder cells, with low or no serum added and with low oxygen tension. It should be noted that the beneficial effects of serum supplementation of IVC media are thought to be due to the nutrients, vitamins, growth factors, hormones and antioxidative compounds contained in the serum. The problem has been that the biological activities of sera are likely to vary greatly from one batch to another and there are potential risks of virus or mycoplasma infections. Inclusion of serum in the culture medium is believed to have direct adverse effects on the embryo itself, resulting in perturbations in metabolism, disruptions in the ultrastructure of the mitochondria and development of large lipid vesicles that displace the normal ultrastructure. For such reasons, a serum-free IVC system without somatic cells, capable of producing high-quality transferable embryos, is a worthy research objective.

In Japan, Hoshi (2003) reported serum-free media which they found effective in the production of good-quality IVP embryos capable of sustained development after transfer; such media are currently commercially available in Japan and have been shown to improve the rates of blastocyst formation during IVC and post-thaw embryo viability. The same author found that calves born from embryos produced in the serum-free media showed a lower incidence of calf mortality and large-offspring syndrome (LOS) than those embryos produced in the presence of serum. In Italy, Lazzari *et al.* (2002) cultured bovine embryos in IVC medium in the presence of high concentrations of serum or bovine serum albumin (BSA) and compared these with embryos cultured *in vivo* in the sheep oviduct; birth weights of calves derived from IVP embryos were significantly higher than those of calves derived from sheep oviduct culture, superovulation or artifical insemination (AI).

On the other hand, there are workers who have used forms of cow serum (CS) which have proved superior to BSA and FCS for the support of embryo development (Vajta *et al.*, 2003); this was ascribed to the markedly superior mitogenic activity shown by some commercial batches of cow serum; the same workers also reported studies in which embryo culture in SOFaaci (aa, amino acid; ci, citrate) medium with 5% CS did not result in increased birth weight or any alterations related to LOS. There are studies indicating that problems ascribed to serum may be limited by restricting serum to a certain period in culture; although there may still be pregnancy complications, co-culture systems can result in good rates of blastocyst development if serum is restricted to the period after the first 72 h of culture.

Embryo density

Among factors known to influence embryo development *in vitro* is embryo density; significantly higher rates of development can be achieved simply by increasing the number of embryos per culture unit in cattle, sheep and pigs. This apparently beneficial effect is believed to be due to an increase in growth-promoting substances that emanate from the accumulated embryos. In pigs, for example, increasing the number of embryos from 10–15 to 25 per 40 µl of medium led to a significant improvement in numbers reaching the blastocyst stage (Yoshioka and Rodriguez-Martinez, 2003).

The oviduct for embryo culture

Several culture systems have been developed to culture the fertilized oocyte (the zygote) to the blastocyst stage; these include an *in vivo* culture procedure using the surrogate sheep oviduct, as well as various co-culture and cell-free systems. The sheep oviduct method is one that yields embryos of a quality almost comparable to those recovered from the cow itself (Fig. 4.9). Although the sheep oviduct procedure is normally used in a research setting, in Italy it has been the basis of IVP cattle embryo production for commercial application over a period of some years (Galli *et al.*, 2003a). In Ireland, a change from the sheep oviduct system to a totally IVC system was associated with a marked decline in embryo quality, especially in the ability of IVP embryos to withstand freeze–thawing procedures.

Sequential media

Encouraging results are found using media in which composition changes according to the embryo's stage of development. Data

presented by Lane *et al.* (2003), for example, using a physiologically based, sequential, serum-free culture system (G1.2/G2.2), showed equivalent pregnancy and cryosurvival rates to those achieved with the traditional BRL co-culture system. The G1.2/G2.2 media have been used for the culture of embryos of many species, the time of medium change from G1.2. to G2.2 varying according to the species: 65–76 h post-insemination in humans, 72 h post-insemination in pigs and 96 h post-insemination in cattle.

These media are based on synthetic oviductal fluid, with the G2.2 portion containing a higher glucose concentration than G1.2. The authors found no difference in embryo development between change of medium at 72 or 96 h post-insemination. The sequential media ISM1/ISM2 (Medicult, Copenhagen) have also been used successfully; a study reported by Heyman *et al.* (2003b) in France assessed the potential for full-term development of IVP blastocysts developed in these sequential media. From a limited number of transfers, no fetal loss was observed, there was no instance of LOS and all calves born survived to 1 month in the ISM group compared with 80% survival in controls developed in a co-culture system.

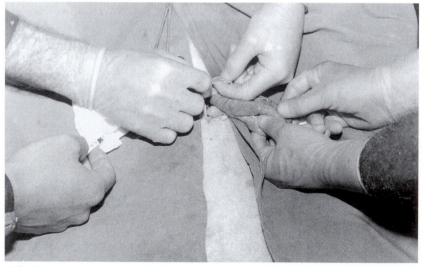

Fig. 4.9.   The sheep oviduct as an incubator.

4.2.8. Cryopreservation of embryos and oocytes

Cattle embryos

Although frozen–thawed IVP cattle blasto-cysts have resulted in many normal healthy calves, it is also clear from numerous reports that such embryos are much more sensitive to freezing than *in vivo*-derived embryos. Pregnancy rates with IVP embryos have been inconsistent and markedly lower than those reported for *in vivo*-derived embryos, which reflects a number of differ-ences at the morphological, ultrastructural, metabolic, biochemical and genomic level between the embryos from the two sources. Various studies have sought to improve embryo culture, using defined media, refin-ing the oxygen concentrations during cul-ture and providing substrates to ameliorate free-radical accumulation, but it is evident that some proportion of bovine blastocysts are incapable of the sustained development found within their *in vivo*-derived counter-parts. On the other hand, culturing the bovine zygote in the ewe oviduct can markedly increase cryotolerance to a level comparable to that found with totally *in vivo*-produced embryos. In Ireland, Rizos *et al.* (2003a) found that the presence of serum in the IVC medium reduced the cryotolerance and quality of the blastocysts produced; culturing embryos in the absence of serum significantly improved the cryo-tolerance of blastocysts to a level intermedi-ate between serum-generated blastocysts and those derived *in vivo*. The same authors found that reduced cryotolerance in the presence of serum was accompanied by deviations in the relative abundance of developmentally important gene transcripts.

Sheep embryos

A report by Rizos *et al.* (2002) demonstrated that the quality of IVP sheep blastocysts was significantly better than their bovine counterparts produced under identical *in vitro* conditions; the evidence suggested inherent species differences between sheep

and cattle affecting embryo quality. At the ultrastructural level, IVP embryos showed a lack of desmosomal junctions, a reduction in the microvilli population, an increase in the average number of lipid droplets and increased debris in the perivitelline space and intercellular cavities, in comparison with *in vivo*-derived embryos; such differ-ences appeared to be more marked in bovine IVP embryos, which also displayed electron-lucent mitochondria and large intercellular cavities.

Pig embryos

Various research groups have sought to develop an efficient cryopreservation proto-col which could be applied to the pig. Data reported by Esaki *et al.* (2003) demonstrated that vitrification, using their mimimum volume cooling (MVC) method, could enable porcine IVP blastocysts to be cryo-preserved; elsewhere in Japan, Kobayashi *et al.* (2003) used Vajta's open pulled straw (OPS) method to vitrify expanded and hatched blastocysts in 7.2 M ethylene glycol; transfer of embryos resulted in three of four recipients becoming pregnant, two of these delivering a total of 12 piglets. A Hungarian–Japanese study reported by Dinnyes *et al.* (2003) tested a vitrification method for dealing with pig embryos pro-duced in a serum-free IVC system; embryos were dropped on to an approximately −180°C metal surface in 2 µl droplets of vitrification solution, the method resulting in relatively high rates of embryo survival.

Cryopreservation of oocytes

The first report on successful freezing of a mammalian oocyte was the work of David Whittingham in London in the late 1970s with mice; it was shown to be possible to obtain live offspring after IVF of mouse oocytes frozen with dimethyl sulphoxide (DMSO) and stored in liquid nitrogen. Despite such success, the history of oocyte freezing during the past quarter-century has revealed difficult problems; certainly, in terms of applications in human assis-ted reproduction, current cryopreservation protocols are regarded as far from optimal (Van der Elst, 2003).

Several workers have reported on their efforts to develop effective vitrification techniques for the cryopreservation of bovine oocytes. In Nottingham, for example, Mavrides and Morroll (2002) suggest that the cryo-loop vitrification technique followed by ICSI could be effective. The cryo-loop is a technique where a thin nylon loop is used to suspend a film of cryoprotectant containing the oocytes before they are immersed in liquid nitrogen. Using bovine oocytes, the Nottingham workers reported a survival rate of 90.5% in comparison with a slow-freezing technique (54.4%).

Cryopreserving ovarian tissue

Some have attempted to preserve oocytes by vitrifying ovarian tissues. A study by Al-Aghbari and Menino (2002), for example, reported that sheep oocytes could be successfully cryopreserved by vitrification of ovarian tissues; they subsequently exhibited IVM rates similar to those of vitrified and non-vitrified oocytes.

4.2.9. Evaluating embryo quality

Attempts to identify factors influencing the viability of IVP embryos have usually involved studies up to the blastocyst stage of development (Fig. 4.10). However, certain crucial developments occur beyond that point (rapid growth and extension of trophoblast) and it is essential to have accurate knowledge of such events, which cannot be simulated in the laboratory. For that reason, it has proved necessary, in cattle, for example, to transfer embryos into a synchrous uterine environment and recover them a week or so later. In the USA, Bertolini et al. (2002) used such a technique to study the growth, development and gene expression of in vivo-produced and IVP day-7 and day-16 cattle embryos. In Canada, Arnold et al. (2003) sought to evaluate expression of certain genes (Mash2 and IFN-τ mRNA) as indicators of trophoblast differentiation and function between IVP

and nuclear transfer (NT) embryos; they did this with IVP-derived embryos transferred on day 8 and recovered from the recipient on day 17. In Florida, Kramer et al. (2003) examined fetal and placental expression of insulin-like growth factor (IGF)-family genes in NT, IVP and in vivo embryos at day 25. They found evidence of differences in growth factor and imprinted gene expression in the bovine conceptuses at the time of implantation; such differences may be associated with losses and abnormalities found in the NT-derived embryos. In Wisconsin, Fischer-Brown et al. (2003) reported on the influence of the uterine environment of the cow on IVP embryo development, transferring on day 7/8 and recovering on day 14/15; the system was seen as useful in evaluating and improving the efficiency of embryo production.

4.2.10. Pregnancy rates with fresh and frozen IVP embryos

A review by Peterson and Lee (2003) in New Zealand dealt with two aspects of research at Ruakura which were based on 2300 embryo transfers using IVP-embryos; on average, 50% survived to day 24, 40% to day 60 and 30% to full-term. They contrasted such results with those for natural mating or for cows becoming pregnant after AI, where embryo loss from day 24 to term was no more than 5%. The New Zealand researchers noted that the high and sustained fetal loss was a major deterrent to the uptake of IVP technology, especially in countries with a seasonal constraint to breeding. The same workers also suggest that commercial and research cattle ET practitioners worldwide are not able to distinguish accurately those recipients that are intrinsically subfertile. Evidence suggests that uterine, rather than ovarian factors may be important in influencing pregnancy rates; it is possible that, using a proteomic approach to identify proteins in the uterine luminal fluid, it may be possible to characterize the uterine environment more accurately.

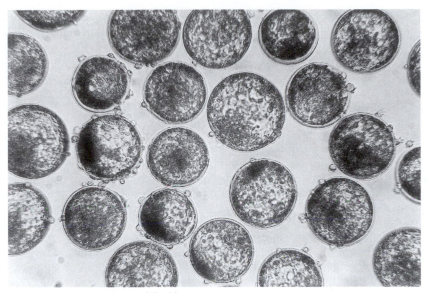

Fig. 4.10.   Evaluating embryos beyond the blastocyst stage.

4.3. IVP Embryos in Commerce and in Research

4.3.1. Current production statistics

Data collected from 25 countries in Europe for 2002 showed that rather more than 8% of the total cattle embryos transferred in that year were IVP embryos, which indicates that a considerable amount of time and attention is currently directed to this new form of embryo production.

4.3.2. *In vitro*- vs. *in vivo*-produced embryos

IVP bovine embryos differ from their *in vivo* counterparts in having darker cytoplasm, reduced buoyant density, a more fragile zona pellucida, higher chromosomal abnormalities, higher abortion/pregnancy failure rates and altered gene expression patterns. Subtle differences exist at the cellular level in terms of metabolic profiles and morphology as well as in gene expression. It is believed that the characterization of early deviations in gene expression will bring a greater understanding of the biology of early embryo development and provide the

means of improving IVC systems (Mohan *et al.*, 2003).

It is known that, during formation of the bovine embryo, differentiation is up-regulated within the outer trophectoderm (TE) layer and suppressed within the inner cell mass (ICM). One critical process in TE differentiation is the construction of tight junctions at the border between cells, which provides the sealing that permits expansion of the blastocoel cavity. It is now well established that IVP cattle embryos are retarded in development and often show a reduced degree of compaction before blastocyst formation compared with their *in vivo* counterparts. In some instances, IVC has been reported to cause decreased pregnancy rates and long-term effects, with higher perinatal death rates in conjunction with malformations and LOS. Studies by Miller *et al.* (2003) of gene expression in cattle have provided evidence that embryos showing a detectable and well-formed compaction period *in vitro* are of similar quality to their *in vivo* counterparts; the same workers showed that *in vitro* and *in vivo* embryo development differed mainly during a critical period before blastocyst formation. It was found that failure of embryos to compact *in vitro* resulted in a significant reduction of specific transcript levels, which depended on culture conditions.

Major developmental events in the early embryo

Several major developmental events occur during the 6-day period between zygote and blastocyst formation in cattle and these events have been examined critically (Fig. 4.11). The events in question include the first cleavage division, the timing of which is of crucial importance in determining the subsequent development of the embryos, the activation of the embryonic genome at the 8–16-cell stage, compaction of the morula on day 5 and blastocyst formation on days 6–7, involving the differentiation of two types of cells, the TE and the ICM. In Ireland, Lonergan *et al.* (2003a) demonstrated that culture of cattle embryos in the ewe oviduct for 6 days resulted in embryos of high quality with cryotolerance similar to that of *in vivo*-produced blastocysts. In contrast, culture *in vitro* (SOF) for 6 days resulted in embryos of low cryotolerance. The data supported the contention that certain windows of embryo development are more predisposed to aberrant programming than others; it was also found that data on embryo quality were supported by observations on the relative abundance of certain gene transcripts in the embryos produced.

In Canada, Dufort *et al.* (2003) found that reduced cryotolerance of bovine blastocysts generated in the presence of serum is accompanied by deviations in the relative abundance of important transport and metabolism gene transcripts. In Belgium, Leroy *et al.* (2003) developed a method to visualize and quantify the differences in lipid content in a single bovine oocyte or embryo by means of a fluorescent dye and a photometer; they confirmed that dark immature oocytes contain more lipid than those with a pale appearance and that the addition of serum to the maturation medium did not increase the lipid content. However, morulae cultured in the presence of serum did accumulate lipid droplets and contained more lipids than embryos cultured in

serum-free conditions; this has implications for embryo cryotolerance and general viability.

Chromosomal abnormalities

A study reported by Knijn *et al.* (2003a) was among many to show that bovine embryos produced *in vitro* differ markedly in quality from those developed *in vivo*; they demonstrated that the lower quality of IVP embryos can be attributed to the ICM having less viable cells because of a lower number of cells and a higher incidence of apoptosis, which appears to be determined before completion of the fourth cell cycle. Some reports show that, in cattle blastocysts, the incidence of mixoploidy may be almost three times greater in IVP embryos compared with those produced *in vivo*. Similarly, in pigs the incidence of chromosomal abnormalities in IVP embryos is much higher than that reported for *in vivo* embryos in this species (McCauley *et al.*, 2003).

4.3.3. Large-offspring syndrome (LOS)

Considerable attention has been given in the literature to problems that may arise with calves derived from IVP embryos. As noted by Galli *et al.* (2003a), reports of extended gestation, dystocia, large calves, increased perinatal mortality and such are usually taken as evidence of LOS; the same authors conclude that most of the calves described have been the result of co-culture systems or high-serum/high-BSA systems involving limited transfers performed under uncontrolled conditions by research laboratories. There is also the point that males used in production systems may not have been selected on the basis of calving ease, an important consideration in the use of bulls in normal farming practice. In commercial programmes, especially where the embryos have been generated by way of the surrogate sheep oviduct, most pregnancies are normal. In Ireland, dealing with a large-scale field trial, a paper in the mid-1990s

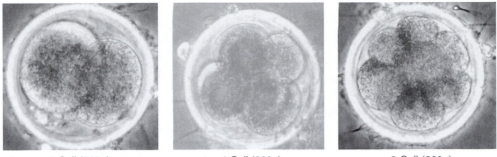

2-Cell (600×) 4-Cell (600×) 8-Cell (600×)

Early bovine embryos from IVF of IVM at 46 h after insemination

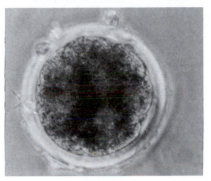

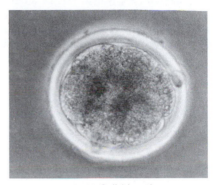

16-Cell (day 3)
(600×)

16–32-Cell (day 4)
(600×)

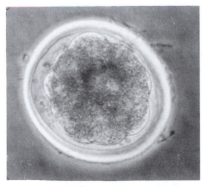

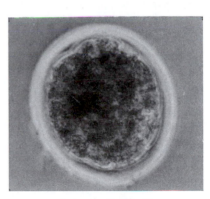

Compact morula (day 5)
(600×)

Late morula (day 6)
(600×)

Fig. 4.11. Early bovine embryos – two-cell to morula.

reported that more than 97% of calves derived from similarly produced embryos were normal. In sheep, although it is clear that the survival rate of IVP sheep and goat embryos is significantly lower than that of their *in vivo* counterparts, there is also information from the main INRA laboratory in France to show that its IVC system (SOF supplemented with 5–10% FCS at 2–3 days post-insemination) has never been associated with abnormal development of sheep or goat embryos as reported by other groups using various forms of serum as a protein source during culture.

4.4. Future Developments

4.4.1. Oocytes from pre-antral follicles

One important area of research in cattle embryo production lies in exploring ways and means of obtaining much greater numbers of oocytes per ovary. Developing a culture system for pre-antral follicles has important biotechnological implications due to its potential to produce large numbers of oocytes for embryo production and ET. Although the development of such a culture system currently remains a long-term goal, it is likely to be achieved at some point in the years ahead. Artificial maturation of small oocytes could provide a new source of mature oocytes for livestock production and assisted reproduction in humans and for use in dealing with endangered species (see Miyano, 2003). It has been demonstrated in mice that oocytes in primordial follicles can be developed *in vitro* to reach maturity and to be capable of fertilization and the production of live young.

Workers at the Jackson laboratory in Bar Harbor in the USA devised a two-step culture protocol which produced murine oocytes capable of undergoing maturation, fertilization and development to live offspring from primordial follicles of newborn mice. The first step was organ culture of intact newborn ovaries and the second step was isolation and culture of oocyte–granulosa-cell complexes isolated from pre-antral follicles that developed in the organ-cultured ovaries. Since the original report in 1996, which led to the birth of a single mouse pup (known as Eggbert), Bar Harbor workers devised a revised two-step protocol which dramatically improved results and led to the birth of 59 offspring (O'Brien *et al.*, 2003).

The bovine ovary, in common with the ovaries of other farm animals, contains a large number of small follicles (Fig. 4.12), of which only a small proportion grow to full size, mature and ovulate; since > 99.9% of ovarian follicles undergo atresia, it would clearly be of great practical benefit if these follicles, destined to become atretic, could be rescued by a suitable culture system, thus providing a large pool of oocytes for embryo production, cloning and transgenesis. It is known that there may be more than 100,000

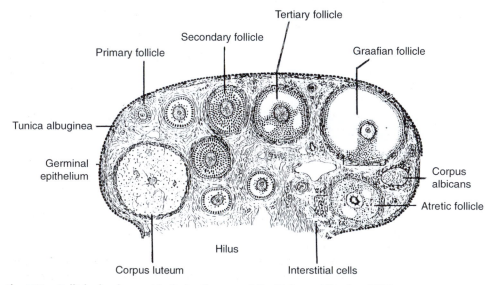

Fig. 4.12.   Follicle development in the bovine ovary (after Hafez and Gordon, 1962).

small, non-growing and growing follicles in the ovaries of the cow; Japanese workers have already produced a calf by growing bovine follicles recovered from early antral follicles. Other workers in that country have also been able to grow bovine oocytes from early antral follicles in appropriate culture media. However, there is a large gap between such success and producing young from oocytes originating in bovine pre-antral follicles.

Developing systems to maintain follicular growth and oocyte development is a highly complex process which requires a comprehensive knowledge of folliculogenesis. It is necessary to develop culture systems capable of simulating the biology of oocyte growth and granulosa-cell multiplication and differentiation as found in the ovary (see Amorin *et al.*, 2003). The literature of recent years contains details of many studies directed towards elucidating factors involved in the growth of pre-antral follicles, including the role of growth factors. It is believed that theca cells secrete growth factors that stimulate granulosa cells to proliferate and differentiate in a paracrine manner, which helps to promote follicular development. Growth differentiation factor 9 (GDF-9) is an oocyte-specific, transforming growth-factor-beta family member which has been implicated in follicle development. The GDF-9 growth factor has been found in several species, including sheep, cattle and humans; working in pigs, Bark *et al.* (2002) found that the same growth factor was an oocyte-secreted factor in that species. In the same species, Wu *et al.* (2001) assessed the outcome of ICSI of oocytes from pre-antral follicles matured *in vitro*, finding that such oocytes could undergo fertilization and subsequent embryonic development.

Fertilized oocytes from primordial pig follicles

Workers in Japan have shown that it is possible to mature and fertilize pig oocytes from primordial follicles by a combination of xenografting and IVC (Kaneko *et al.*, 2003); this was done using ovarian tissue taken from 20-day-old piglets and transplanted to ovariectomized mice, which were then treated with gonadotrophin over a period of 2 months. It should be noted that the pig apparently differs from most other farm animals in its pattern of oogenesis. Oogonia and oocytes are formed during the first half of fetal life in the cow and ewe; in the sow, evolution is apparently slower, with some reports showing oogenesis continuing for some days after birth.

4.4.2. Gene expression studies

The success or otherwise of bovine embryo production systems has often been based on the percentage of embryos that reach the blastocyst stage. While it has been useful to employ such developmental end-points as markers of embryo production efficiency, it is probable that the effects of a culture system may not reveal themselves during the relatively short period of IVC. Production efficiency is likely to be a matter of quality rather than embryo yield. In Ireland, Rizos *et al.* (2003a) combined measurement of developmental competence (cleavage and blastocyst rates) with qualitative measurements, such as cryotolerance and relative mRNA abundance, to give a more complete picture of the consequences of modifying IVC medium composition for the embryo. It seems likely that the study of gene expression will be the method of choice for evaluating bovine embryo quality in the years ahead. However, there are still many problems to be solved before it can be used in routine research in the IVF laboratory; decisions have to be made about which gene should be studied and the normal level of expression of this gene in normal animals. Studies reported by Wrenzycki *et al.* (2003) have identified genes that are exclusively expressed in the ICM and TE cells of the early developing bovine embryo; expression of these genes is affected by the timing of blastocyst expansion. It is known, for example, that the transfer of day-7 expanded blastocysts, whether produced naturally or in embryo production systems, result in higher pregnancy rates than those from day-8 expanded blastocysts.

In Ireland, Lonergan *et al.* (2003b) compared gene expression in bovine embryos cultured *in vivo* and *in vitro* from the zygote to the blastocyst stage; this pattern was strongly influenced by the culture environment, with some transcripts showing differences as early as 10 h after initiating culture. The authors note that such information may have implications for human assisted reproduction, in which there are moves towards culturing embryos to the blastocyst stage, which calls for prolonged culture under potentially deleterious conditions. Elsewhere, in France, Dalbies-Tran and Mermillod (2003) analysed gene expression in bovine oocytes before and after IVM; they reported that some 300 identified genes were expressed in the oocyte.

In Canada, McGraw *et al.* (2003) investigated the precise expression patterns of seven genes involved in chromatin structure during early bovine embryo development. A report by Knijn *et al.* (2003b) found that the relative abundance of Glut-3, -4 and -8 transcripts was significantly different between *in vitro*- and *in vivo*-developed cattle blastocysts. In this regard, it is known that glucose transport is critical for mammalian blastocyst formation and subsequent development and that there are differences in the mRNA expression of glucose transporter genes. A study reported by Corcoran *et al.* (2003) in Ireland sought to identify differentially expressed genes in bovine day-7 blastocysts derived from either *in vivo* or IVC treatments; they were able to identify a number of genes that were linked to blastocyst quality. Further work in Ireland, reported by Fair *et al.* (2004), identified a gene, *Ped*, previously described in mice and known to regulate embryonic development; expression of this gene was considerably greater in bovine embryos cultured *in vivo* (sheep oviduct) than *in vitro* (SOF medium).

As with the evaluation of cattle embryos, studies are being carried out in the buffalo on gene expression in bubaline IVP embryos. In India, the temporal expression of genes for several growth-factor ligands and receptors was examined in buffalo embryos by researchers using reverse-transcription PCR; the identity of the resulting PCR products was determined by their expected size, restriction analysis, Southern blot hybridization and nucleotide sequence analysis.

4.4.3. Development of microfluidic technology

Currently appearing on the horizon are oocyte/embryo culture systems based on microfluidic technology, which offer the advantage that medium can be readily and constantly changed to meet the precise requirements of the oocyte or the developing embryo. Already, sequential media are being applied in embryo culture; microfluidic technology is a more advanced form of such technology. Microfluidic devices would enable medium around the oocyte or embryo to be gradually changed, rather than subjecting the oocyte/embryo to the sudden changes in microenvironment that are part of conventional *in vitro* embryo production. Already, microfluidic devices have been shown to be capable of maturing bovine and porcine oocytes (Beebe *et al.*, 2002). The objective with these devices is to provide conditions which closely resemble those found in the live animal, with small volumes of fluid and, in the case of sperm, much reduced total numbers. The eventual outcome could be a system which enables maturation, fertilization and early embryonic development all to flow automatically after introduction of oocytes and sperm into the microfluidic device. As noted by Suh *et al.* (2003), microfluidics might eventually provide an alternative to ICSI for some human oligospermic patients, by simulating *in vivo* conditions, using low fluid volumes and sperm in numbers much the same as occur in the human oviduct; even with extemely low sperm numbers, it may be possible to effect fertilization. In view of the concern that is occasionally expressed about ICSI, it could be useful to have an alternative. In farm animals, it may be possible to view the future of laboratory embryo production in terms of simply adding gametes to the micro-fluidic device at one end and recovering, a week or so later, high-quality viable embryos ready for transfer or cryo-preservation – by no means an impossible scenario.

5

Controlling Oestrus and Ovulation

5.1. Oestrus and Its Detection

The phenomenon of oestrus in farm animals is now recognized as being more complex than was at one time thought. The behaviour shown by farm animals during the period of sexual receptivity can vary markedly (Table 5.1). Evidence in sheep and pigs, for example, shows that the presence of the male can reduce the duration of behavioural oestrus, and advance the time of ovulation, relative to the onset of oestrus,

by advancing the timing of the preovulatory surge of luteinizing hormone (LH). In the absence of the male, presumably the pre-ovulatory LH surge must await the build-up of follicular oestradiol to a certain level before it is triggered. In sheep, ewes can display a strong ram-seeking behavioural pattern, so that contact between the sexes is by no means wholly dependent on the ram's activity; studies have shown that many ewes in oestrus seek out and remain near rams. The conditions in which animals

Table 5.1. Behavioural characteristics of sexual receptivity in farm animals.

Species	Behaviour
Horse	Urinating stance repeatedly assumed; tail frequently erected; urine spilled in small amounts; clitoris exposed by prolonged rhythmical contractions of vulva; relaxation of lips of vulva. Company of other horses sought; turns hindquarters to stallion and stands still
Cow	Restless behaviour; raises and twitches tail; arches back and stretches; roams bellowing; mounts or stands to be mounted; vulva sniffed by other cows
Pig	Some restlessness may occur, particularly at night, from pro-oestrus into oestrus. Sow stands for 'riding test' (the sow assumes an immobile stance in response to pressure on its hindquarters). Sow may be ridden by others; some breeds show 'pricking' of ears
Sheep	During pro-oestrus there may be a short period of restlessness, courting ram. During oestrus, the ewe seeks out the ram and associates with it; may withdraw from flock. Remains with ram when the flock is 'driven'; may show tail-wagging on occasion
Goat	Restless in pro-oestrus. During oestrus, the most striking behaviour includes repeated bleating and vigorous tail waving; poor appetite
Deer	Short chases with male in pursuit usually precede mating; these chases may be in small rings or figures-of-eight
Bactrian camel	Crosses neck with male; adopts sitting posture with male

©I.R. Gordon 2004. *Reproductive Technologies in Farm Animals*
(I.R. Gordon)

find themselves may be a strong factor influencing their oestrous behaviour. In cattle, for example, the duration of standing oestrus is known to be significantly shorter for cattle kept on concrete surfaces compared with animals on straw bedding, dirt pads or pasture (Diskin and Sreenan, 2003); data from recent Irish studies are in Table 5.2.

5.1.1. Need for accurate oestrus detection

Cattle

One of the most important technical problems facing the cattle artificial insemination (AI) industry since its inception has been the detection of oestrus; accurate detection is essential if there is to be successful application of AI in cattle herds, although the methods commonly used in detection have remained largely unchanged through the 60 and more years that this breeding technology has been employed. The problem has not been eased by the fact that an increasing proportion of those dealing with cattle may have an urban rather than a rural background; for that reason, they may not always be as familiar with the subtleties of cattle behaviour as formerly was the case.

Many attempts have been made to develop technological methods to detect oestrus in cattle and obviate the need to check the animals several times daily. Such methods include: tail paint, vasectomized teaser bulls, pressure-activated heat mount detectors, pedometers, electronic pressure-sensitive count devices, computerized radiotelemetric pressure sensors, devices for measuring vaginal electrical resistance and hormone assays for determining progesterone changes in milk. All these are valuable research tools and many have proven their worth in the farmyard.

Much of the literature dealing with oestrus detection relates to cattle, particularly dairy cattle; this largely reflects the extensive use of AI in such animals. Behaviour during oestrus and the oestrus-to-ovulation interval are essential for estimating the optimum time to inseminate cows. The economic importance of traits such as longevity, health and reproduction has increased in the past two decades in comparison with milk yield; effective oestrus detection is important for improved reproduction. In most farm situations, oestrus detection is performed by visual observation, although this is difficult on large dairy farms due to the short observation periods that occur during feeding and milking; this has led to attempts to monitor cows by various forms of automatic oestrus detection.

In The Netherlands, some studies have tested automated models for detection of oestrus and mastitis in dairy cows; a system of sensors was employed, alerting operators to changes in milk yield, temperature,

Table 5.2. Effect of underfoot surface on oestrus in cattle. Data clearly demonstrate that cows dislike being mounted when standing on concrete and have a preference for softer underfoot surfaces, such as grass, dirt or straw-bedded yards. Detecting oestrus when cows are held on concrete slats is much more difficult than when they are on softer undersurfaces. (From Diskin and Larkin, 2003.)

Incidence of false heats (%)	Average duration of oestrus (h)	Average no. mounts received during oestrus	Incidence of silent heats (%)
Concrete slats (32)	8.6	20	8
Straw bedding (17)	14.2	30	0
Plastic-covered slats (28)	12.2	40	4
Pasture (8)	14.7	48	3

electrical conductivity of milk, cow activity and concentrate intake. Results were good for detection of oestrus but variable for the detection of mastitis. It was concluded that automated detection of these two conditions can be carried out under commercial conditions. Traits useful for oestrus detection and their suitability for automatic detection have been the subject of some studies (Firk *et al.*, 2002).

There may be unexpected relationships between oestrus detection methods and the ability of the cow to become pregnant. A study by Yaniz *et al.* (2003), in Spain, for example, demonstrated that pedometer measurements of increased walking activity at oestrus could enable an accurate prediction of the probability of pregnancy in dairy cows to be made; a clear relationship was found between increased walking activity during oestrus and fertility. The possibility of pregnancy increased progressively as locomotion at oestrus rose to five times the basal level at inter-oestrus and declined thereafter. Dramatically decreased fertility was observed in cows showing the highest increase in walking activity at oestrus (> 500% activity); this may reflect an endocrine imbalance or ovarian disorder in such animals.

Zebu cattle

Oestrous behaviour and oestrus-to-ovulation interval are parameters that have not been well characterized in zebu cattle. In Nelore cattle, the predominant beef breed in Brazil, ovulation was recorded about 26 h after the start of oestrus. It was also found that oestrus was shorter in comparison with European breeds, with a high incidence of oestrus at night, making detection difficult and leading to impaired fertility after AI.

Sheep

Features displayed by sheep during oestrus are likely to include the ram nuzzling the genital region of the ewe and showing a particular form of behaviour known as the 'flehmen response' (Fig. 5.1). This behaviour enables the vomeronasal organ, located

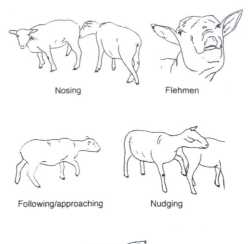

Fig. 5.1. Some features of ram sexual behaviour (after Pelletier *et al.*, 1977).

in the roof of the mouth, to examine fluid-borne pheromones. The 'flehmen response' is characterized by the ram elevating its head and curling its upper lip; this closes the nostrils and enables fluids to be aspirated into the vomeronasal organ, via the nasopalatine duct, where they can be analysed by the organ's sensory neurones. Similar activities are found with the bull and stallion and, less commonly, with the boar.

5.1.2. Oestrus detection rates

Oestrus detection rates in dairy herds are often typically below 50% and this represents a major limiting factor in optimizing reproductive performance. Some have suggested a new approach to an old problem, in which they composed a scale based on the frequency of behavioural patterns during heat and between heats. Use of the scale and twice-daily observation periods of 30 min each gave a detection rate of 74%, with no incorrect assessments. Studies in

The Netherlands showed that two 30 min observation periods per day in lactating Holstein cows resulted in an efficiency of oestrus detection of 74% and an accuracy of 100%; time of observation during the day was very important, those made before milking giving a lower detection rate than observations after milking and feeding. The way in which the efficiency of oestrus detection interacts with the conception rate shown in cattle during the early months after calving is shown in Table 5.3.

The importance of accurate oestrus detection was emphasized in a survey of 45 dairy herds in the USA; the studies indicated that a high proportion of cows are artificially inseminated following the visual method of observing behavioural changes associated with heat. However, certain behavioural characteristics, such as nervousness and mounting other cows, used in commercial herds as indications of oestrus, can be wrongly interpreted to indicate that a cow is acceptable for insemination. The American studies showed that oestrus detection was a problem in 30% of the 45 herds examined, with up to 46% of cows inseminated when milk progesterone levels were high, a procedure which can only result in poor conception rates. Many studies show the importance of inseminating cows when progesterone levels are low, which is normal when the animals are in oestrus.

Workers at Cornell (Bob Foote and associates) have shown that many non-pregnant cows in the luteal phase of an oestrous cycle or while pregnant are submitted for AI in the USA, and no doubt elsewhere. For such reasons, the training of herd managers, and especially inseminators, to evaluate many criteria of oestrus and to reject for AI cows not in oestrus can improve the efficiency of semen use, reduce abortion of pregnant cows, increase overall breeding efficiency and reduce calving interval, components vital to an economically successful dairy farm operation.

Failure to detect oestrus remains a major factor, resulting in costly delays in the breeding of cattle; detection rate is highly correlated with calving interval. In the UK, it has been estimated that average detection

Table 5.3. Herd pregnancy rate as determined by heat detection and conception rates. For dairy herds using AI, it is the combined effect of heat detection rate and conception rate that determines pregnancy rate and the compactness of calving. For example, with both heat detection and conception rates optimal, it is possible to have 96% of the herd calving in a 90-day period. (From Diskin and Larkin, 2003.)

Heat detection rate (%)	Conception rate (%)			
	60	50	40	30
90	96	91	83	71
70	89	82	73	61
50	76	68	59	48
40	67	59	48	40

rates are around 55%, having remained at this level for the last 25 years. The methods and aids currently available, e.g. behavioural observations, milk progesterone, mount detectors and suchlike, have been the subject of many reports but none is completely reliable and they often suffer from problems of detection failure and false positives. Those who believe that detection in cattle is ready for major new initiatives point to exciting developments in sensing technology which may offer such possibilities as detection of oestrous pheromones by 'electronic noses' and/or online milk hormone detection.

Up to the present time, standing behaviour has been the symptom used to determine the right time for insemination. However, standing behaviour may not be observed in more than 50% of cows actually in oestrus in many herds and there is no guidance to relate the time of ovulation to other oestrous symptoms. In a study reported from Holland by Van Eerdenburg et al. (2002), 100 cows were detected in oestrus visually, using a scoring system; the time of ovulation was estimated by ultrasonic scanning of the ovaries. Standing oestrus was observed in 50% of the animals; cows that ovulated 0–24 h after AI scored almost three times the number of oestrous behaviour points compared with those ovulating 24–48 h after AI.

5.1.3. Aids to detection

In the UK and Ireland, as in other countries, such as the USA, herd fertility is clearly declining, despite improved knowledge and herd health monitoring by animal health professionals and farmers. One of the most important factors in herd fertility is oestrus detection. For good oestrus detection, two major requirements must be met; a skilled observer must spend sufficient time making observations and cows must show overt signs of oestrus during these observations. To make visual detection of oestrus more sensitive, some workers have used a scale covering the different signs of oestrus.

In the use of devices to detect oestrus in cattle, their relevance will depend on the particular circumstances, and cost is a critical consideration determining their commercial applicability. Electronic heat mount detector systems and activity meters (e.g. pedometers) may be used where cattle are housed but are much less appropriate in a pasture-based setting. In countries such as New Zealand, relatively simple detection measures, such as using tail paint (Fig. 5.2) and heat mount detectors, have become the standard aids and are inexpensive (Verkerk, 2003).

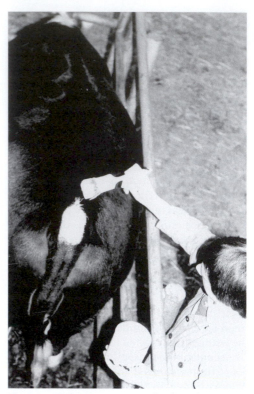

Fig. 5.2. Tail-painting as an aid to heat detection.

Buffaloes and pigs

The average duration of oestrus in the buffalo has been given as 15 h in some reports and 19 h in others; ovulation has been estimated to occur 10–18 h after the end of oestrus. Data for 15,186 sows and gilts on 55 Dutch farms gave the average duration of oestrus as 48.4 ± 1.0 h, ranging from 31 to 64 h among farms; in most herds, gilts exhibited a significantly shorter heat period than sows (40.8 ± 1.1 vs. 48.5 ± 1.0, respectively).

Sheep, goats and horses

In sheep, many studies have shown the mean duration of oestrus as approximately 24 h. In Uruguay, researchers working with Corriedale, Polwarth and merino sheep recorded a mean value of 24 h, with no differences between breeds, season, young ewes and old ewes; both body weight and body condition score were significantly correlated with the duration of oestrus. In sheep, some workers have classified oestrous cycles as normal (15–19 days), short (< 14 days), long (20–26 days) and multiple (> 27 days). The goat has a duration of oestrus somewhat longer than that of the ewe, with ovulation occurring some 30–36 h after the start of sexual receptivity; such characteristics appear to be common to goats in both the temperate and the tropical regions.

The mare's oestrous cycle is on average 21 days long, consisting of 14 days of di-oestrus (luteal phase) and 7 days of oestrus, when the mare is sexually receptive. Several aspects of the mare's reproductive endocrinology were so surprising to early workers that they hesitated to publish their findings. First, the mare's ovulatory LH surge is prolonged, with levels gradually rising throughout oestrus to reach a peak on the day after ovulation; this is in sharp contrast to the large surge of LH that occurs a day prior to ovulation in sheep, goats and cattle; secondly, there is a distinctive FSH surge during the mid-luteal phase of the cycle.

5.1.4. Measures of genuine oestrus status

Milk progesterone testing may be employed to confirm oestrus in the cow. Several reports have shown that a suboptimal conception rate in cattle bred by AI may be due to insemination not being carried out at a time coinciding with ovulatory oestrus; in some reports, the incidence of this problem has sometimes reached 20%.

5.1.5. Future developments in oestrus detection technology

Many reports have appeared over the years on the use of vaginal probes as a method of oestrus detection, particularly in dairy cattle and pigs; the probe provides information of changes in the electrical resistance of vaginal fluids at the time of oestrus. However, quite apart from the time and labour involved in the use of such devices, their accuracy has often left much to be desired. Eventually, it is possible that miniature implantable devices with transponders sending information directly to computers may appear. From an animal welfare viewpoint, this may be valuable, enabling the individual to be identified, not only for oestrus, but for its health status as well.

In the age of the microchip and increasingly sophisticated electronic technology, it seems likely that miniature sensing devices, implanted subdermally in the animal to detect changes in impedance, temperature or activity at oestrus, may become a practical reality in future years; allied to new electronic methods of identifying cows, there would seem considerable scope for developments in this area. The combination of accurate sensing and identification devices could greatly assist in future developments in oestrus detection technology.

Relatively inexpensive closed-circuit televisions (CCT) are now available for monitoring cows in calving boxes and the same equipment can be used to monitor cows for oestrus; using time-lapse and fast play-back, the oestrous activities of the night can be viewed in a short time. With CCT systems specifically designed for oestrus detection, the camera could be switched on automatically for a few minutes when mounting occurs to identify the cows involved. The normal way of using pedometers in dairy cattle is to read them in the milking parlour, which can mean a time-lag of 8 h between the time of maximum activity and the attendant becoming aware of this. In The Netherlands, Van Eerdenburg (2003b) used pedometers that record cow activity in 2 h periods, which enabled detection rates to be improved; the author concludes that, by combining pedometer data with the history of the cow, accuracy could be improved still further. Various of the aids to heat detection in cattle are noted in Fig. 5.3.

5.2. The Oestrous Cycle

Many names are associated with publications dealing with the oestrous cycle in farm animals and it is worth mentioning a few. In 1903, Marshall in the UK published papers on the oestrous cycle of the ewe and this was followed by papers by Cole and Miller in 1935 and McKenzie and Terrill in 1937, all working in the USA. In cattle, John Hammond in Cambridge published his monograph on the cow and its oestrous cycle in 1927. Around the same time, Fred McKenzie in the USA dealt with the pig's cycle in work published in 1926; at a later stage, Nishikawa in Japan was to publish his monograph on the horse in 1959.

5.2.1. Physiology and endocrinology of the oestrous cycle

With the advent in the 1960s and 1970s of sensitive immunoassays (radioimmunoassays (RIAs)), enzyme immunoassays

(EIAs) and other methods for measuring steroid, polypeptide and other hormones involved in the reproductive processes of farm animals, a considerable body of information was rapidly accumulated about endocrine events in the oestrous cycle and at other stages of the animal's reproductive life. Some hormonal characteristics of oestrous cycles in farm animals are summarized in Fig. 5.4., taking the pig as the example. Such information has guided the thoughts of those involved in the control and manipulation of reproduction. All the hormones mentioned act at extremely low levels, from nanograms to picograms per millilitre of blood. Just how low these

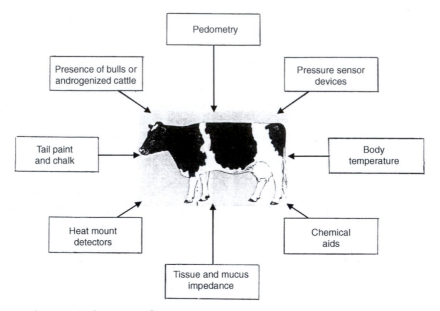

Fig. 5.3.   Aids to oestrus detection in the cow.

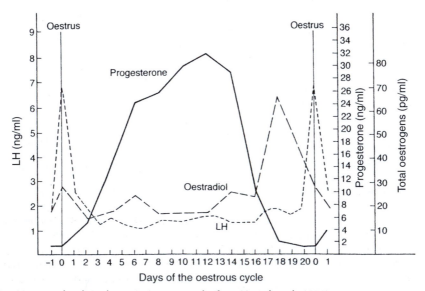

Fig. 5.4.   Hormone levels in the pig's oestrous cycle (from Hansel *et al.*, 1973).

levels really are can be appreciated with reference to Table 5.4. One of the great advances in the 1960s was the development of assay procedures enabling such low levels to be detected and measured.

Phases of the oestrous cycle

The phases of the oestrous cycle have been described by some as comprising pro-oestrus, oestrus, metoestrus and dioestrus; certain events occur in these stages which characterize them. During oestrus, the animal stands for mounting by the male; in pro-oestrus certain changes may be evident that alert stockmen to the approach of oestrus. In metoestrus, a blood discharge may be evident on the tail of the cow; in dioestrus the female refuses to stand for mounting and the male no longer shows interest in her. More commonly, the oestrous cycle is divided into two phases, the follicular and luteal phase. The follicular phase is characterized by preovulatory follicle growth, the exhibition of oestrous behaviour, the preovulatory LH surge and ovulation. The transition from the follicular to the luteal phase is marked by ovulation; the transition from the luteal to the follicular phase is more complex and is marked by a rapid regression of the corpus luteum (CL), in which prostaglandins (PGs) play a prominent role.

An understanding of mechanisms involved in the control of the oestrous cycle has been influenced by several major discoveries made over the past half-century. Two of the most important of these were the discovery and development of the hormone-receptor concept and the discovery that the hypothalamus and brain regulate secretion of anterior pituitary hormones by way of several small peptides of neurosecretory origin. One of these hypothalamic peptides is gonadotrophin-releasing hormone (GnRH), now known to stimulate the synthesis and secretion of LH and follicle-stimulating hormone (FSH) by specific cells located in the animal pituitary (gonadotrophs). Some of the factors influencing reproductive activity in the farm animal, hormonal and otherwise, are indicated in Fig. 5.5.

One of the most recent additions to the list of hormones known to be involved in reproduction is leptin, discovered in 1994. Leptin is secreted as a hormone from adipocytes and is relevant to the interests of those dealing with farm animals, where food intake, body composition and reproductive performance are of considerable economic importance. Although much of the information about leptin relates to its action on the brain, one site of action of the hormone is the ovary. Since its discovery a decade ago, much research effort has focused on the role of leptin in regulating growth and reproduction in rodents, humans and farm animals; it is clear, for example, from many studies that leptin acts centrally to inhibit feed intake and stimulate growth hormone (GH) secretion in the pig.

Management of reproduction in dairy and beef cattle requires a thorough understanding of the changes in physiology and endocrinology that occur during the oestrous cycle and, in the case of prepubertal heifers and cattle after calving, the transition from anoestrus to cyclicity. Such an understanding will improve reproductive management and facilitate the successful application of oestrus control measures; in turn, this can facilitate the application of AI and speed up the genetic improvement of cattle production systems. A review by Okuda et al. (2002) in Japan summarized current understanding of the endocrine mechanisms that regulate the timing and pattern of uterine PG (PGF$_{2\alpha}$) secretion during the oestrous cycle and early pregnancy. Luteal regression is caused by a pulsatile release of PGF from the uterus in the late luteal phase in most mammals. Although it has been proposed in ruminants that pulsatile PGF secretion is generated by a

Table 5.4. From gram to picogram.

Units of weight	Name
1.0	Gram
10^{-3} 0.001	Milligram
10^{-6} 0.000,001	Microgram
10^{-9} 0.000,000,001	Nanogram
10^{-12} 0.000,000,000,001	Picogram

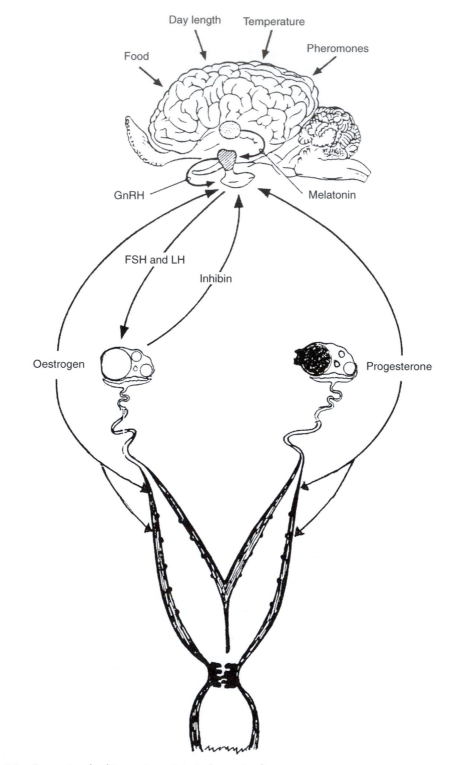

Fig. 5.5. Factors involved in ovarian activity in farm animals.

positive-feedback loop between luteal and/or pituitary oxytocin and uterine PGF, the bovine endometrium is believed to possess other mechanisms for the initiation of luteolytic PGF secretion. It is now evident, for example, that the growth factor tumour necrosis factor-alpha (TNF-α) stimulates PGF output from bovine endometrial tissue not only during the follicular phase but also during the late luteal phase, indicating that TNF-α is a factor in the initiation of luteolysis in cattle. Growth factors, of which there are many that influence reproduction, are hormone-related substances that are produced and secreted in minute quantities by cells of various tissues and exert their action by diffusing into their target cells.

Pigs

Although the expected average interval between heats in the pig is normally 21 days, several reports from the USA have shown that, when a large number of oestrous cycles is recorded in mated gilts, there are two peaks in the frequency of these intervals; one occurs near 21 days and the other at 26 days. The occurrence of two phases of return to oestrus in sows was also recorded in commercial units in the UK; one phase peaked at 20.7 days and the second at 26.5 days. It is believed that two mechanisms could be operating to account for these phases; the first peak is derived from a failure in fertilization and the second from embryonic loss at a later stage.

Buffaloes

Those reviewing studies on the oestrous cycle, oestrous behaviour and endocrinology of the buffalo record the length of the cycle as averaging 21 days; there are, however, marked variations in cycle length, which is known to be influenced by climate, photoperiod, temperature and nutrition. Buffaloes are seasonal breeders, particularly under tropical climatic conditions; the majority of animals exhibit oestrus during winter and fail to show signs of heat during the summer. The hormone profile during the buffalo's oestrous cycle is similar to that observed in cattle, but hormone levels are lower and the difference between the two species is greater in summer than during winter. Concentrations of most of these hormones appear to be lower in the swamp than in the river buffalo. The hormone prolactin has been implicated in the ovarian inactivity that occurs during the hot summer season, with concentrations of prolactin becoming very high during this period compared with winter.

5.2.2. Monitoring ovarian activity

Studies with beef cows in the USA have shown that such cows ovulate about 30 h after the onset of oestrus; this was determined by transrectal ultrasonography at 4 h intervals, starting 20 h after heat onset. The onset of oestrus corresponds closely to, but precedes, the preovulatory LH surge by several hours. Other American studies examined oestrous behaviour and time of ovulation during the four seasons in beef cattle; cows had longer heat periods, fewer mounts and longer intervals between mounts in summer than in winter. The conclusion was that season can alter oestrous behaviour of beef cows but time of ovulation relative to heat onset is not influenced; during all seasons, more cows are mounted between 6 a.m. and noon than during other times of the day (White and Wettemann, 2000). In the UK, this time with dairy cows, there have also been studies, which agree with the beef animal reports, showing that ovulation in Holstein–Friesian cattle occurs about 30 h after oestrus onset.

5.2.3. Follicular dynamics

The Finnish researcher Erikki Rajakoski, back in 1960, coined the term 'follicle wave' to describe the follicular events he observed in the ovaries during the cow's 3-week oestrous cycle; follicular waves are now known to occur periodically during various reproductive states in cattle. Much information on follicular dynamics in the cow has come from the laboratory of Ginther and associates in Wisconsin. The focus of attention in much of this work is factors controlling the emergence of a single dominant follicle from the cohort of follicles which commence development at the start of a follicle wave (Ginther *et al.*, 2002). Knowledge of follicle development is important in devising methods to manipulate the fertility and productivity of farm animals (Evans, 2003). A follicle wave emerges approximately every 10 days and each wave contributes a large (> 10 mm) dominant follicle; the remaining follicles regress and are known as subordinate follicles. A different number of follicle waves may occur each oestrous cycle; opinions vary on whether two or three follicle waves occur more frequently; it is known that an oestrous cycle with three waves is of longer duration than one with two cycles, and that the wave from which the ovulatory follicle is generated is shorter than earlier waves. It is not certain whether two or three waves reflect environmental conditions,

a chance event or occur as a specific characteristic of an individual animal.

Among farm animals, pigs are the exception in that follicle development is apparently continuous, with follicles developing to intermediate diameters before the follicles destined to ovulate are selected from this pool and continue growing to ovulation. The follicles destined to ovulate increase growth between days 14 and 16 of the cycle; follicles grow from about 4–5 mm diameter on day 15 to an ovulatory diameter of 9–11 mm by the end of the cycle (Fig. 5.6).

Follicle recruitment and selection

Two processes are involved in the growth and development of the follicle destined for ovulation in the cow. The first is follicle recruitment, which results in the development of a cohort of follicles from which the dominant follicle emerges; the second is follicle selection, in which one follicle becomes dominant and continues towards

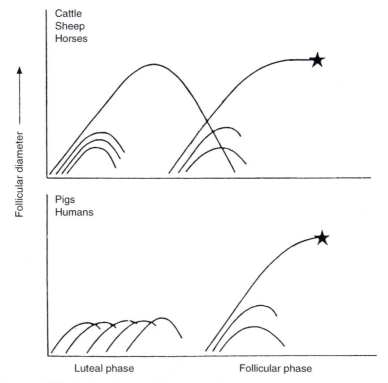

Fig. 5.6.   Patterns of follicular development in farm animals.

ovulation while the others regress. Much research effort has been directed to identifying factors leading to the dominance of a single follicle in the cow and the mechanisms that suppress the growth of the subordinate follicles.

Value of real-time ultrasonics

The availability of equipment permitting real-time ultrasonic scanning of the reproductive organs *per rectum* provided a valuable new approach to the study of follicular dynamics in farm animals. Ultrasonography is now the basis of an extremely accurate technique for the estimation of the antral follicle population within the limits of resolution imposed by the scanning device; follicles as small as 2–3 mm diameter can be visualized, measured and sequentially monitored. When required in research studies, the number of CL in farm animals can also be accurately assessed by way of transrectal ultrasound.

Zebu cattle

Despite a similarity in patterns of follicular dynamics, European (*Bos taurus*) and zebu (*Bos indicus*) cattle apparently differ in several aspects of their reproductive physiology; the dominant follicle and CL are smaller and the duration of oestrus shorter in Nelore than in European breeds. A report by Viana *et al.* (2003) in Brazil noted that follicular dynamics in zebu cattle is characterized by a lower maximum diameter and persistence of dominant follicles, when compared with taurine breeds; these workers showed that follicular divergence occurs earlier in zebu cattle. The reproductive organs of the zebu show several morphological and physiological differences from European taurine cattle; ovaries of zebu cows are smaller than those of taurine animals and the CL may be embedded deep within the ovary.

Buffaloes

Ovarian follicular dynamics during the oestrous cycle in Murrah buffaloes have been reported by workers in Brazil; 3.3, 63.3 and 33.3% of animals showed one, two and three waves, respectively. Although follicular dynamics in buffaloes have been studied by several authors, information is more limited than in cattle. Workers in Italy showed that the majority of buffalo calves displayed a typical pattern of follicular development; Presicce *et al.* (2003) reported an average of four follicular waves among the calves examined.

Sheep

Follicular growth and steroidogenesis have been examined in sheep largely by way of direct observation of the ovaries by laparoscopy, by real-time ultrasonics and by measuring hormone levels. Some studies suggest that follicular growth in the ewe is continuous and independent of the stage of the oestrous cycle, while others maintain that follicular growth occurs in two or three waves. One study showed that follicular growth in the ewe started at three distinct phases of the oestrous cycle, with two waves occurring in the luteal phase and one wave in the follicular phase; similar follicular waves were observed in the anoestrous period as well as in the breeding season. Canadian workers concluded that growth of antral follicles to an ovulatory size is maintained throughout anoestrus in sheep, with a transient shift in the number of small and medium-sized follicles during mid-anoestrus; they also found that the periodic emergence of waves of large follicles occurred in synchrony with an endogenous rhythm of FSH secretion.

Studies in Ireland by Evans *et al.* (2000) used ultrasonography to determine whether follicle development occurs in a wavelike manner; they concluded that it did, with a predominance of two or three waves per cycle. The same workers also found evidence for the phenomenon of follicular dominance in sheep; however, this appears to be a matter of co-dominance rather than single follicle dominance where ovulatory follicles assume the growth profile of a dominant follicle while subordinate follicles regress. In Canada, Duggavathi *et al.* (2003) concluded that follicle wave emergence in sheep involves the recruitment of one to three follicles from a relatively stable, small follicle pool, in contrast to the six to nine follicles from the analogous follicle pool in cattle.

Prepubertal and pregnant sheep

Ultrasonography has been used in many studies to examine antral follicular development in prepubertal sheep. A report by Bartlewski *et al.* (2003) showed that antral follicle recruitment and growth increased after the first 2 months of age and just before puberty in ewe lambs; however, the rhythmic pattern of follicular wave emergence, as seen in adult ewes, was not established in pre- and peripubertal ewe lambs. Studies in sheep have examined follicular activity during pregnancy. A report by Bartlewski *et al.* (2000) showed that ovarian antral follicle development was suppressed in ovaries containing CL in early pregnancy; it was believed that this suppression was exerted primarily by the developing conceptus and was restricted to CL-bearing ovaries.

Goats

Ultrasound has been used to determine the nature of follicular dynamics in goats; reports show several follicular waves during the cycle. It appears that follicular dominance is expressed more weakly and less frequently in goats than in cattle. Workers in Uruguay scanned the ovaries of Saanen goats ultrasonically on a daily basis and recorded animals showing two, three and four follicular waves per cycle. In Spain, again in dairy goats, evidence has been presented showing follicular growth occurring in waves during the oestrous cycle. In Poland, Schwarz and Wierzchos (2000) reported data supporting the existence of a wave pattern of folliculogenesis, recording the average waves per cycle in Polish white goats as 4.8.

Horses

Much has been reported from the USA on follicular dynamics in the horse. In Wisconsin, data were reported supporting the view that the future dominant follicle has an early size advantage over future subordinate follicles and indicate that the advantage is present as early as 6 days before deviation (deviation = beginning of a marked difference in growth rate between the two largest follicles).

Camelids

Camelids of the New World (e.g. llama and alpaca) and the Old World (dromedary and Bactrian camels) are induced ovulators and do not exhibit any clearly defined reproductive cycle. They do, however, show evidence of follicular waves. Studies on the follicular wave pattern in dromedaries and llamas have shown waves emerging every 11–20 days; in Bactrian camels, studies have shown an average follicular wave duration of 19 days (range, 14–21 days).

5.2.4. Growth and regression of the corpus luteum

The CL of the cow, as in other farm mammals, is a differentiated follicle in which the theca and granulosa cells give rise to small and large steroidogenic luteal cells, respectively; the bovine CL grows in 11 days to weigh 4 g in the typical beef cow and still larger in the dairy animal. Such explosive growth arises from the hypertrophy of the differentiating theca-derived small and granulosa-derived large luteal cells that make up some 70% of the mass of the bovine CL. In the cow, the CL usually protrudes above the surface of the ovary, as shown in Fig. 5.7.

Luteotrophic hormones

LH is the major luteotrophic hormone in sheep and cattle, although local intraovarian regulation of luteal development and function may also be important. There is also evidence that GH may be required as well as LH for normal luteal function, and the fact that luteal cells contain GH receptors indicates a direct role for GH in luteal

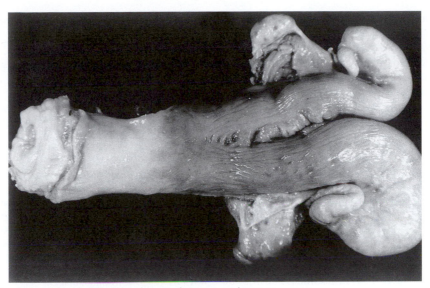

Fig. 5.7. Cow tract with fully formed corpus luteum in right ovary.

function and development. During the growth of the pig CL, which is extremely rapid, most of the proliferating luteal cells are associated with vascular endothelial cells; this is consistent with the high vascularity and blood flow of the mature porcine CL and implies a critical role for angiogenesis in luteal development. Paracrine regulation of luteal function in the pig has been reported by some workers; during luteolysis, macrophages that invade the CL secrete TNF, a growth factor which inhibits luteal oestrogen production, allowing $PGF_{2\alpha}$ to become luteolytic.

Prostaglandin release

$PGF_{2\alpha}$, released from the uterus in a pulsatile fashion, induces regression of the CL in the cow. In addition to the uterus, the CL has been shown by Shirasuna *et al.* (2003) in Japan to be a site of $PGF_{2\alpha}$ production, although the physiological relevance of CL-derived $PGF_{2\alpha}$ remains poorly understoood. It is becoming increasingly evident that luteal regression is a process depending to some extent on oxidative mechanisms. $PGF_{2\alpha}$, an established luteolysin in farm mammals, has been shown to trigger accumulation of reactive oxygen species within luteal tissues; these superoxide and peroxide radicals

cause lipid peroxidation, decrease membrane fluidity and markedly attenutate luteal steroidogenesis. It is well accepted that luteal regression in the ewe is initiated by $PGF_{2\alpha}$ of uterine origin, although the exact mechanism of this regulation is not fully understood. Functional regression is apparently directly stimulated by PGF via activation of its membrane receptor, but whether structural regression is also initiated by PGF is at present unclear. One view of the factors leading to luteolytic release of $PGF_{2\alpha}$ involves the release of uterine platelet-activating factor (PAF), which creates a local loop that causes the pulsatile release of $PGF_{2\alpha}$, which is augmented by oxytocin and inhibited by interferon-tau (IFN-τ).

Large and small luteal cells

It is known that the sheep's CL contains two morphologically and functionally distinct steroidogenic cell types, designated small and large cells. Receptors for PGF are located exclusively on the large cells, and the signal for regression is received in these cells.

Effect of dietary lipids

It is believed that, in cattle, dietary lipids can improve reproductive functions by way of their effects on steroidogenesis; this may have a favourable effect on embryonic survival. Studies in Scotland have shown similar evidence in sheep; results reported in that country showed that dietary lipids enhanced postovulatory ovine luteal function.

Premature and delayed regression

It is well established that progesterone given early in the luteal phase of the oestrous cycle causes premature luteolysis in cattle, sheep and goats. It is also known that certain factors can prolong the life of CL and that this depends on the species. Certain events during the oestrous cycle and early pregnancy appear to differ markedly between pigs and ruminants. During the luteal phase of the sow's cycle, oestrogens are very luteotrophic; an injection or oral administration of an oestrogen can prolong the lifespan of porcine CL for several weeks.

Camelids

Results presented by workers in the late 1990s showed that, as in other large domestic animal species, release of $PGF_{2\alpha}$ controls luteolysis in camels; the embryonic signal for maternal recognition of pregnancy must be transmitted before day 10 after ovulation if the CL is to be maintained. In contrast to events in ruminants, however, release of endometrial $PGF_{2\alpha}$ in non-pregnant camels does not appear to be under the control of oxytocin.

5.3. Advantages of Oestrus Control

There are many reasons why farmers and researchers are likely to be interested in oestrus control in farm animals; some of these are detailed in Table 5.5.

5.3.1. Cattle and buffaloes

Oestrus control in cattle has played an important part in reproduction control during the past three decades in making AI,

Table 5.5. Advantages of oestrus control in farm animals.

1. Saves time and labour in detecting oestrus
2. To permit and simplify the use of AI, particularly in beef cattle where observation of oestrus poses problems
3. To permit fixed-time AI, where animals are inseminated at a particular hour to suit the farmer
4. Makes for most efficient feeding of animals in groups according to their stage of pregnancy
5. To permit compact lambings and predictable farrowings and keep births within specified time-limits
6. To avoid losses of young at birth by providing care and supervision
7. To provide for more uniform groups of young, which can be reared to market weight more easily than those born in uncontrolled births
8. To provide for stricter control of disease hazards; batch farrowing can allow buildings to be cleaned thoroughly between batches
9. Essential for synchronizing donor and recipient animals in embryo transfer programmes
10. The technique can be used as part of superovulation protocols in embryo transfer

superovulation and embryo transfer (ET) procedures much easier to apply under farm conditions. Oestrus control allows breeding by AI to be planned according to a strict timetable. It also permits batched calvings, the start of a breeding period at a specific date and inseminating groups of cattle at fixed times rather than detecting heat periods and inseminating animals individually. When dealing with cattle in groups, good handling facilities on the farm are essential for a smooth operation; this means appropriate holding pens, races and crushes.

Although the possible advantages resulting from effective regulation of the oestrous cycle in cattle have been the subject of numerous reports, it was only in the mid-1970s that commercially acceptable forms of oestrus control emerged and became available to the farmer. For oestrus control to be effective, clearly it should solve more problems than it creates, and it should do so in a cost-effective way; regardless of what form treatment may take, care is necessary to ensure that stress-free handling of animals is ensured. Oestrus control can only work if the cost of materials is reasonable and the protocol can be applied by professionals and farmers without adding substantially to that cost. In Ireland, for example, oestrus control measures applicable to beef suckler cattle, which would be useful in making full use of the best bulls, have apparently been abandoned because of legal and financial constraints.

Dairy cattle

Reproductive performance is a major factor affecting the profitability of dairy production; this is especially true for pasture-based, seasonal dairy production systems which require not only the maintenance of a 12-month calving pattern but also a compact calving period to enable peak milk production to coincide with maximum pasture growth. For seasonal dairy herds, as exist in a country such as New Zealand, oestrus control could improve reproductive performance and shorten the calving interval by allowing for the breeding of most cows in the herd in the first few days of the breeding season rather than over an entire oestrous cycle. Many oestrus control protocols have been devised, some more complex than others.

Zebu cattle

In the Nelore breed in Brazil, the short oestrus associated with such cattle, together with the high incidence of oestrus at night, has led to low reproductive efficiency in AI programmes. To overcome such difficulties, there is a need to use timed AI, using hormonal treatments that can synchronize ovulation sufficiently accurately in zebu cattle. In countries with large dairy and beef cattle populations, such as Brazil, in which a small proportion of cattle are bred by AI (< 5%), fixed-time AI by way of oestrus control may have particular attractions; an increase in the use of semen from the best bulls in the production of milk and meat in that country could be valuable. There is little doubt that AI of zebu cattle such as the Nelore kept under extensive management conditions would be greatly facilitated by single, fixed-time insemination strategies if they could achieve pregnancy rates equivalent to those in cows exhibiting typical oestrous cycles and mated by natural service.

Buffaloes

$PGF_{2\alpha}$ and its synthetic analogues have been used to synchronize oestrus in cyclic and suboestrous buffaloes. Fertility after such control is known to be affected by factors such as body condition, nutritional level and season; it is also known that the cost of synchronization made farmers reluctant to adopt oestrus control measures.

5.3.2. Sheep and goats

There are several reasons why oestrus control in sheep can be useful on the farm and in research. The technique can be important in saving time and labour, especially with flocks that are small, by ensuring that lambings occur in a short period of time. For the application of AI, oestrus control can be essential in permitting sheep to be bred to genetically superior rams according to a planned and predictable timetable. The same principles apply to goat breeding.

5.3.3. Pigs and horses

In contrast to what is possible with farm ruminants, a considerable measure of oestrus control is possible in pigs by management. In sows, the post-weaning oestrus can be expected to occur 4–5 days after weaning; among postpubertal gilts, on the other hand, oestrus occurs at random throughout a 3-week period and management alone is no more likely to be effective in achieving its synchronization than it is in cattle and sheep.

Horses

In dealing with the mare, in view of its long oestrous period, it is more relevant to talk about ovulation rather than oestrus control;

certainly, there are good practical reasons for controlling the time of ovulation in the mare. A delay of ovulation in mares showing normal oestrus can be a significant cause of infertility in the mare. Various authors have drawn attention to the need for an effective hormone protocol that could be applied to mares to induce fertile ovulations at a predictable time, with sufficient accuracy to enable single inseminations to be made without the need for veterinary examination of ovaries, so as to be within 3 days (fresh semen), 2 days (cooled semen) or 1 day (frozen semen) of ovulation. In seeking to control the time of ovulation in the mare, it may occasionally be better to use hormones to postpone ovulation rather than induce it.

5.3.4. Camelids

Factors controlling ovulation in camelids are still not well understood but it is known that in Bactrian camels it can be induced by deep intravaginal deposition of whole semen or sperm-free seminal plasma, as well as by the intramuscular injection of semen or seminal plasma. In dromedaries, ovulation can be induced by mating, but manual stimulation of the cervix or intra-uterine injection of whole semen and seminal plasma apparently does not stimulate the release of sufficient LH to bring about ovulation. However, in view of growing interest in using reproductive technologies such as AI and ET, there is need for effective methods of inducing and synchronizing ovulation in this species, without using natural mating with sterile males, which carries with it certain health risks.

Like camels, female llamas and alpacas ovulate in response to mating; unmated females remain in a follicular phase, characterized by almost constant sexual receptivity and regular emergence of an ovulatory follicular wave. However, despite virtual constant sexual receptivity, the ability to ovulate in response to a mating stimulus is influenced by the developmental status of the dominant follicle at the time of mating.

5.4. Growth and Development of Oestrus Control Technology

5.4.1. Historical

The technique of intravaginal steroid administration, first demonstrated in the early 1960s in sheep, was subsequently developed in the 1970s for use in cattle (Fig. 5.8) by way of the progesterone-releasing internal device (PRID) and later by the controlled internal drug release (CIDR) device. A method permitting continuous administration of progesterone/progestogen obviously eliminated many management problems associated with earlier treatments, which had involved either tedious injections or oral treatments. One obvious practical advantage of the intravaginal devices is that they involve no breakage of skin or the strict aseptic protocol required with implants that have to be inserted and removed.

In 1964, Terry Robinson in Sydney devised the polyurethane sponge, impregnated with a suitable progestogen, which was inserted into the ewe's vagina and left *in situ* for 12 days; during this period, progestogen was absorbed through the vaginal wall in physiologically effective amounts. The Searle pharmaceutical company, who prepared the progestogen fluorogestone acetate (FGA), undertook the commercial development of the product and the Upjohn company was licensed to market a sponge pessary carrying an alternative progestogen, medroxyprogesterone acetate (MAP). In the early 1970s, the progestogen-impregnated sponge was to become the cornerstone of almost all controlled breeding applications in sheep and goats, whether in the breeeding or non-breeding season, in France, Ireland, the UK and many other countries. In the mid-1970s, the PRID, devised by Russ Mauer in Abbott Laboratories in Chicago, became a widely used means of controlling oestrus in cattle. Later, in the 1980s, a further intravaginal device, the CIDR device, based on progesterone, was shown to be highly effective by New Zealand researchers (Table 5.6).

Fig. 5.8. Oestrus control using intravaginal devices.

Table 5.6. Devices used to administer progesterone/progestogens.

Method	Luteolytic agent	Duration of treatment (days)	Onset of oestrus (days)[a]	Fertility
PRID	10 mg capsule of oestradiol	10–12	2–3	Normal
Ear implant of norgestomet	3 mg norgestomet + 5 mg oestradiol valerate at start of treatment	9–10	2–3	Normal
CIDR	10 mg capsule of oestradiol	10–12	2–3	Normal

[a]Days after removal of the device.

Prostaglandins

During the 1970s. much research was conducted to examine the biological properties of PGs; these substances were first detected in the seminal fluid of rams and were initially thought to be secreted by the prostate gland: hence the term 'prostaglandin'. A great deal is now known about the distribution and biological effects of the PGs; first reports of the use of $PGF_{2\alpha}$ as a luteolytic agent in cattle date from the early 1970s. By the mid-1970s PGs had arrived on the farm, where they proved particularly useful in horses and heifer cattle. Much less work was conducted with PGs in oestrus control in sheep and goats, part of the explanation being that PGs are not relevant to oestrus control during the sheep anoestrus. In pigs, the arrival of the same

PGs heralded a new era in the control of farrowing.

Prolonging or shortening the luteal phase

In early attempts at oestrus control in cattle, two main approaches were followed in devising treatments that would give acceptable fertility levels: it was either a question of prolonging the luteal phase of the cycle artificially using progesterone/progestogens or shortening the cycle by means of the luteolytic action of $PGF_{2\alpha}$ (Fig. 5.9).

5.4.2. Fertility at the controlled oestrus

Because progesterone treatment suppresses LH release, preventing oestrus and

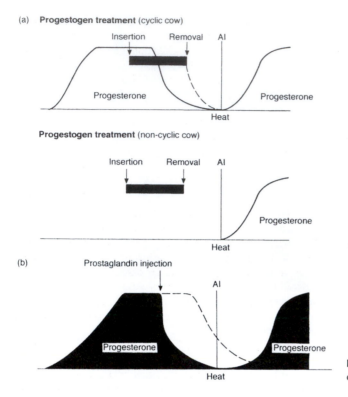

(a) **Progestogen treatment** (cyclic cow)

Progestogen treatment (non-cyclic cow)

(b)

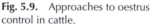

Fig. 5.9. Approaches to oestrus control in cattle.

ovulation, as mentioned earlier, various progesterone/progestogen delivery systems have been used in cattle oestrus control. However, it became clear at an early stage of oestrus control in cattle that prolonged use (14–21 days) of progestogen resulted in reduced fertility at the subsequent oestrus, apparently due to the development of persistent follicles and reduced oocyte competence. Shorter periods of progesterone usage (8–10 days), however, resulted in a much more fertile oestrus; with the shorter treatment it became necessary to induce luteolysis in those cattle treated in the early luteal phase of their cycle. This could be done using progestogens or oestrogens, or combinations of the two, to ensure that the CL regressed before the oestrus control measure terminated.

5.4.3. Accurate control of ovulation

Research in oestrus control in cattle has been directed towards providing the farmer

with a method whereby his cows can be inseminated at a predetermined time (fixed-time AI); by this means, so the theory goes, the farmer is saved the time and labour involved in heat detection. The fact remains, however, that, despite advances in the development of short-term progestogen treatments and in combination treatments involving progestogens, oestrogens, prostaglandins and GnRH, there may still be some way to go in meeting the needs of farmers seeking an optimally high calving rate after a single, fixed-time insemination.

5.5. Practical Applications of Technology

It has long been recognized that, to achieve precise control of the oestrous cycle in cattle, it is essential to control both the life-span of the CL and the follicle wave status. Studies in Ireland and elsewhere show that a better understanding of the hormonal control of follicle growth is necessary to achieve sufficiently precise control of the cycle to permit a single AI at a predermined time without the need for oestrus detection treatment (Diskin *et al.*, 2002). Factors affecting the uptake of controlled breeding technologies in New Zealand seasonal dairy herds have been discussed by various workers; the consensus appears to be that most oestrus control treatments had only a marginal effect on the percentage of cows pregnant in the first 6 weeks of the breeding season. The greatest advancement achieved in the mean day of conception was less than 1 week, leading to the conclusion that direct financial benefit from oestrus control was questionable. It was also concluded that oestrus control saved little labour compared with daily examination of cows to detect oestrus.

The future of beef herds in the USA and elsewhere is likely to depend on production systems producing a uniform, quality product making maximum use of modern technology as appropriate; effective methods of oestrus synchronization may also be the key to the uptake of much new technology. The extensive use of oestrus synchronization will depend upon research providing producers with more predictable control programmes. Fine-tuning of ovulation synchronization and timed AI programmes, where the aim is to reduce the need for heat detection and animal handling and the shortening of the breeding season, should eventually lead to an increased use of AI in the beef industry.

5.5.1. Control measures currently available

Melengestrol treatment

Melengestrol acetate (MGA), a synthetic progestogen, has been used in various regimens for oestrus synchronization in cattle; advantages of MGA include low cost, negligible toxicity, oral administration (via feed) and induction of cyclicity in anoestrous, prepubertal and post-partum animals. The duration of MGA feeding varies among the different protocols that have been devised, but the level of feeding (0.5 mg per cow/day) is regarded as critical to success. The

decreased fertility in cattle after long-term MGA treatment or after short-term MGA treatment initiated late in the cycle may be overcome in various ways. In one protocol, for example, MGA is fed for 14 days and PG given 17 days afterwards; such a protocol has been shown to result in a well-synchronized, fertile oestrus. However, for many beef farmers, the lengthy duration of such treatment is likely to be considered impracticable. Attempts have been made to improve the synchrony of oestrus and ovulation within an MGA–PGF$_{2\alpha}$ oestrus control protocol. In Missouri, for example, Wood *et al.* (2001) showed that injection of GnRH within a 14–19 day MGA–PGF protocol increased the synchrony of oestrus during the synchronized period and concentrated the period of detected oestrus; they suggest that this protocol may be useful in the fixed-time insemination of replacement beef heifers.

Feeding MGA for 14 days, followed by administration of PGF 17–19 days after MGA withdrawal, has been developed as an effective method of oestrus control in beef heifers; studies with post-partum beef cows have shown an improved outcome among cows that received MGA prior to the injection of PGF in comparison with animals receiving PGF only. It has been shown that improved synchrony of oestrus could be achieved in beef cows when animals were pretreated, either short- or long-term, with MGA prior to GnRH and PGF.

A review by Patterson (2002) suggested that MGA treatment prior to the GnRH–PG oestrus control protocol could be useful in: (i) inducing ovulation in anoestrous beef cows; (ii) reducing the incidence of a short luteal phase among anoestrous cows induced to ovulate; (iii) increasing oestrous response, synchronized conception and pregnancy rates; and (iv) increasing the possibility of successful fixed-time AI. According to Kesler *et al.* (2002), the MGA–PGF protocol has emerged in the USA as the procedure of choice for beef heifers. However, the original protocol was based on the detection of oestrus for breeding and this requires time and labour. For such reasons, most farmers prefer timed AI. Although an interval of 17 days to PGF was originally used, a 19-day interval is known to reduce the variability in the timing of oestrus. Obviously, such an oestrus control procedure is relevant only to feeding conditions where beef heifers are accustomed to meal feeding rather than to pasture grazing.

Ovsynch protocol

For fixed-time AI to be successful in cattle, both the functional lifespan of the CL must be regulated and the follicle wave status synchronized in all animals, irrespective of the stage of cycle or follicle wave. New oestrus control techniques have largely succeeded in synchronizing follicle development with the occurrence of CL regression, thereby controlling the timing of ovulation much more precisely and enabling AI to be carried out within certain fixed time limits. For research purposes, follicle growth can be synchronized in various ways with hormones such as progesterone, oestradiol and GnRH. Oestrogen treatments on the farm, however, with or without progesterone, are prohibited in lactating dairy cows in the USA and severely frowned upon in the countries of the European Union (EU). This makes GnRH the hormone of choice to synchronize follicle growth in such animals; unfortunately for farmers, one major consideration with GnRH is its cost. There is little point in waxing lyrical about oestrus control measures that are beyond the pockets of those wishing to use them.

A relatively recent advance by workers in the USA enabled them to induce synchronized ovulation in dairy cattle, within a predictable 8 h period, using a GnRH–PG–GnRH sequence, which came to be known in that country as the Ovsynch protocol (Fig. 5.10). It was suggested that this protocol

could permit effective management of AI in lactating dairy cows without oestrus detection; the schedule was apparently less effective in dealing with heifers. In some countries, the protocol may be favoured if heat detection is particularly difficult. French workers, for example, concluded that, in herds in which oestrus detection is good, a PG protocol (one or two injections) appeared to be the more appropriate, using AI after an observed oestrus; when oestrus detection is poor, then GnRH–PG–GnRH + fixed-time AI may well be appropriate.

Studies in the UK have shown that progesterone levels in the early luteal phase (3–4 days after ovulation) are higher when the second GnRH dose is given after PG. Such evidence is interpreted as suggesting that the GnRH stimulated the production of a better-quality CL; this could be a reason for using a GnRH–PG–GnRH rather than a more simplified GnRH–PG protocol. As with other oestrus control measures, the Ovsynch treatment is initiated in all cows without regard to stage of cycle (i.e. all cows start on the same day of the week). However, a study by Cavestany *et al.* (2003) has shown the importance of stage of the oestrous cycle at the start of the Ovsynch protocol. Although the Ovsynch protocol has proved to be an effective planned breeding method that permits dairy cows to be inseminated without reference to oestrus detection, there are several reservations about the effectiveness of the treatment under some conditions. While the protocol is designed to induce synchronized ovulation from 24 to 32 h after the second injection of GnRH, failure of the cow to respond to the initial injection of GnRH may result in lower pregnancy rates due to asynchronous ovarian responses. Some workers have detected premature oestrus in 5–10% of cows between the first GnRH and $PGF_{2\alpha}$

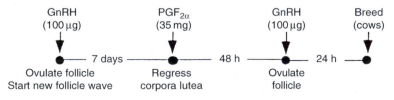

Fig. 5.10.   Ovsynch oestrus control protocol.

injections. As well as premature oestrus, incomplete luteal regression following the injection of $PGF_{2\alpha}$ is also associated with conception failure in the Ovsynch protocol. In Wisconsin, Cordoba and Fricke (2002) reported that lactating cows in a grazing-based dairy operation synchronized poorly to Ovsynch, resulting in reduced conception to timed AI (27.3%) compared with AI after tail-paint removal (47%); the problems were thought to lie in incomplete luteal regression after PGF and poor ovulatory responses to GnRH.

Prostaglandin treatments

Oestrus control in cows is often based on the synchronization of luteal regression using PG. The administration of PG in two treatments 10–14 days apart results in oestrus being shown over a 5-day period; such variability in the occurrence of oestrus is due to the maturity of the ovulatory follicle at luteolysis, which in turn reflects the stage of follicular growth. However, it is now apparent that improved synchrony of oestrus may be achieved by the synchronization of the follicular waves, in addition to the synchronization of luteolysis. With this in mind, GnRH treatment in the luteal phase, a few days before PG, can reprogramme and thereby synchronize pre-ovulatory follicle development. A second GnRH treatment after PG can further improve the synchrony of ovulation; it is on this basis that the GnRH–PG–GnRH Ovsynch form of treatment is believed to result in reasonably normal morphological and hormonal changes at the ovarian level.

Ear implants

Ear implants have been employed as a means of achieving long-term progestogen treatment in cattle, sheep and goats. Oestrus control measures in goats, for example, have occasionally involved the use of ear implants containing the highly potent Searle-produced progestogen, norgestomet. In France, researchers have used whole (3 mg) or half implants (1.5 mg) in a comparison with 45 mg FGA sponges; progestogens were left in place for 11 days, with pregant mare's serum gonadotrophin (PMSG) and $PGF_{2\alpha}$ given 48 h prior to progestogen withdrawal. Results were less favourable with the implants than with the FGA sponges.

5.5.2. Oestrus synchronization and resynchronization

Farmers using oestrus control in cattle commonly employ a 'sweeper' bull to pick up cows that fail to become pregnant at the controlled oestrus (Fig. 5.11). However, it is likely that the genetic merit of such 'sweeper' bulls may often leave much to be desired and there is the cost of keeping the animal. For such reasons, oestrus control may be used, not only for first inseminations, but for the inseminations that have to be carried out on animals that 'repeat'. It is known, from various surveys, that unnecessarily high culling rates of dairy cattle may occur because too many non-pregnant cows are not detected promptly in oestrus after first and subsequent inseminations. Such poor rates of detected oestrus may be due to limited expression of oestrus, a physiological problem, or insufficient observation by stockpersons, a management problem.

In Scotland, the efforts of Basil Lowman in the Scottish Agricultural College (SAC) and Colin Penny in the Edinburgh Veterinary School have led to a protocol which could make the 'sweeper' bull redundant in a beef cattle herd, all matings being by way of AI. However, the success of such a venture requires appropriate handling facilities and careful planning to minimize stress and to ensure that animals are in optimum body condition. The main features of the triple synchronization programme devised in Scotland over a 4-year period are in Table 5.7. The CIDR devices, which still contain adequate levels of progesterone after their initial use, are employed in resynchronization to reduce treatment costs. Cows in poor condition (< 2 body score) are given a low

Fig. 5.11. Declaring the 'sweeper bull' redundant.

Table 5.7. Triple synchronization protocol (based on protocol devised by Lowman and Penny in Scotland over a 4-year period).

Day	Action
0	Insert CIDR + 2.5 ml Receptal
7	Inject prostaglandin PGF$_{2\alpha}$
9	Remove CIDRs at 9 a.m.
	Inject 400 iu PMSG (optional)
11	First insemination (a.m.)
12	Repeat insemination (a.m.)
27	Reinsert CIDRs
32	Remove CIDRs (a.m.) and tail-paint cows
33	Observe carefully for heats
34	Second insemination (a.m.) for cows observed in heat
35	Repeat insemination (a.m.)
46	Scan cows assumed pregnant and insert new CIDRs in non-pregnant animals
50	Insert CIDRs in second AI animals
53	Inject prostaglandin (scanned non-pregnant cows)
55	Remove CIDRs and tail-paint cows
56	Observe carefully for heats in second AI group
57	Third insemination (a.m.) any animals observed in heat
58	Repeat insemination (a.m.) any cows in heat

dose of PMSG (400 iu) as a means of enhancing fertility. The practical outcome of the programme on a test farm saw 75% of the suckler cows calving over a 4-week period and 90% over 46 days.

5.5.3. Cost–benefit calculations

Workers in Florida compared cost–benefits of the Ovsynch/timed artificial insemination (TAI) protocol with AI at a detected oestrus; using economic models, they showed that the protocol could be a profitable alternative for managing dairy herds in which oestrus detection rates are suboptimal. Various methods have been used to evaluate the cost–benefit of different treatments aimed at improving reproductive performance in herds with low oestrus detection efficiency. In Germany, using simulation and modelling techniques and dealing with a dairy herd with an oestrus detection efficiency of 55%, it was found to be cost-effective to spend up to US$10 per dose of PG, US$9 per milk progesterone test and US$6 per rectal palpation. The

Intercept regime developed in the UK is the same protocol as Ovsynch in the USA. Calculations based on costs of £3 per extra day of calving interval and £770 per extra non-pregnant cow 150 days after calving showed the economic advantage of the Intercept regime compared with conventional breeding to be £91 per cow.

Reducing treatment costs

Workers in the USA examined ways of reducing costs in the Ovsynch-type treatment and reduced GnRH doses from 100 to 50 µg per injection; using 50 rather than 100 reduced costs from US$20.27 to US$6.40 per cow and it was concluded that a reduced GnRH dose could be employed without compromising the efficacy of the synchronization protocol. Replacement of the second GnRH (of the Ovsynch protocol) with oestradiol benzoate (ODB) may be another way of reducing costs because the steroid is much cheaper than GnRH analogues. There may, however, be restrictions on the use of oestrogens, as noted earlier. Workers in Brazil have shown that both GnRH–PG–GnRH and GnRH–PG–ODB protocols were effective in synchronizing ovulation in cyclic Nelore cattle and allowed a 45% pregnancy rate after TAI; the protocols, however, were not effective in non-cyclic animals. The Brazilian workers concluded that the modified oestrogen protocol was an attractive alternative to Ovsynch due to the low cost of the oestrogen; the same workers successfully used a half-dose of PG (12.5 mg vs. 25 mg) injected *intra vulvo* in the submucosa and found that the half-dose achieved pregnancy rates comparable to the full dose injected intramuscularly (i.m.). A South African study led workers to the conclusion that using halved sponges (60 mg MAP) in the breeding season was justified by their fertility results.

5.5.4. Animal handling considerations

The insertion of norgestomet ear implants (Synchromate-B and Crestar treatments), in those countries where they are commercially available (e.g. UK and USA), generally calls for a two-person operation. Devices such as the intravaginal PRID and CIDR device, on the other hand, can be inserted by a single operator without undue difficulty. With implants, as well as with intravaginal devices, the majority of cattle can be expected to exhibit oestrus 24–48 h after their removal. In terms of the percentage of animals in oestrus after a progestogen-synchronizing treatment, that is likely to be markedly lower in cattle than in sheep, where an oestrous response in excess of 95% is by no means uncommon; such results reflect endocrine differences between sheep and cattle. One factor, when vaginal inserts (PRIDs and CIDR devices) are used in oestrus control in cattle, is the loss of the device during treatment. One Irish study of factors affecting loss involved 1559 dairy cows on more than 40 farms; a loss rate of ~3.6% of CIDR inserts was recorded. In the USA, other workers recorded a comparable loss (3%) in cattle on open pasture but a much higher loss (9.6%) in cows in feedlot-type pens, presumably a reflection on behavioural differences between the two sets of cattle.

Management procedures may influence the way in which animals respond to oestrus control measures. In the USA, for example, Romano (2002) found that exposure to female Nubian goats in pro-oestrus/oestrus hastened the onset of oestrus in does synchronized for oestrus with progestogen-impregnated sponges (FGA/MAP) during the breeding season; there was a highly significant decrease in the interval of sponge withdrawal to oestrus in goats exposed to oestrous does at the termination of treatment (26.0 ± 7.1 h vs. 45.3 ± 13.5 h).

6

Control of Post-partum Ovarian Activity

6.1. Factors Influencing Post-partum Ovarian Activity

All farm animals undergo a period following parturition when they are not in a position to become pregnant. In a species such as the horse, this period may be no more than a week; in others, such as the ewe kept in conventional sheep farming, there may be many months between lambing in the spring and mating once more in the autumn. However, for the farmer, time is money and there are likely to be many occasions when he wishes to have a degree of control over the date when his animals are able to start their next pregnancy. Of course, the delay in resuming reproductive activity is nature's way of ensuring that the mother has sufficient resources to meet her own needs and those of her newborn offspring. The process of lactation can be a serious drain on bodily resouces and adding pregnancy to the load may not be at all helpful. In bygone days in the Western world, the same reasoning applied to humans, where nursing the baby was a fairly certain way of keeping a sensible interval between successive births.

6.1.1. Cattle and buffaloes

Dairy cattle

The fertility of the cow in the months following calving depends on the satisfactory involution of the uterus and the re-establishment of cyclical breeding activity. The interval between successive calvings, the calving interval (CI), is usually regarded as one of the important factors determining profitability of dairy herds (Fig. 6.1). A long CI can be particularly damaging in strictly seasonal farming systems, such as those found in New Zealand, where concentrated, well-timed calvings are essential to ensure optimal economic returns from the dairy enterprise. For those who regard a 1-year interval as the ideal, this can only be achieved if the conception rate and efficiency of oestrus detection are high and the interval between calving and first service is less than about 90 days; despite some trend towards earlier breeding, 60 days after calving is usually quoted as the earliest that a cow can be mated and show an acceptable conception rate. It is a different story in zebu dairy cattle. The average duration of gestation in such cattle has been reported as 295 days; for such reasons, it is not surprising that CIs of 14 months or more are commonplace in such cattle.

CI is primarily a function of the number of days from calving to the initiation of the next pregnancy – days open – and the effect of gestation length; numerous management and feeding protocols have been designed in an effort to reduce this interval. Workers in the USA have emphasized that to be effective any programme of post-partum

Fig. 6.1. The post-partum period: a difficult time for the dairy cow.

oestrus control must be consistent with well-established physiological concepts.

Twelve-month calving intervals

For the past half-century and more, dairy cows in the UK and elsewhere have been expected to produce increasing amounts of milk; this has been achieved by way of effective breeding programmes and improved feeding and management systems. These high-producing cows are expected to produce a calf every 12 months, since that interval has usually been considered to be the most profitable in terms of milk prices, feeding and housing costs and other factors. This demand for a short CI (12 months) in many instances will call for treatments with both hormones and antibiotics. In terms of breeding requirements, the recommendation is that cows should be inseminated at the first oestrus appearing 50 days after calving; at this time, however, some cows have not yet regained normal reproductive functions and are still in an anoestrous state.

There is therefore a demand for hormonal treatment to make the cattle regain cyclical activity and conceive within predictable time-limits. Even when cows have

regained reproductive activity, often, due to a negative energy balance at this stage (Fig. 6.2), they may suffer from health problems and show impaired fertility, resulting in a low conception rate. High levels of milk production and a 12-month calving interval may involve a risk of problems during the dry period; drying off at milk yields of 10 kg or more, for example, may well increase the risk of mastitis during the dry period, resulting in a need for antibiotic treatment. Furthermore, high yield at drying-off means increased intramammary pressure, which causes discomfort to the animal.

Increasing the calving interval in high yielders

If the CI is voluntarily increased from 12 to 15–18 months, this would mean postponing artificial insemination (AI) for 3–6 months, which would certainly make drying-off easier and increase the time available for the cow to recover from puerperal diseases and return to normal reproductive function, leading to better fertility. Workers in Sweden have confirmed that an extended calving interval can be beneficial for the fertility of high-producing cows; the animals had longer to recover after calving and to restart their normal ovarian function. The

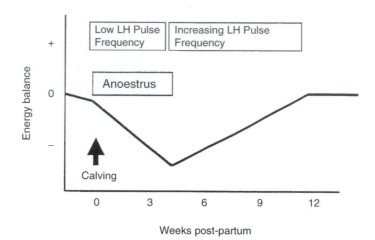

Fig. 6.2. Negative energy balance in the post-partum cow (source: Roche and Diskin, 2001).

extended period reduced the need for hormonal treatments for anoestrus and more than one insemination; fewer cows were culled due to fertility problems and there was a reduced incidence of ovarian cysts.

Many researchers in North America, dealing with year-round calving herds, suggest that a calving interval of 12.5 months (100 days open) provides optimal financial benefits for commercial milk producers; estimates of the cost of non-pregnant cows to the producer have ranged from US$1 to US$10/day for every day over the 100 days. In commercial herds, detection of oestrus is the primary factor influencing reproductive efficiency; for such reasons, prostaglandin (PG)-based reproductive programmes have become important tools because they reduce the need for constant oestrus detection. The use of PG at 14-day intervals, combined with detection of oestrus, has been a successful strategy for improving reproductive efficiency, but the method is limited due to variations in the time of ovulation; the success also depends on the stage of the oestrous cycle when treatment starts.

Beef cattle

In beef farming, the aim is to obtain and wean one calf per cow per year; thus, after a period of pregnancy of about 280 days, beef cattle have to establish the next pregnancy within 80–85 days of calving. However, cyclicity is absent during the pregnancy and early post-partum period. Although cyclic activity may resume within 2–3 weeks of calving in dairy cows, it does not often resume before 35–60 days or even longer in suckled beef cattle; usually only 30–50% of cows resume cycling by 2 months post-partum. Prolonged post-partum acyclicity or anoestrus in suckler cows influences the efficiency of oestrus control procedures.

Post-partum anoestrus can be regarded as a transition period during which the functional hypothalamic–pituitary–ovarian–uterine axis recovers from the previous pregnancy. The efforts of many researchers have resulted in a reasonable picture of endocrine events in the post-partum cow; this enables various practical measures to be taken to deal with the problem. In beef cattle nursing calves, the physiology of post-partum acyclicity and resumption of cyclic ovarian activity post-partum has been well mapped out. In brief, follicle-stimulating hormone (FSH) release and the development of dominant follicles resume soon after parturition; however, these dominant follicles fail to undergo terminal maturation, a prerequisite for ovulation, due to the absence of appropriate luteinizing hormone (LH) pulses and the dominant follicles become atretic. Absence of LH pulses early in post-partum is due to depletion of LH stores in the anterior pituitary and is independent of suckling. After the replenishment of LH stores between days 15 and 30 post-partum, the absence of LH pulses is dependent on suckling; suckling inhibits LH pulses by inhibiting gonadotrophin-releasing hormone (GnRH) discharges.

Involution of the uterus

The first 2–3 weeks are necessary for uterine involution to begin, for anterior pituitary LH stores to be replenished and for waves of follicular growth to resume; thus, within 3 weeks, the beef cow becomes ready to resume cyclic activity. However, the inguinal perception of the calf by the suckler cow during suckling increases sensitivity of the putative hypothalamic GnRH pulse generator to the negative-feedback effect of oestradiol via release of endogenous opioid peptides; this results in suppression of pulsatile release of LH, failure of ovulation and prolonged anoestrus. Thus, following replenishment of the anterior pituitary LH stores, post-partum anoestrus resembles the luteal phase of the cycle, with waves of dominant follicles which fail to ovulate, even though they have the potential to ovulate. In buffaloes, uterine involution after calving is reported to be complete by about 3 weeks post-partum; some studies have shown the time of completion of uterine involution to be significantly correlated with calf birth weight.

6.1.2. Sheep and goats

If a ewe is to be bred twice yearly or three times in 2 years, the interval between lambing and rebreeding will be markedly shorter than usual; for such reasons, events in the post-partum ewe become of real importance in any consideration of more frequent lambings. Ewes producing young in the spring give birth at a time when they normally enter the seasonal anoestrus, even had lambing not occurred. The spring-lambing ewe, with lambs at foot, represents the most difficult category of ewe in which to induce pregnancy; she has both

lactational and seasonal anoestrus to deal with in the early months after giving birth. As shown in data from extensive Irish trials in the 1970s conducted in early anoestrus, with lambs running at foot the conception rate in ewes after controlled breeding was little more than half the value recorded among similar sheep several months later in late anoestrus, when lambs were no longer with them (Table 6.1).

Goats

In goats, it is known that the time elapsing before resumption of sexual activity after parturition depends on several factors, including nutrition, nursing of offspring and season of parturition. Post-partum anoestrus is likely to be longer in undernourished than in well-nourished goats; anoestrus is also likely to be longer in does with suckling kids than in non-nursing animals. In sheep and goats originating from high and mid-latitudes that exhibit seasonality in their reproductive activity, the length of post-partum anoestrus also depends on the time of year of parturition. The results of a study by workers in Mexico, for example, with goats in a subtropical region, showed the duration of post-partum anoestrus to be significantly longer for goats kidding in January (200 days) than for those kidding in May (100 days) or October (50 days); such data demonstrated that, in a subtropical environment, season of parturition, rather than food availability, is the factor having a marked influence on the time of reinitiation of sexual activity after parturition.

Table 6.1. Lambing outcome in spring- and summer-mated ewes.

	Season	
	Spring	Summer
Groups	83	594
Ewes treated	2,508	21,594
In oestrus (%)	93.0	97.0
Ewes lambing	871	13,795
Conceptions (%)	37.0	66.0
Lambs/conception	1.58	1.62

6.1.3. Pigs

Sow productivity is closely related to the number of weaned piglets produced per sow per year; this, in turn, is a function of the number of young born and reared per litter and the number of litters per sow per year. One major way in which sow productivity can be influenced is by reducing the age of weaning. Decreasing the length of the lactation period by early weaning can increase the number of litters per sow from 2.05 when piglets are weaned at 8 weeks to 3.0 when they are weaned at birth, assuming a week or so between weaning and conception. Although, in theory, it may appear possible for the sow to be capable of producing three litters per year, there are, in practice, several serious difficulties facing the producer who attempts this. In contrast to the cow, the sow is normally in a state of anoestrus during lactation and for that reason cannot be bred until some days after the piglets are weaned.

There is commercial interest in the use of segregated early-weaning management systems in pig production systems; this interest is primarily due to the role of segregated early weaning in breaking the pathogen transfer cycle from the sow to the piglet. There is also the practical point that early weaning can offer the potential for increasing the farrowing frequency and increasing the annual reproductive output of the sow. According to studies in pigs conducted in the 1960s, uterine involution is completed histologically after a lactation length of 3 weeks, but less information has been available for modern sow genotypes and pig production systems that are used today.

Lactational anoestrus

Methods of dealing with the lactational anoestrus are important in any attempt to increase the frequency of farrowings. It is believed that the inhibition of the hypothalamic–pituitary axis during lactation in the sow is mainly due to suckling-induced neuroendocrine reflexes, although nutritional deficit becomes relatively more important during the third and fourth weeks after farrowing (Fig. 6.3).

In wild boar herds, or where domestic sows are living under semi-natural conditions, weaning of piglets is a slow and gradual process extending over many weeks. In commercial pig production, however, lactation is interrupted at 3–5 weeks after farrowing, which is about the time of peak milk production.

Stress factors

In current commercial pig production, the mixing of recently weaned sows is a common practice; with this, a rank order is immediately established and sows obtaining a low rank often have less access to food. Food deprivation is associated with increased cortisol, indicating stress, and with changes in the levels of prostaglandin and progesterone concentrations; there are clear indications that stress, induced by mixing of sows and/or lack of food during the early stages of pregnancy, can reduce reproductive performance in the sow. There is evidence that elevated $PGF_{2\alpha}$ release might be one cause for disturbances in

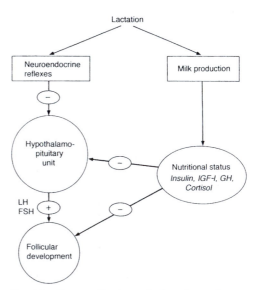

Fig. 6.3. Factors affecting secretion of gonadotrophins in the lactating sow. IGF, insulin-like growth factor; GH, growth hormone. (From Quesnel and Prunier, 1995.)

embryo transport through the oviducts of sows subjected to such stress events as food deprivation or mixing. In Sweden, for example, Mwanza *et al.* (2000) investigated the effect of postovulatory food deprivation on the hormonal profiles and activity of the oviduct; it was concluded that food deprivation is associated with changes in the hormonal profiles, activity of the oviduct and a delay in the transport of oocytes.

6.1.4. Horses

In view of the relatively long gestation period in the mare (more than 11 months), if she is to produce a foal each year, she must obviously become pregnant within a few weeks of foaling. Among the farm animals, the mare has the unique ability to return quickly to cyclicity and fertility after giving birth; it is estimated that about 90% will show a foal heat. After a long period of ovarian inactivity during pregnancy, the mare's ovaries are capable of developing follicles immediately after parturition; by a week post-partum, a preovulatory follicle is present and the animal is showing signs of oestrus (foal heat).

Foal heat

The management of the post-partum period is of great importance in the practice of stud farm medicine because the time from parturition to the resumption of cyclic ovarian activity is short. The first fertile oestrus, together with the first ovulation, occurs as early as 6–20 days after foaling. This phenomenon is the result of the long gestation period and the evolutionary pressure to maintain an annual interval. It is known that several factors influence ovarian activity in mares after foaling. In the period known as the 'lactational anoestrus', mares fail to ovulate and/or show overt oestrus during the post-partum period. This condition appears to be more closely related to photoperiod than to lactation itself; the interval from foaling to first ovulation and the incidence of ovarian inactivity after the first ovulation is significantly increased in

mares foaling early in the year. Although short-term removal of the foal has no effect on post-partum ovarian activity or LH secretion, removal of the foal on the day of parturition or restricted suckling may alter follicular growth or time to first ovulation. Body condition is also known to play an important role in influencing post-partum ovarian activity. Work in Hungary led to the conclusion that the resumption of ovarian activity is strongly influenced by the season of parturition and parity.

Fertility at foal heat

In horses, because of the long 11-month gestation period, it is of practical interest to breed the mare soon after foaling. Many studies have examined the fertility of mares at the first post-partum oestrus and compared this with the fertility observed at later times. Studies in Italy in the 1990s found that the foaling rate following conception at the foal heat (72.1%) was no different from the foaling rate after conception at any other cycle; based on the absence of differences, it was concluded that breeding at the foal heat is advisable.

6.2. Development of Control Measures

6.2.1. Physiology and endocrinology of the post-partum animal

Strategy for survival

It has been suggested that delayed resumption of ovarian cycles after calving could be regarded as a strategy for survival on the part of the cow, designed to avoid initiating pregnancy during periods of environmental or physiological stress. With this in mind, there are obvious possibilities of conflict between the commercial objectives of the dairy farmer and the mechanisms involved in the restoration of the cow's cyclical breeding pattern. It is clear that, in most lactating cows, there is a suppression of follicular development immediately after calving but that changes occur quite rapidly within 2 weeks of parturition. Although the physiology of the hypothalamic–pituitary–ovarian axis is yet to be fully understood, it is clear that the pulsatile release of LH and GnRH and the pituitary sensitivity to GnRH gradually increase after calving; it is also evident that the first luteal phase is shorter and progesterone concentrations are lower than in subsequent cycles.

Need to sensitize brain receptors

The failure of the post-partum cow to express oestrus prior to the first ovulation is thought to be due to the absence of prior progesterone action; it is believed that the high concentrations of placental oestrogen in the cow's circulation during late pregnancy may induce a refractory state in which the brain cannot respond, in the absence of prior progesterone action, to the oestrus-inducing action of follicular oestrogen. The corpus luteum formed at first ovulation provides the progesterone required for the sensitization of receptors in the brain, which are then able to respond to ovarian oestrogen by eliciting the psychic phenomena associated with oestrus. It is known that corpora lutea of shorter than normal lifespan are common after the first post-partum ovulation; it is also evident that there may be considerable irregularity in both the apparent lifespan of the corpus luteum and the level of progesterone associated with it.

Post-partum interval and conception rate

An increasing first-service conception rate and a declining number of services per conception as the interval between calving and breeding increases has been shown in many research reports and from cattle AI organizations. Data have shown conception rates varying from 5% to 35% in matings occurring within 2 weeks of parturition; the conception rate usually improves as the post-partum interval lengthens, but there is relatively little change after about 3 months. New Zealand Dairy Board figures for 69,000 inseminations in dairy cattle showed that, of cows mated within 30 days of calving, 31.3% conceived to first service, 42% conceived within 40 days, 49% within 50 days and 54% within 60 days; a conception rate of 62% within the 60–90-day period post-partum was increased only fractionally in cows bred beyond that time.

For those attempting to maintain a calving interval of 1 year by breeding cows as soon as they show oestrus after calving, there is no reason why this should not be done, although the low conception rates and associated insemination costs of rebreeding are factors to keep in mind, particularly if expensive semen straws are being used. The earlier that ovarian activity is re-established after calving, the better; cows failing to exhibit oestrus in the first 30 days after calving are likely to require more services than those that do exhibit oestrus.

Suckling effects

First ovulations usually occur later after calving in beef cattle than in dairy animals; it is also well recognized that the frequency and intensity of suckling can affect the duration of the post-partum period of anoestrus. For that reason, complete, partial or short-term weaning has been employed by various workers in their efforts to reduce the length of the post-partum anoestrus. The mechanisms by which suckling delays the resumption of ovarian activity after calving are still a matter of some speculation. Prolactin released in response to nipple stimulation may be part of the story and the suppressive effects of β-endorphins on the release of GnRH may be another.

Components of the post-partum period

The various components of the post-partum period in cattle include uterine involution, regeneration of the uterine epithelium, elimination of bacterial contamination and the return of ovarian cyclicity (Fig. 6.4). After calving, major changes occur in the structure and function of the uterus and ovaries as well as their hormonal control by the hypothalamus and pituitary. Cows are usually inseminated from 6 weeks or so after calving and, for that reason, a short post-partum period is important.

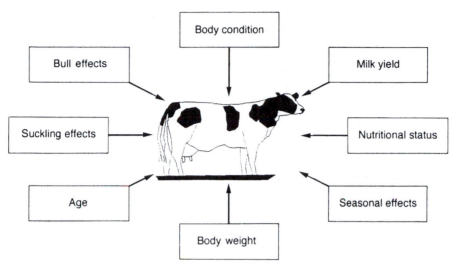

Fig. 6.4. Factors affecting duration of the post-partum interval.

Negative energy balance

During early lactation in high-yielding dairy cattle, the amount of energy required for maintenance of body tissue functions and milk production is well known to exceed the amount of energy cows obtain from dietary sources. Most glucose is used directly in milk synthesis, while glucose utilization elswhere is reduced and cows utilize body fat as an energy source. A negative energy balance post-partum not only contributes to increased ketogenesis but also delays the onset of ovarian function, especially if energy deficiency is prolonged. During the period from about 3 weeks before to 3 weeks after calving, high-yielding dairy cows do not eat enough feed to supply their nutritional requirements, falling into negative energy balance and sometimes negative protein balance; body reserves are mobilized and body condition decreases. Hormone and metabolic profiles are changed, resulting in alterations of the hypothalamic–hypophyseal–ovarian axis, which compromise reproductive physiology and adversely influence oestrous behaviour and ovulation.

Transition period

The transition period in dairy cows covers the 3 weeks before calving to 3 weeks after parturition; this period is characterized by marked physiological and metabolic changes and calving-related disorders, such as milk fever, retained fetal membranes, metritis and fatty liver. These calving-related disorders can be expected to have a serious negative effect on the economics of dairy farming. Efforts to decrease the incidence of calving-related disorders should have a positive effect on both milk production and reproductive responses, and a considerable literature is devoted to such work.

The primiparous sow

Various workers have pointed out that the reproductive performance of the primiparous sow is a key factor in pig production; because the gain to be expected is always higher than that expected from a replacement gilt, the culling of such sows should be avoided. Studies in The Netherlands have shown that about 14% of primiparous sows are culled and 50% of this culling is for reproductive reasons. It is known that primiparous sows show a prolonged weaning-to-oestrus interval, require greater use of oestrus-inducing hormones and have a smaller litter size and a lower farrowing rate to first insemination (Table 6.2). Australian workers have reported results suggesting that, with the exception of parity-1 sows,

Table 6.2. Breeding performance of primiparous and multiparous sows after weaning (from Hughes, 1998).

	Oestrus (%)			Farrowing rate (%)		
	3–9 days	> 10 days	No. of sows	3–9 days	> 10 days	No. of sows
Primiparous sows	76	24	9,557	82	78	10,368
Multiparous sows	89	11	27,369	83	73	30,087

extended weaning-to-oestrus intervals are not major problems; results also indicated that, where the weaning-to-oestrus interval was extended, regular boar contact post-weaning was unlikely to stimulate an earlier return to oestrus.

Goats

Studies in Brazil examined post-partum ovarian activity in goats as influenced by suckling effects; for goats suckling young continuously, twice daily and once daily, workers recorded intervals from kidding to first oestrus averaging 46.4, 33.9 and 30.0 days, respectively.

6.2.2. Strategies for inducing resumption of ovarian activity

Avoiding early resumption

Getting the post-partum cow pregnant again shortly after calving may not necessarily be a question of inducing an early resumption of ovarian activity. In fact, there is evidence that the occurrence of ovulation within 21 days of calving may compromise the fertility of dairy cows. As a means of initiating cyclical ovarian activity in the early weeks after calving, several research groups in the 1970s examined the use of GnRH, given as early as 14–20 days after calving; however, the responsiveness of the pituitary and the levels of LH released were inconsistent. It is now known that uterine involution, which should normally be complete by 25 days post-partum, can be prolonged because of the association between early ovulation and an increased incidence of uterine pyometra; it is thought that extending the interval to

first ovulation may reduce the incidence of abnormal cycles and uterine pathology.

Australian workers, using a GnRH implant (Ovuplant) within 48 h of calving, found they could extend the interval to first ovulation to at least 24 days after calving. They concluded that such GnRH treatment could be useful in inducing anoestrus during periods of severe negative energy balance in early lactation in high-producing Holstein cows; as well as influencing events in the uterus, the animals were given a chance to pass through this period without the need to divert resources to ovarian activity, which is known to be especially sensitive to nutritional effects.

Once-daily milking

Dealing with a dual-purpose milk/beef production system commonly practised in Mexico, where cows are milked once daily in the presence of the calf and then the calf suckles for the next 1–8 h, Perez-Hernandez et al. (2002) attempted to separate inhibitory factors by delaying the start of suckling after milking; this was on the understanding that this would reduce the impact on the hypothalamic–hypophyseal axis and allow it to recover more readily. They were able to demonstrate that delaying suckling for 8 h after milking increased the proportion of cows ovulating within 100 days of calving, shortened the calving to first ovulation interval and improved calf performance without reducing cow milk yield or body condition.

'Shang' treatment

It is all too clear that the suckling of calves can markedly affect the resumption

Fig. 6.5. Suckling as a factor influencing post-partum interval.

of ovarian activity in beef cattle (Fig. 6.5); for that reason, management routines have been devised to temporarily suspend suckling in order to facilitate reproductive function. The frequency of LH pulses, and hence ovarian activity, is known to be influenced by several factors, acting either alone or in conjunction with others. In Scotland, Sinclair *et al.* (2002) are among those who have shown interactions between suckling and body condition in beef cows, suggesting that calf restriction (suckling once daily rather than unlimited access to dam) could alleviate the suppressive effects of undernutrition on episodic LH release.

In beef cattle there have been reports showing that oestrus may be induced in anoestrous cows nursing calves by way of short-term progestogen treatment allied to 48 h calf removal; this procedure, known as the 'Shang' treatment, was suggested by a Texas rancher. For the 'Shang' method to work, the beef cows must be in reasonable body condition and on an adequate diet; a 24 h separation period was found to be less successful than 48 h. Removal of the calf resulted in a 5% reduction in calf weight at weaning, a relatively small price

to pay for getting the cow pregnant on schedule.

Pheromonal effects

Workers in Scotland a decade ago found evidence of pheromonal influences on the duration of the post-partum anoestrus in beef suckler cattle; cervical mucus from cows in oestrus was found to contain agents (pheromones) capable of assisting in the re-establishment of breeding activity, particularly in cows with extended anoestrous periods. The same workers suggested that this finding may have implications for the design of housing for suckler cows; lack of opportunity to mix with other cows or keeping cattle in small herds may exacerbate the problem of extended post-partum intervals. In the USA, other workers reported that beef cows exposed to possible pheromonal influences (bulls or androgenized–treated cows) in the early post-partum period returned to oestrus earlier than animals isolated from such forms of stimulation. The effect does not necessarily depend on having an intact bull; some report that the presence of a teaser (vasectomized) bull can stimulate ovarian activity. In the USA and the UK, it was found that teaser bulls significantly reduced the post-partum interval, but there was no benefit when cows were in good body condition at calving.

Boar effects

After weaning, contact with the boar can induce the onset of ovarian activity and advance oestrus in sows by days; boar stimuli are probably important in activating neuroendocrine pathways involved in the regulation of ovarian activity. Although boar presence can stimulate oestrus, excessive boar stimulation is also known to reduce the expression of oestrous behaviour; it is believed that increased boar contact can reduce the responsiveness of the sow to such stimuli.

Lactation length

The effects of lactation length on the weaning-to-service interval, first-service farrowing rate and subequent litter size were studied by workers in the USA; they analysed 178,519 records from 13 commercial herds to show that third-parity and older sows could be weaned after only 9 days of lactation and still retain the ability to recommence oestrous cycles in an average of 7 days or less with a subsequent first-service farrowing rate of > 70%. Second-parity sows could be weaned after 12 days of lactation or less. In first-parity sows, however, those weaned earlier than 14 days required > 10 days to recycle. The study showed that reducing lactation length from 20 to 15 days reduced litter size by 0.2 and from 15 to 10 days by a further 0.2.

The influence of feed intake during different weeks of lactation on the reproductive performance of sows was investigated in the USA, using data from 18,243 farrowing records on 30 commercial farms in Minnesota and Iowa; data dealt with sows with lactation periods between 7 and 22 days. It was found that increased feed intake during early and mid-lactation reduced the weaning-to-service interval more than it did during late lactation; greater feed intake during mid- to late lactation increased litter weaning weights more than it did during early lactation.

6.2.3. Assessing nutritional status of animals

Numerous studies have examined the nutritional aspects of post-partum anoestrus in cattle and related nutrition and body condition to ovarian activity after calving. Target condition scores for autumn- and spring-calving beef cattle are indicated in Fig. 6.6.

It is well accepted that, as high-yielding dairy cows enter a state of negative energy balance during early lactation and that as lactation progresses, the animals regain energy balance. However, the consequences of the previous imbalance for fertility have been difficult to evaluate because factors other than nutrition may be involved (e.g. metabolic or infectious diseases). In Spain, Lopez-Gatius *et al.* (2003) conducted a meta-analysis to evaluate the effects of body condition score at calving and time of first AI on reproductive performance in dairy cattle. They record that pregnancy rate at first AI significantly decreased by about 10% in cows calving in poor condition; animals with a high body condition score at parturition showed a shorter interval to conception than animals with an intermediate or low body condition. The Spanish workers concluded that special attention should be paid to the cow's nutritional needs during the periparturition period, between late pregnancy and early lactation. This may be important in preventing metabolic disorders. Overfeeding during the dry period has been associated with a predisposition to accumulate fat in adipose tissue, increased lipolysis after calving and a reduced ability of adipose tissue to re-esterify mobilized fatty acids.

6.3.4. Current treatment protocols

Ovsynch treatment

According to several observers in the USA, there is interest in the Ovsynch oestrus

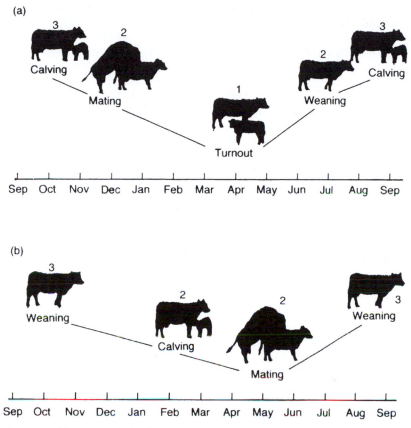

Fig. 6.6.   Target condition scores for beef cows.

control protocol and dairy operations are utilizing this programme to inseminate lactating cows; for such reasons, it is important to understand the mechanisms that influence pregnancy rates so that fertility may be improved. Studies have shown a positive association between body condition score and fertility to the Ovsynch protocol, which suggests that body condition should be considered when determining the time post-partum that cows are inseminated. It is still unclear how low body condition may result in reduced pregnancy rates.

Suckler cows

In suckled beef cows, post-partum anoestrus is the main factor which increases the interval between calving and conception, particularly in primiparous cows. Efforts to improve fertility have usually involved

treatments using progestogen, $PGF_{2\alpha}$ and pregnant mare's serum gonadotrophin (PMSG) to induce ovulation and synchronize oestrus. For the induction of ovulation and cyclicity after the third week post-partum in suckled beef cattle, a suitable protocol was that proposed by Yavas and Walton (2000); this involved the use of progesterone-releasing internal device (PRID) or controlled internal drug release (CIDR) intravaginal devices to establish ovulations and normal cycles at 3–4 weeks post-partum. In France, Mialot *et al.* (2003) compared the effectiveness of an oestrus control and AI programme based on GnRH + $PGF_{2\alpha}$ + GnRH (Ovsynch) with the standard PRID + $PGF_{2\alpha}$ + PMSG protocol; the study involved Charolais and Limousin beef cows under French field conditions after autumn calving, without calf removal and in herds with a high percentage of cyclic animals. They found that pregnancy rates were similar in both beef breeds and after both oestrus control treatments when most cows were cyclic; however, it was concluded that the Ovsynch protocol should not be used in beef herds with a low proportion of cyclic animals, which is a common feature of French beef herds.

Buffaloes

Measures to induce breeding activity in buffaloes that show anoestrus after calving are similar to those formulated for cattle. In India, workers reported that combined treatment with melengestrol acetate (MGA) (0.5 mg daily for 14 days) + oestradiol benzoate (400 µg) 48 h after the last MGA feeding was effective in inducing fertile oestrus in anoestrous buffaloes during the normal breeding season.

Fertility prophylaxis

In the former East Germany in the 1970s, a novel concept termed 'fertility prophylaxis' was used in trying to reduce the post-partum interval in dairy cattle. This involved daily oral doses of a progestogen (chlormadinone acetate; CAP) for a 20-day period (days 15–35 after calving) to clinically healthy dairy animals; over a 4-year period they gathered evidence showing an earlier than usual resumption of cyclical activity, a consequent decrease in the CI (by 10.5 days on average) and a reduction in the level of infertility in the cattle herds so treated. In the USA at that time, attempts to stimulate resumption of ovarian activity used the progestogen MGA, which was fed for a 14-day period, starting on day 21 after calving; as with CAP and the German work, this was found to be effective in reducing the post-partum interval, although no firm conclusions were reached as to the mechanisms involved. Such data are of historical interest in supporting the concept that chronic progestogen treatment may have a positive influence on endocrine events in the early weeks after calving. As well as such effects

recorded with progestogens, there have been studies, in both dairy and beef cattle, suggesting that $PGF_{2\alpha}$ treatment in the early post-partum period might decrease the calving interval; again, the mode of action was not clear, but was seen to be of sufficent interest to justify research to determine the way in PGF might achieve such an effect.

Progestogens in pigs

The use of a progestogen (Regumate) at the weaning of primiparous sows was examined by workers in France; this took the form of a 3-day treatment with the progestogen starting on the day of weaning or a 5-day treatment starting 2 days after weaning. The French workers recorded an improvement in fertility and productivity after such treatments.

Pig gonadotrophin 600 and PMSG

The effect of exogenous gonadotrophins on the weaning-to-oestrus interval in sows has been the subject of many studies; one such involved Yorkshire sows injected with pig gonadotrophin 600 (PG600), which is a mixture of 400 iu of PMSG + 200 iu human chorionic gonadotrophin (hCG); it was shown that PG600 increased the percentage of sows in oestrus within 7 days of weaning, the treated sows also showing oestrus significantly earlier than controls. The strategic use of PG600 in first-litter sows after weaning has been described by Hughes *et al.* (2000) in Northern Ireland; they recorded 94% of the sows receiving their first service on either the fourth or fifth day after weaning compared with 77% of the controls. In Brazil, workers administered PMSG (750 iu) to primiparous sows on the day after weaning piglets (18-day lactation length); it was concluded that the use of PMSG was associated with a more precise prediction of oestrus duration and allowed optimization of breeding management of the pigs. Some studies, such as those recorded above in Brazil, have used PMSG alone rather the PG600 product.

Prostaglandins (PGs)

Various attempts have been made to enhance the reproductive performance of sows in the post-partum period using PG therapy. In the USA, for example, workers studied the effect of injecting sows with $PGF_{2\alpha}$ immediately post-partum on subsequent performance; they recorded positive effects on weaning weights of progeny of sows receiving PG injections and in some instances a reduction of 19% in neonatal mortality after injections of PG on the day of farrowing and 24 h later. In Canada, other studies evaluated the effect of a single injection of $PGF_{2\alpha}$ given post-partum to sows 24–48 h after farrowing; this work demonstrated numerical improvements in the number of piglets weaned and milk yield per sow and a more rapid decline in post-parturient serum progesterone levels in PG-treated sows.

Split weaning

The effect of split weaning in first- and second-parity sows on sow and piglet performance was the subject of work in The Netherlands; in split weaning the heaviest piglets were weaned at 3 weeks, leaving six piglets with the sow to be weaned at 4 weeks (Fig. 6.7). It was found that second-parity split-weaned sows showed a higher farrowing rate compared with similar sows that were weaned at 4 weeks; it was concluded that split weaning in first- and second-parity sows should be considered in current sow farming to improve production. In a thesis dealing with the causes and consequences of variation in the weaning-to-oestrus interval in the sow, among the facts reported is the finding that, by avoiding mating at the first oestrus after weaning in first-parity sows, farrowing rate could be improved by 15% and litter size by an average of 1.2 live-born piglets per litter. In Italy, workers reduced litters to the four or five heaviest piglets some 3–6 days prior to the date of the weaning of the remainder of the litter; they reported a significant reduction in the interval from final weaning to oestrus and a tendency for more sows to show oestrus after final weaning.

Fig. 6.7.   Split weaning in sows.

7

Control of Seasonal Breeding

7.1. Advantages of Control Measures

7.1.1. Factors affecting seasonal breeding

Knowledge of the important bearing that seasonal changes in day length have on the behaviour and function of farm livestock has steadily accumulated over the past half-century. It was in the 1920s that attention was first drawn to the phenomenon of photoperiodism in plants, showing how their reproductive phase may be related to length of day. In this same period, it was realized that the annual rhythm of breeding in birds was not due to seasonal changes in temperature, as had been commonly assumed up to that time, but to changes in light. In the 1930s, investigations were extended to mammals; in the farming world, attention was focused on the observation by the reproductive biologist Marshall in Cambridge that, when British breeds of sheep were transported across the equator to New Zealand, they changed the time of year at which they mated, so conforming with the seasons of their new environment. This was taken as strong evidence that the seasonal breeding habits of the ewe must be the result of an environmental factor, now known to be day length. It became clear in the 1940s that farm animals fell into two categories, those termed 'short-day breeders', such as sheep and goats, and those termed 'long-day breeders', such as horses and chickens. In the early 1950s, it became possible to use

hormonal therapy to overcome the ewe's seasonal anoestrus, opening the way to new possibilities in sheep farming (Fig. 7.1).

7.1.2. Sheep and goats

Ewe as the model

Our understanding of the mechanisms controlling reproductive seasonality in female farm animals has largely been derived from studies in ewes (Rosa and Bryant, 2003). It is believed that the seasonal anoestrus in the ewe is the result of light-induced changes in the sensitivity of the hypothalamic–pituitary axis to the negative-feedback action of ovarian oestrogen. Once the last corpus luteum of the breeding season starts to regress in the ewe, tonic luteinizing hormone (LH) secretion rises, leading to an increase in follicular oestrogen. However, unlike events in the cyclic ewe, this elevated oestrogen inhibits secretion of tonic LH by its negative-feedback action. The events that characterize anoestrus in ewes are basically thought to be a failure in the normal preovulatory development of ovarian follicles due to inadequacies in the LH pulse frequency.

Seasonal changes in ram activity

In Ireland, all breeds of sheep show a well-defined breeding and non-breeding season;

Fig. 7.1. Autumn lambs by progesterone–PMSG (1953). Autumn-born lambs with their Half-bred mothers on a Cambridgeshire farm in 1953. These were the first such lambs born in the UK after treatment with the newly devised progesterone–PMSG treatment for the induction of pregnancy in anoestrous ewes.

for most lowland breeds, the season spans the 6-month period running from September to February; a similar story would hold true for many British breeds, as shown by Hafez in Cambridge half a century ago. The seasonal breeding found in ewes, however, is not evident in rams to the same extent. Although reports at one time suggested that rams were seasonal breeders, this is certainly not true of the main breeds in the UK and Ireland (Fig. 7.2). That is not to say that seasonal fluctuations in semen quality and sex drive do not occur, but simply that, if they do, they do not materially reduce the efficiency of the mating process. While there is evidence of considerable seasonal variation in testis size and semen production in primitive sheep breeds such as the Soay, such is not the case with rams used in normal commercial practice. Among environmental factors (day length, feed, climate) likely to reduce the effectiveness of the ram, elevated temperature would seem to be the most important. In both Australia and the USA, there are plenty of studies showing that high summer temperature

may be responsible for temporary infertility in rams.

Increased lambing frequency

The short gestation length in sheep (~147 days) enables the production of more than one lamb crop per ewe per year. However, the seasonal breeding habits of the ewe limit the farmer's ability to capitalize on this opportunity, although numerous strategies have been developed to facilitate out-of-season breeding, with varying levels of effectiveness and cost efficiency. Authors in the 1960s, reviewing attempts up to that time to increase the frequency of lambing in sheep, concluded that there was no way in which this could be achieved consistently on a flock basis with the techniques then available.

All this was to change in the early 1970s when John Robinson in Aberdeen demonstrated, using appropriate sheep and management, that it was possible to achieve two lamb crops every 13 months on a flock basis. Not only that, but it was possible to produce an average of twins each time,

Fig. 7.2. Suffolk rams remain sexually active in all seasons. At one time it was thought that Suffolk rams became sexually inactive during the spring and summer months. However, when effective treatments for the induction of oestrus and ovulation became available in the early 1950s, it was found that the rams were quite capable of mating at all times of the year, although their sperm output might be somewhat less in the spring/summer period.

giving a remarkable figure of 3.7 lambs per ewe per year, when compared with the figure of 1.28 lambs achieved in the UK under conventional sheep-farming conditions. The ingredients of the Aberdeen story included the Finn × Dorset ewe, fluoro-gestone acetate (FGA)-impregnated intra-vaginal sponges (Fig. 7.3), carefully controlled nutrition and the weaning of lambs at 1 month of age. The results achieved in Aberdeen made it possible to draw up a specification for an intensive lamb production system suitable for commercial exploitation in the UK and elsewhere.

Goats

In temperate regions, the goat can be regarded as an autumnal breeder, like the sheep, with sexual activity occurring in northern latitudes between September and January. There are, however, goat breeds in which, like the sheep, breeding activity never stops completely when kept in a suitable environment. This is true for breeds in tropical and subtropical regions, but may also be evident in other breeds which are located further from the equator.

Fig. 7.3. Progestogen-impregnated sponge.

South African studies with the Boer goat showed a peak in breeding activity during autumn and low activity in late spring to mid-summer; however, periods of complete anoestrus were never observed. In Japan, there is the Shiba goat, which is a continuous breeder, showing no seasonal variation in its reproductive activity.

7.1.3. Horses

Using ultrasonography and rectal palpation, seasonal changes in the reproductive
activity of mares in England were recorded
by workers in the October to January
period. The number of mares diagnosed as
anoestrous increased from 6.5% in October
to 68.9% in January; no distinct transition
phase between dioestrus and anoestrus
was evident. In other studies, dealing with
mares in the January to May period, 78%
of mares examined in February were
anoestrous compared with 8% in April and
0% in May. It was concluded that the first
ovulation of the breeding season generally
occurred in mid-April.

Natural and arbitrary breeding seasons

Some problems encountered in assisted
horse reproduction differ in nature from
those experienced in cattle, sheep and pigs;
certain difficulties arise from the unique
features of the animal's reproductive physiology and endocrinology (e.g. reluctance to
superovulate) but others are human-made.
One such human-made problem in racehorse
reproduction arises from the fact that the
physiological reproductive season does not
necessarily coincide with the artificial breeding season imposed on the animal (Fig. 7.4).

The reproductive physiology of the
horse differs from that of other farm animals
in several ways. In cyclic mares, the gradual
preovulatory increase in LH secretion begins
about 4 days before ovulation and the
duration of oestrus is considerably longer
than in farm ruminants and in pigs. In
contrast to cows and sows, mares have
a seasonal pattern of breeding and a prolonged transitional period between winter
anoestrus and cyclical ovarian activity.
Although sexual activity may be influenced
to a minor extent by the season of the year,
stallions will usually mate with mares in any
season. Sperm production in the stallion is
influenced by several factors, including season, size of testes, frequency of ejaculation
and age. Studies in the USA and elsewhere
have recorded better sperm quality in the
spring and summer than in autumn and
winter.

Stallion shuttle service

The demands made on males may differ
among the several farm species that have
breeding seasons. In horses, the stallion
may find himself being shuttled from the
UK or the USA to the southern hemisphere
to enable him to be used for two breeding
seasons in the one year. Such practices and
costs, of course, can only be countenanced

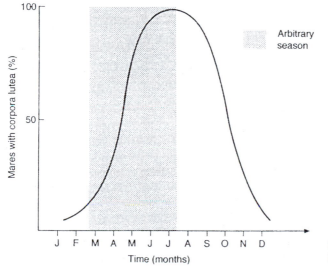

Fig. 7.4. The mare's arbitrary
breeding season (from Allen, 1978).

with stallions that are among the élite and the very élite.

Seasonal changes in follicular activity

The annual reproductive cycle in mares is characterized by minimal follicular activity during short days (anovulatory season) and maximal activity during long days (ovulatory season). A sound understanding of the processes involved in the growth and development of follicles is essential for those working in oestrus and ovulation control in this species. Many studies have been conducted, using real-time ultrasonics to determine the sequence of events in follicular development in the mare since the early 1980s. There is a clear size difference between ovarian follicles in the mare and those in the other farm mammals; during the oestrous cycle, the ovaries contain several follicles greater than 10 mm in diameter and a preovulatory follicle that eventually grows to 35–55 mm diameter. Although follicle waves are evident during the cycle, they are categorized as major and minor waves; in major waves there is divergence of follicles into dominant and subordinates, while in minor waves there is no such divergence and no dominant follicle develops. Major waves are further subdivided into primary and secondary waves, the former giving rise to a follicle that ovulates at oestrus and the latter to a follicle which does not ovulate or ovulates after the primary wave to give a secondary ovulation. Secondary and minor waves are believed to occur more frequently in spring than in autumn.

Evidence presented by Irvine *et al.* (2000) indicated that gonadotrophins continue to regulate ovarian function in the autumn as at other times of the year and showed that inadequate gonadotrophin in early dioestrus may be a critical event leading to suboptimal follicular and luteal development and eventually the cessation of oestrous cycles. Follicular waves have been demonstrated during the mid-anovulatory and the transitional periods; the higher follicular activity during the transitional period results in waves being detected at

a later stage of development than during the mid-anovulatory period (Donadeu and Ginther, 2003).

Year-round breeders

Although categorized as a seasonal breeder, a small proportion of mares do not exhibit anoestrus and continue to exhibit oestrous cycles throughout the winter months. There is some evidence that this phenomenon is a characteristic of mares which exhibit high body-fat scores and the implication is that the timing and occurrence of anoestrus may be causally related to body fat stores. It is known that leptin concentrations reflect the level of fat-stores and may provide a metabolic signal of nutritional well-being.

7.1.4. Deer

Red deer (*Cervus elaphus*) are highly seasonal breeders, with polyoestrous females exhibiting the onset of ovulatory activity in autumn and a high conception rate (~85%) to the first mating of the season. Red deer hinds in the UK can be expected to exhibit oestrous cycles of 18–21 days' duration at the beginning of October. In animals that do not become pregnant, cyclical breeding activity continues for about 6 months. Hinds are mated in the 'rut', usually in October–November (Fig. 7.5). The 'rut' is the term used to describe the period of concentrated sexual activity in wild deer, when individual stags appropriate a group of hinds. Calvings occur in the mid-summer period after a gestation period of ~233 days, which involves a misalignment of pasture growth with the high energy demands of the milking hind in pastoral countries such as New Zealand and Great Britain. For such reasons, there is an obvious case for using hormonal and other means of realigning the calving season with the grass-growing season.

It is clear from many reports that red deer, given adequate nutrition and management, are capable of exhibiting high fertility, with more than 95% of mature hinds bearing single calves each year. In the absence of pregnancy, however, red deer hinds are

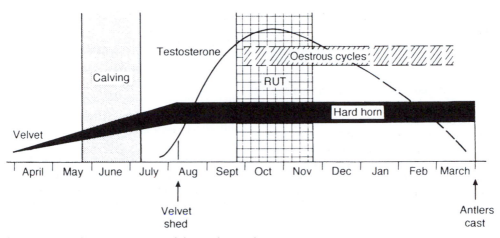

Fig. 7.5. Reproductive activity in red deer in the UK (from Perry, 1971).

polyoestrous and are capable of exhibiting between four and nine continuous 17–19-day oestrous cycles over a 3–6-month period between autumn and spring. New Zealand studies have demonstrated that antral follicle growth and regression continues during both the breeding and non-breeding seasons. Oestrous cycles are characterized by a variable number (one to three) of follicular waves, from which a single large (> 6 mm) follicle emerges; it is evident that anoestrus in red deer hinds represents a period of dynamic changes in follicular activity.

A further consideration in deer farming is the influence of season on growth. In contrast to what the farmer expects with his/her sheep and cattle, growth in male deer is seasonal, which can impose limits on farmed venison; red deer grow rapidly in the spring and slowly in winter, despite *ad libitum* high-quality feed. There is evidence that such seasonal changes are regulated by variations in growth hormone levels. Results from New Zealand studies have shown that exposure of red deer calves to an extended photoperiod (16 h light, 8 h dark) could result in growth rates comparable to those normally associated with spring and summer.

7.1.5. Buffaloes and camelids

Water buffaloes, like cattle, are polyoestrous and capable of breeding throughout the year. The characteristic seasonal breeding patterns in this species, which have been reported in many countries, particularly under tropical climates such as in India, are attributed to the ambient temperature, photoperiod and feed supply. Although authors may often refer to the buffalo as a seasonal breeder, this is because most animals are concentrated in the tropical and subtropical regions of the world; the same holds true for countries in the southern hemisphere, such as Brazil, where buffaloes are reported to have a higher conception rate during March to August, when the temperature is lower and day length is shorter. In India, studies have shown a majority of buffaloes exhibiting oestrus between October and March, when days are shorter and the temperature relatively cool.

The breeding season in New World camelids in South America is usually confined to the autumn months, although both males and females are capable of showing sexual activity throughout the year. Seasonal changes in semen characters were observed by workers in Germany; sperm concentration was highest in November–December and lowest in summer. It was concluded that male llamas have adapted satisfactorily to European climatic conditions. Camels are sexually active for a limited number of months each year, the period of the 'rut'; the time of their breeding period is likely to differ from region to region.

7.1.6. Poultry

Wild birds in the temperate zone generally display a well-defined and restricted breeding season, regulated primarily by length of day. As with farm mammals such as sheep, the influence of day length and of other factors is coordinated by nerve centres in the hypothalamus, which, acting by way of the pituitary, regulate ovarian activity. The avian breeding season starts in the spring as days are growing longer, but terminates while days are still long, due to a phenomenon known as refractoriness, which is a delayed response to long day length.

7.2. Breeding and Non-breeding Seasons

7.2.1. Short- and long-day breeders

The seasonality of reproduction is an adaptive physiological process utilized by wild animals to deal with seasonal changes in their environment and in particular the availability of food. Through centuries of domestication, there has been an almost complete loss of this adaptation in cattle and pigs but it remains in most breeds of sheep, goats and horses originating from temperate latitudes. At these latitudes, photoperiod is clearly the main environmental factor determining the onset and duration of breeding seasons. It is known, however, that, since sheep deprived of photoperiodic stimuli exhibit an endogenous rhythm of reproduction, the main role of day length may well be that of synchronizing this internal rhythm. The timing of the breeding season depends on the duration of the gestation period to ensure that parturition occurs when pasture is available in the spring. For such reasons, sheep and goats (5-month pregnancy period) are short-day breeders, with conception occurring in the autumn and winter, whereas horses (11-month pregnancy period) breed during the increasingly long days of the spring.

Refractoriness to light

Short days stimulate sexual activity in the ewe but prolonged exposure results in a refractory condition, with subsequent cessation of reproductive activity. Refractoriness can be broken by exposing ewes to long days; thus, alternation between long and short days is essential for the photoperiodic control of seasonal breeding. In fact, as shown later, short-term alternation between long and short day length can be employed to maintain rams and male goats in optimal sperm-producing ability.

7.2.2. Physiology and endocrinology of seasonal breeding activity

Sheep

Anoestrus in the ewe is known to be the result of a marked reduction in gonadotrophin secretion; during the transition back into the breeding season, levels of LH and follicle-stimulating hormone (FSH) increase substantially. Ovulation does not usually occur in anoestrus, but, although there is decreased gonadotrophin secretion during the non-breeding season, the size and number of large antral follicles are similar to those observed in the cyclic ewe; the only difference in follicular dynamics between anoestrous and cyclic sheep is in the increased number of small antral follicles found in the anoestrous animal. It is believed that, during the transition to anoestrus in ewes, the endogenous rhythm of FSH release remains strong; the season-related alterations in the normal pattern of ovine ovarian cycles appear to be due to reduction in ovarian responsiveness to gonadotrophins or diminished secretion of LH.

Melatonin

Photoperiodic information is conveyed through several neural relays from the retina to the pineal gland, where the light signal is translated into a daily cycle of melatonin secretion; levels are high at night, low during the day. Melatonin levels show considerable variation among individuals, but the length of the nocturnal secretion of melatonin reflects the duration of the night and regulates the pulsatile secretion of gonadotrophin-releasing hormone (GnRH) from the hypothalamus (Fig. 7.6). Changes in GnRH release result in corresponding changes in LH secretion, which are responsible for the alternating presence or absence of ovulation in the ewe and varying sperm production in the ram. Melatonin does not act directly on GnRH neurones but apparently involves a complex neural circuit of interneurones.

Thyroid hormones

It is now well established that thyroid hormones are essential for the seasonal suppression of LH secretion by oestradiol in the ewe. Several studies have investigated the action of thyroid hormones in the development of anoestrus in sheep. In the USA, studies led to the suggestion that thyroid hormones need only be present for a brief period of time near the end of the breeding season for neuroendocrine changes to occur that lead to anoestrus. It appears that the reproductive neuroendocrine axis is not equally responsive to thyroid hormones at all times of the year.

Evidence from several studies indicates that the relationship between thyroxine concentrations and the timing of the transition to anoestrus is quantitative; oral administration of drugs to induce hypothyroidism can lead to a significant delay in the transition to anoestrus, whereas administration of thyroxine can cause premature anoestrus. It is believed that a major factor responsible for seasonal breeding in sheep is a change in the responsiveness of GnRH neurones to

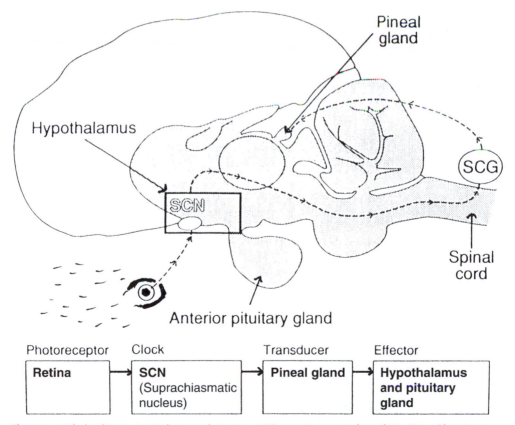

Fig. 7.6. Light leads to action in brain and pituitary. SCG, superior cervical ganglion. (From Chemineau and Malpaux, 1998.)

the inhibitory effects of gonadal steroids; because the majority of GnRH neurones lack oestrogen receptors, seasonal changes in responsiveness to oestradiol are probably conveyed by other factors.

First ovulations of season

Just prior to the start of the breeding season, there is a transient increase in progesterone in the ewe's circulation, thought to originate in luteal tissue; the first ovulation of the breeding season follows this transient increase in progesterone, although first oestrus is usually delayed until the second ovulation. Daily ultrasonography has been used during the transition between non-breeding and breeding season to observe how ovarian antral follicle dynamics change in the period leading up to the first and second ovulations of the season. It is evident that the cessation of anoestrus in the ewe involves an increase in LH secretion, with a preovulatory surge of LH apparently preceding the transient period of progesterone secretion; it is thought that the short-lived period of progesterone secretion is necessary to obtain adequate follicle development synchronized to an LH surge for normal ovulation and the start of regular cycles in the breeding season.

Horses

The physiological mechanisms governing transition from the breeding season into the anovulatory season in horses are not well understood; transition may be due to a failure of LH secretion, since both follicle development and FSH secretion continue after the final ovulation of the breeding season, whereas LH secretion does not. It is known that, during the anovulatory season, pituitary LH secretion and content decline and total hypothalamic GnRH content and secretion are decreased; pituitary FSH secretion, but not pituitary content, is also reduced. During anoestrus, the lack of gonadotrophin secretion results in little follicular development, and no corpora lutea are formed; the lack of ovarian activity in anoestrus results in undetectable levels of oestradiol and progesterone in the circulation at this time.

Endogenous opioid peptides are known to be involved in the suppression of gonadotrophin secretion in the mare's seasonal anoestrus. Several groups have reported on the dopaminergic control of seasonal reproduction and use has been made of such knowledge in devising protocols for inducing oestrus in anoestrous mares. Results from work in France led to the conclusion that dopamine plays a role in the control of seasonal reproduction but its effect, or the expression of its effect, depends on the stage of anoestrus. It was also concluded that, using certain criteria, dopamine antagonists (e.g. sulpiride) can be successfully used to induce cyclic ovarian activity in anoestrous mares.

The onset of the mare's reproductive activity in the spring involves a series of events which include renewed GnRH secretion, FSH secretion, follicle development, oestrous behaviour, oestradiol secretion, LH secretion and finally ovulation. Although there is evidence suggesting that seasonal regulation of gonadotrophin secretion is largely independent of the ovary, ovarian feedback may contribute to the dynamics of LH secretion during the re-establishment of reproductive activity.

Deer

Thyroid hormones are required for the neuroendocrine processes involved in the onset of the non-reproductive state in red deer. Studies in New Zealand have shown that, as in sheep, thyroxine is required for the termination of the breeding season in red deer hinds and that thyroid gland secretions block reproductive activity during the non-breeding season.

7.2.3. Environmental and genetic factors

Seasonal changes in ovulation rate

Sheep from temperate latitudes exhibit seasonal variations in breeding activity that are controlled by annual changes in day length. It is well established from studies in several countries that there are significant differences in the duration of the breeding season among sheep breeds (Fig. 7.7). In Britain, this was clearly demonstrated by workers such as Saad Hafez, working in Cambridge in the 1950s; in Mexico, in more recent times, workers showed that Rambouillet and Criollo ewes had a significantly higher number of oestrous periods per year (11.5 and 13.0, respectively) than Romney, Corriedale and Suffolk sheep (8.8–8.7). It is also known that the ovulation rate in sheep can be influenced by changes in day length that occur during the autumn–winter breeding season. In Scotland, workers found evidence that the seasonal decline in ovulation rate may result in a lower lambing percentage in sheep mated towards the end of their natural breeding period; this may have practical implications for farmers who may wish, in an effort to reduce labour and feed costs, to gear the lambing season to greater pasture availability in the month of May.

Food intake and diet

Many mammalian species show marked seasonal changes in their food intake and body weight as well as reproductive activity; appetite and body weight increase in spring–summer and decrease in the autumn–winter period. Both light and diet modulate reproductive activity. Several of the farm mammals (sheep, goat, red deer) breed in the autumn and many studies in sheep suggest that improved nutritional status achieved during the long summer days may be a factor influencing reproduction in the short autumn days. The hypothalamus is the key site where the photoperiodic and nutritional signals are integrated, although a full understanding of the mechanisms involved is still lacking; continuing research in this area is likely to provide new insights into the mechanisms involved in photoperiod–nutrition interaction. Some workers have conducted trials to compare follicular activity in mares during the spring transition and to determine whether green pasture would hasten the onset of the ovulatory season. In the USA, for example,

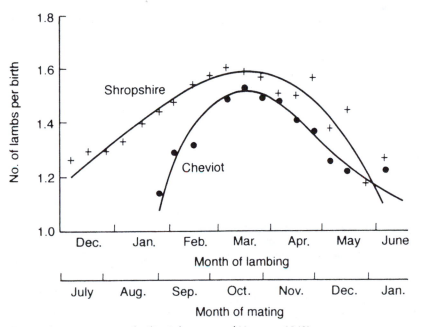

Fig. 7.7. Seasonal variation in ewe fertility (Johansson and Hansson, 1943).

it was found that the interval from development of a follicle > 30 mm to ovulation was significantly shorter for mares on pasture than for those on dry feed.

Goats

Many reports have dealt with the annual cyclic ovarian activity shown by goat breeds in various countries and in different latitudes. In Italy, for example, workers dealing with Mediterranean goat breeds noted that all such goats showed reproductive seasonality, with a spring–summer period of anoestrus and a well-defined period of cyclic ovarian activity in autumn–winter, with the highest percentage of cycling females in the November–February period; significant breed differences were also observed in the start and duration of the breeding season, as well as in the incidence of non-seasonal episodes of luteal activity.

Food availability

Studies with male goats under subtropical conditions have shown that local breeds exhibit seasonality in their reproductive activity similar to that shown by breeds from temperate latitudes, in which sexual activity throughout the year is controlled by marked changes in daylight length. However, because the amplitude of daylight changes is much lower in tropical and subtropical latitudes compared with those of temperate areas, it has often been assumed that nutritional conditions are the main modulators of sexual activity. Workers in northern Mexico, however, demonstrated that male Creole goats, fed constantly throughout the year, exhibit large seasonal variations in sexual activity, with the most intense activity occurring in the May–December period; the authors concluded that food availability was probably not the main factor influencing activity but that this was probably synchronized by changes in the photoperiod.

Condition scoring in sheep

The effect of body condition score (BCS: 1 = emaciated; 9 = obese) on the onset of the breeding season in sheep has been examined in some studies. Condition scoring is assessed by handling over and around the backbone, in the loin area immediately behind the last rib and above the kidney, using the fingers along the top and sides of the backbone (Fig. 7.8). In the USA, workers have found evidence suggesting that poor body condition (BCS < 3) inhibits the secretion of LH, which is associated with the onset of the breeding season, by altering the relative amounts of insulin-like growth factor (IGF)-binding proteins within the hypothalamic–pituitary axis.

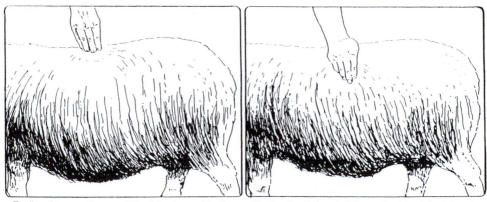

Feeling the smoothness of the ridge formed by the vertical processes of the backbone – the first step in condition scoring.

Feeling the fullness of the underside of the loin – the final step in condition scoring.

Fig. 7.8.   Condition scoring in sheep.

7.3. Development and Application of Control Measures

7.3.1. Practical applications of technology

The practical management of reproductive seasonality in sheep may involve the use of progesterone, progestogens, pregnant mare's serum gonadotropin (PMSG) and melatonin. Various authors have dealt with the use of melatonin as the means of manipulating reproduction in farm livestock. It is known that about 40 days are required for melatonin to stimulate pulsatile GnRH activity; although the sites and mode of action of melatonin are not fully understood, its action is believed to be on a region of the hypothalamus. In practice, the most frequently used method of melatonin administration is the subcutaneous implant, which can be used to advance the cyclical ovulatory activity of ewes and goats.

Melatonin can be used alone, in association with other hormonal treatments or in conjunction with an artificial photoperiodic treatment. It is clear that melatonin can be used to produce a quantitative and qualitative increase in out-of-season sperm production in rams and male goats. Various reports have shown evidence of seasonal patterns in the production and quality of ram semen. In New Zealand, it was found that semen density, volume and number of sperm per ejaculate varied significantly with season and were lowest in the spring and highest in the autumn.

7.3.2. Animal management

The ram effect

The seasonality of breeding activity in sheep represents an important constraint in the breeding programme of commercial flocks. The commercial advantages of the 'male effect' has been thoroughly explored in sheep during the past two decades, although the evolutionary significance of this male effect has received less attention.

Originally, the effect was demonstrated in the sheep by Australian workers in the mid-1940s but has since been shown to operate in a wide range of species, including goats, deer and wild pigs. Various studies have dealt with the physiological basis of the ram effect, including the roles of pheromones and hormones, the capacity of different rams to induce the effect, variation in the response of ewes, the control of reproductive activity in rams, the regulation of sexual behaviour and the use of hormones in combination with the ram effect.

The ram effect is among several possible approaches employed to manipulate reproduction during the anoestrous season but has the obvious advantage of being inexpensive, easy to apply and free from raising concerns among food consumers. Those studying the effect of ram introduction found that there were often two peaks of oestrous activity in response to the ram effect, the first at about 18 days after male introduction and the second after about 22–24 days (Fig. 7.9); New Zealand workers were to show that many ewes ovulated twice before exhibiting oestrus, with corpora lutea (CL) established initially regressing after 6–8 days. It was evident that a period of progesterone priming was necessary for many ewes to maintain CL for the full 14-day luteal phase.

A review by Rosa and Bryant (2002) has dealt with many aspects of the ram effect, including the nature of the ram stimuli involved, the behavioural and physiological events elicited in the ewe as a result of the presence of the ram, the neuroendocrine basis of the phenomenon and the various factors associated with variation in response. The authors conclude that the ram effect is an inexpensive technique that can be used to stimulate ovulation in ewes during the non-breeding season in extensive sheep management systems.

Priming with progesterone

The ewe responds to the ram effect with an increase in LH pulsatility and an LH surge; this results in a silent ovulation

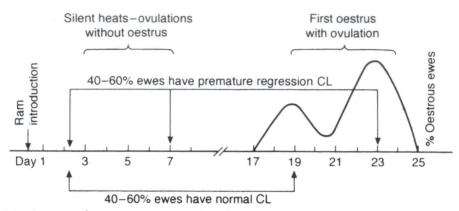

Fig. 7.9. Response of anoestrous ewes to ram introduction. CL, corpora lutea. (From Knight, 1983.)

(ovulation unaccompanied by oestrus). In some ewes, however, the CL formed after the first ovulation regresses after 4–5 days; this leads to a second ovulation, again without oestrus. For oestrus to accompany ovulation, it is first necessary for the ewe to be adequately 'primed' with progesterone/progestogen; the use of such 'priming' prevents the occurrence of short luteal phases and allows the CL to be maintained for the full 14-day luteal phase. Results reported by Ungerfeld *et al.* (2003) in Uruguay showed that short-term (6-day) medroxy-progesterone acetate (MAP) priming was as effective as longer-term treatment; the workers concluded that progestogen priming allowed a significant percentage of ewes to display oestrus during the first day after the introduction of rams and prevented short luteal cycles.

Testosterone-treated castrates

The rams employed in trying for the male effect do not necessarily have to be intact. In Australia, for example, results of studies indicate that the reproductive efficiency of ewes bred early in the breeding season can be improved by having a more concentrated lambing following the introduction of castrated rams treated with a subcutaneous injection of 375 mg testosterone oenanthate before the entire rams were joined with them; the duration of the breeding period was reduced to 5–6 weeks compared with the 8 weeks used for unteased flocks.

Male effect in goats

The male effect has been demonstrated by various workers in goats. Angora goats in Spain, for example, were exposed to males at the end of September, after a separation period of > 7 months; oestrus was recorded 7–16 days after exposure to bucks in 97.4% of goats. The teasing induced a normal oestrous cycle in 5% of does, a short cycle of around 7 days in 79% and two short cycles in 16% of females. Studies in goats have examined whether the presence of oestrous females can improve the response of seasonally anovulatory goats to the introduction of bucks in the group; their results indicated that this was not sufficient to induce an adequate stimulation of seasonally inactive males. The same workers noted that the use of sexually active bucks is necessary to induce reproductive activity in anovulatory females; bucks could be prepared by long-day treatment and melatonin implants to gain the required sexually active status.

7.3.3. Melatonin treatment

Diurnal production of the hormone melatonin by the pineal gland mediates the effects of the environmental photoperiod to regulate seasonal cycles in reproduction in several of the farm species, particularly sheep and goats. It remains unclear, however, which of the characteristics (amplitude, duration or phase of secretion) of that rhythm conveys the photoperiodic informa-tion to the reproductive axis. Various work-ers in the USA have observed the absence of a nocturnal rise in serum concentrations of melatonin in prepubertal and postpubertal gilts; they also found that gonadal steroids failed to influence melatonin secretion in pigs. It is evident that sensitivity of the pineal gland to light intensity differs markedly among mammals. In the 1970s, it was shown that the secretion of this indoleamine has a distinctive diurnal rhythm, with increased levels of melatonin coinciding with darkness; in the 1980s, it became clear that exposure to short day length or administration of melatonin by daily treatment or by implants significantly advanced the sheep breeding season and could also favourably influence the ovulation rate.

Treatment of ewes and does with melatonin to advance the breeding season has taken several forms; daily oral adminis-tration can be effective, but subcutaneous implants containing the agent, inserted at the base of the ear, have been the most widely used. Treatment with the implant in spring advances onset of the breeding season, the treatment being particularly effective in breeds showing the least seasonality of natural reproductive activity. In France, several reports have described research in sheep and goats using Melovine, a melatonin implant registered in France in 1995; inserting the implant in ewes 30–40 days before mating increased output by 16 lambs per 100 ewes mated and was effective in bringing forward the lambing date. Field trials in New Zealand and elsewhere using Regulin, a melatonin implant, have been described by several workers, who have also dealt with factors that may need consider-ation before embarking on an early lambing programme. The mechanism of action of melatonin in seasonal breeding has been discussed in various articles; it is held to be unlikely that the indoleamine acts directly within the pituitary gland to mediate the effects of light.

Melatonin treatments are not confined to female sheep; various studies have exam-ined effects in the ram and how these may be of practical relevance. In France, workers controlled sperm production in rams by light and melatonin treatments. Their most effective treatment involved 2 months of supplementary light followed by melatonin treatment; this resulted in a 30% increase in the number of artificial insemination (AI) doses per ejaculate in June and July, compared with untreated rams. In Scotland, studies with Soay rams suggested that mela-tonin acts in the pituitary gland to affect prolactin secretion and in the hypothalamus to influence gonadotrophin secretion. In Turkey, it was found that melatonin implant treatment in merino rams improved semen quality and resulted in significantly higher scrotal circumference and testis volume measurements than in untreated rams.

In the UK, workers advanced seasonal changes in rams with melatonin and exam-ined the possibility that allowing rams prior sexual experience would induce ovulation in anoestrous ewes in late anoestrus; they showed that significantly more ewes ovu-lated after exposure to rams with mating experience and melatonin than when simply exposed to untreated rams (56 vs. 24%). For practical purposes, such studies are largely academic and would have limited applica-tion; the farmer's need is for a treatment which can achieve a consistent and high response (e.g. progestogen and PMSG).

Some workers have combined mela-tonin treatment with the conventional progestogen–PMSG regime. Greek workers treated anoestrous ewes with a melatonin implant 5 weeks before insertion of 60 mg MAP sponges (in situ for 14 days), with 500 iu PMSG at sponge withdrawal; they recorded a significantly higher litter size in ewes conceiving at the induced oestrus (1.52 vs. 1.32). In the previous year, a similar melatonin–progestogen combination had also proved successful, but without the need for PMSG at sponge withdrawal in the melatonin-treated sheep. In Belgium, Suffolk ewes were given melatonin (by injection) 3 weeks ahead of treatment with 40 mg FGA sponges and 800 iu. PMSG; the melatonin/FGA/PMSG resulted in a significantly increased ovulation rate compared with controls.

Effects on the fetus

In pregnant sheep, maternal levels of melatonin may have certain implications for the fetus. In Australia, studies led workers to conclude that there are a number of different mechanisms for the neuroendocrine transmission of information about the time of day and duration of the external photoperiod to the sheep fetus during gestation. Evidence shows that the photoperiodic history of a ewe commences *in utero*, the fetus being receptive to the maternal melatonin signal from early pregnancy; the photoperiod experienced by the dam is apparently transmitted to the fetus, modifying the subsequent response of the lamb to the photoperiod.

Horse

Melatonin has been used in attempts to overcome anoestrus in horses. Although it is known that continuous administration of melatonin during anoestrus will not advance the onset of the breeding season in mares, it is known that melatonin treatment during the summer but not the winter may advance the start of the following breeding season in mares. If short-term treatment with melatonin could advance the timing of the re-establishment of reproductive activity, it might enable horse breeders to breed their animals during the late winter without the need for artificial lighting regimens.

Deer

The summer calving season and subsequent period of lactation of farmed red deer in the UK and New Zealand usually coincide with deteriorating pasture quality. For that reason, there is interest in methods of advancing the onset of the breeding season by up to 2 months to bring a closer alignment of feed requirements and pasture production. It is known that the season can be advanced by various forms of melatonin administration, including oral administration, injection and subcutaneous implants. Although melatonin treatment can be applied in prepubertal red deer without problems, in adult hinds the treatment

must be superimposed on the animal during the final period of pregnancy, which may have negative effects; New Zealand workers found that melatonin implantation initiated 80 days before parturition could have an adverse effect on calf viability.

7.3.4. Light control in rams and stallions

Sheep and goats from temperate latitudes display seasonal variation in reproductive activity that is controlled by annual fluctuations in day length. Short daily photoperiods stimulate reproductive activity, whereas long photoperiods have the opposite effect. Out-of-season reproductive activity can be achieved by providing long photoperiods in winter, followed by implantation of melatonin; this treatment advances puberty in rams and increases production of sperm outside the normal breeding season.

It is clear that the photoperiod, by way of its effect on the pineal and melatonin secretion, is the main cue controlling the gonadotroph activity of the hypothalamo-hypophyseal system, which in turn drives gonadal function. It is evident from various studies that a light regime with alternating 1-month periods of increasing and decreasing day length stimulates the gonadotrophs in alternate months and maintains testicular weights at a permanently high level. Using such a regime, plasma testosterone never reaches the levels considered to be inhibitory at the end of the breeding season. Similar results have been found in goats; in both sheep and goats, an alternating light regime can maintain the quantity and quality of semen at the levels normally observed during the breeding season. According to some reports, rams and male goats exhibit permanent sexual activity when submitted to alternation of 1–2 months of long daily photoperiods and 1–2 months of short photoperiods.

Stallions

Of the various environmental factors involved in the horse's breeding season, daylight is known to be the most important.

Studies in Cambridge in the 1940s showed this to be true for the stallion as well as the mare. Currently, large numbers of Thoroughbreds and sports horses are routinely maintained under artifical lighting during the late winter and early spring months in North America, Europe and Australasia. This is part of the breeders' efforts to try to mate their mares as early as possible in the stud season.

7.3.5. Use of progestogen–PMSG treatments

A great deal of research was directed towards overcoming the problem of anoestrus in sheep during the 1960s, mainly by Terry Robinson and associates in Sydney. It became clear that a highly effective treatment involved a 12-day period of treatment with intravaginal sponges (containing FGA or MAP), with a single dose of gonadotrophin (PMSG) given at removal of the device (Table 7.1).

Four decades later, reports are still appearing in the literature as workers in

different countries apply this form of treatment. In Poland, for example, recent studies concluded that PMSG injection was required to induce a satisfactory response in Leine sheep treated with FGA sponges during anoestrus. In Tunisia, work with Thibar ewes and FGA sponges showed that PMSG at sponge withdrawal significantly increased conception rate and litter size. In Syria, workers induced synchronized oestrus and twinning using MAP sponges and 400 iu PMSG at progestogen withdrawal in Awassi sheep. In Uruguay, Ungerfeld and Rubianes (2002) compared the effectiveness of different progestogens and dose levels when used over much shorter treatment periods than the conventional 12–14 days in anoestrous ewes; they concluded from their data that controlled internal drug release (CIDR), FGA and MAP sponges were equally effective, with 6 days' priming for induction of oestrus.

Breeding-synchronized ewes

Much work was done more than 30 years ago in devising ram management routines to be used in sheep flocks synchronized by way of progestogen and PMSG; rams introduced 48 h after progestogen withdrawal achieved significantly greater success than when introduced earlier (Table 7.2). The explanation apparently was largely a matter of ram mating psychology, as detailed in reports at the time. Presented with a bunch of ewes in oestrus, the ram's natural inclination is to mate once and then move on to a 'fresh' ewe. If the male is present from the time of sponge withdrawal, the first ewe to come into oestrus occupies his full attention and depletes his sperm reserves.

Table 7.1. Lambing outcome after progestogen treatment (fluorogestone acetate (FGA) and medroxyprogesterone acetate (MAP)).

Progestogen	FGA	MAP
No. ewes treated	878	1176
No. ewes lambed	628	846
No. lambs born	1157	1150
% ewes lambing	71.5	71.9
Lambs born per ewe lambing	1.80	1.80
% ewes with singles	35.2	34.7
% ewes with twins	48.8	50.3
% ewes with triplets or more	16.0	15.0
Lambs per 100 ewes treated	131	131

Table 7.2. Effect of ram mating management on lambing outcome. Trials with 16 pens of sheep – ten ewes and one Suffolk ram per paddock ($P < 0.01$).

	Ram in at progestogen pessary withdrawal	Ram in at 48 h after pessary withdrawal
Number of sheep	79	80
Bred at the induced oestrus	76 (96.2%)	77 (96.3%)
Conceptions at the induced oestrus	30 (39.5%)	52 (67.5%)
Number of lambs born	43 (1.43)	80 (1.54)

Goats

Treatments initially designed for use in sheep have been successfully employed in goats to control oestrus, both out of season and during the breeding season. In view of the longer oestrous cycle in goats (21 days vs. 17 days in sheep), the intravaginal sponge treatment is likely to be longer in goats than in sheep. In Greece, Lymberopoulos *et al.* (2002) used intravaginal sponges in the month of July, this month being considered to be the transitional period suitable for oestrus control in both Swiss and indigenous dairy goats; they used FGA sponges, PMSG (at sponge withdrawal) and prostaglandin $F_{2\alpha}$ ($PGF_{2\alpha}$) (48 h prior to sponge withdrawal), and bred the goats by AI 42–44 h after sponge removal. French workers compared a long (21 days) with a short (11 days) treatment of progestogen; in the short treatment, $PGF_{2\alpha}$ (cloprostenol) was given in addition to PMSG 2 days before sponge removal. The French research went on to show that satisfactory levels of fertility could be achieved with only one insemination performed on a fixed-time basis after short-term FGA treatment, PMSG and PG. The development of this protocol enabled the French to develop an AI-based genetic improvement programme in goats.

Long-acting GnRH

In several farm species, such as sheep and cattle, continuous exposure to high levels of GnRH rapidly reduces gonadotroph responsiveness in the anterior pituitary to further GnRH stimulation. Mares, on the other hand, can be stimulated to ovulate by such continuous GnRH administration. It is believed that the difference between the species may involve structural differences in the GnRH receptor, distinct second-messenger systems or variations in receptor synthesis and recycling. Many GnRH analogues with different potencies and durations of action have been developed and most are available in long-acting continuous-release formulations that can be implanted subcutaneously (e.g. Ovuplant); their efficacy in stimulating ovarian activity in mares is supported by many clinical trials. There is evidence from work in New Zealand, for example, that GnRH agonist implants can induce follicular development, ovulation and the establishment of CL in approximately 75% of seasonally acyclic mares.

8

Controlling Multiple Births and Litter Size

Ovulation rate is the major factor determining the incidence of multiple births in sheep, goats, cattle and horses and in determining litter size in pigs; multiple ovulations are influenced by genotype and environmental factors and are the result of folliculogenesis. Control of folliculogenesis lies with the gonadotrophins and local regulatory factors in the ovary such as steroids, cytokines and growth factors

Normally, when talking about litters, the pig is the farm-animal species in mind (Fig. 8.1). However, there are situations in which the term 'litter' may be applied to sheep or goats, especially in breeds such as the Finn Landrace and Romanov in sheep, where it is not uncommon for ewes to give birth to four or five offspring. In cattle, it is rare for multiples in excess of two to occur, so the term 'litter' would seldom need to be applied in that species. In zebu cattle and buffaloes and in deer and camelids even twins would be something of a rarity, so

Fig. 8.1. Large White sow and her litter of 14 young. The piglets were reared successfully on the Herefordshire farm of Charles Coxon at Milton, Pembridge, in 1957. Although litter size in pigs is not strongly inherited, it can differ according to breed and whether the sow herd is purebred or crossbred. Apart from genetic considerations, several other factors can enter the picture, not only feed but also the boar and system of mating.

©I.R. Gordon 2004. *Reproductive Technologies in Farm Animals* (I.R. Gordon)

multiples would be the term used in dealing with them.

8.1. Advantages of Control Measures

8.1.1. Sheep and cattle

Sheep

Many economic studies in lowland sheep over the years have clearly shown the importance of high fertility as a major determinant of profitability in the enterprise; a small difference in the proportion of ewes carrying multiples can make a considerable difference to the net income from the flock. In farming situations where the full genetic potential of a particular breed is being achieved and a further improvement in litter size is considered desirable, then the introduction of a more prolific breed, selection of certain ewes within a breed and the artificial control of litter size (by the use of appropriate exogenous gonadotrophins or immunization treatments) are among several of the options available.

Beef cattle

In some farming conditions, the availability of a reliable technique for inducing twin calving in beef cattle could be a valuable means of increasing the biological and economic efficiency of beef production systems. Lifetime efficiency of the beef cow is closely related to reproductive rate, and increasing the twinning rate could be one means of increasing the efficiency of the beef enterprise; the usual estimate is that twinning could increase biological and economic efficiency of beef production by 20–25%. That said, it must also be emphasized that a high level of management is required for twin-producing dams and their calves to achieve the increased productivity.

Dairy cattle

In contrast to what might be achieved by twins in beef animals, there are likely to be many dairy farming conditions in which producers are anxious to reduce twinning or eliminate it entirely. In this context, for countries such as the USA, it is worth noting that the administration of recombinant bovine somatotrophin (rBST) as a means of increasing the milk yield of dairy cattle has usually been accompanied by a significant increase in the incidence of twins in such animals. Twinning is undesirable in milking herds because it can reduce reproductive efficiency by adversely affecting conception rate in the subsequent lactation. Dairy cattle giving birth to twins have a higher risk of serious problems such as stillbirths, retained placenta, metritis, displaced abomasum, ketosis and aciduria than cows calving singletons (Table 8.1).

There is evidence showing a direct relationship between milk production and the incidence of double ovulations in lactating dairy cows. It is believed that high milk production near the time of ovulation can increase the incidence of double ovulations and that this may result in an increased twinning rate; the practical implications of this relationship are important, for current dairy management systems are usually geared towards maximum milk yield per cow. Although the specific physiological mechanisms that may predispose higher-producing dairy cows to exhibit an increased twin ovulation rate are currently unknown, evidence suggests that high feed intake in milking cows may increase hepatic metabolism of ovarian steroids, thereby altering the endocrine environment sufficiently to allow the development of two rather than one follicle during the selection

Table 8.1. Some effects of twins in Irish dairy cattle: effect of natural twinning on dairy cow and calf performance (taken from over 10,000 calvings in the Moorepark herds) (from O'Farrell *et al.*, 1991).

	Single	Twin
Abortion (%)	2	1
Perinatal mortality (%)	6	16
Calving difficulty (%)	8	7
Malpresentation (%)	3	11
Retained placenta (%)	3	19

period of a follicular wave. Once more is known of how increased milk production affects follicle development, it may be possible to devise strategies to avoid or reduce the negative effects of twinning on dairy operations without compromising the animal's ability to become pregnant with a single offspring.

8.1.2. Pigs

It is generally accepted that, when the sow's ovulation rate is increased well beyond the normal physiological limits (e.g. using exogenous gonadotrophins), the benefits of any small increase in the average litter size will almost invariably be outweighed by the variability in the size of the litters and in the birth weights of piglets in these litters. The conclusion reached by many is that, although ovulation rate should be kept as high as possible by natural means, attempts to produce increases artificially are unlikely to be of commercial interest. However, crossbreeding studies, dating from the 1930s, have shown that hybrid vigour can lead to useful increases in litter size; it has become a widespread commercial practice in many countries to use crossbred or hybrid sows as the breeding females in commercial units. Some pig breeding companies have adopted strategies that are designed to increase litter size, with group nucleus breeding schemes using best linear unbiased prediction (BLUP) and artificial insemination (AI) to connect several herds to provide a large nucleus population. The Cotswold Group Nucleus Breeding Scheme (Fig. 8.2) was designed to permit litter size to increase by 0.2 live piglets per litter annually when it was set up in the 1990s.

The question of litter size in pigs involves not only ovulation rate but factors affecting the incidence of embryonic mortality. Since the days of John Hammond in the Cambridge of the 1920s, it has been recognized that the sow differs from the other farm animals inasmuch as litter size depends less on the number of oocytes released from the ovaries and much more on the number of fetuses that survive in the uterus.

Tropical regions

One of the serious problems in pig production in tropical areas is the comparatively low litter size at birth. According to Tantasuparuk *et al.* (2000), the sources of genetic material in Thailand originate from temperate and subtropical areas, such as countries in Western Europe and North America, where the climate is quite

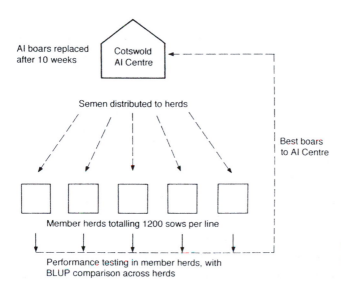

Fig. 8.2. Cotswold group nucleus pig breeding scheme (from Webb, 1995).

different from that in Thailand, which has higher temperatures and humidity. The same workers showed that sows that were weaned/mated during the cool season had the highest litter size whereas those weaned/mated in the hot or early rainy season had the lowest; it was concluded that high temperature plays an important part in explaining the seasonal variation in reproductive efficiency in tropical areas.

Meishan pigs

The high litter size of the Meishan breed has attracted much attention from those interested in factors influencing litter size. In the USA, workers reporting on the prenatal development of Meishan and Yorkshire piglets concluded that, because uterine capacity sets the upper limit on litter size, the decreased endometrial surface area required per conceptus in Meishans compared with Yorkshire pigs may well explain the increased litter size.

8.1.3. Horses

As mentioned earlier, with reference to dairy cows, control measures are not necessarily aimed at increasing the number of offspring born to farm animals. In horses, it is usually much more a question of decreasing or eliminating the incidence of multiple births. In the USA, it is recorded that twinning is a major cause of abortion in mares; workers have employed various techniques in early pregnancy because it is difficult to deal with twins that remain after the 40th day of gestation.

Incidence of multiple ovulations

The incidence of double ovulations and twin births in mares is the subject of several reports. In The Netherlands, the incidence in Dutch Warmblood mares was recorded as 26%; workers concluded that ultrasonographic examination is essential for the accurate detection of multiple ovulation. In Germany, the records of 11,614 inseminations of German Thoroughbred mares,

carried out between 1973 and 1996, were analysed; it was found that the incidence of twin pregnancies increased significantly and continuously from 0.9% for breedings in February/March to 1.34% and 1.76% for breedings in April/May and June/July, respectively; it was concluded that environmental factors increased the chance of two embryos surviving. In the UK, when the oestrous cycles of 1136 Thoroughbred brood mares were monitored over a 3-year period, it was found that multiple ovulations occurred in 22.4% of cycles. It was also found that ovulations, both multiple and single, were equally distributed between the left and right ovaries and that the incidence of multiple ovulations increased significantly with age. In Poland, a relatively high rate of multiple ovulation and resulting multiple pregnancies was among the major problems in Thoroughbred reproduction recorded by Gorecka and Jezierski (2003) in that country.

Dioestrous ovulations

The presence of multiple corpora lutea in the mare's ovaries may not necessarily mean that they originated from simultaneous ovulations. In the UK, it was found that 10.3% of 350 mares examined showed evidence of dioestrous ovulations, most of these occurring early in the cycle (days 4–5); although many mares with dioestrous ovulations had mated at the preceding oestrus, only one became pregnant with twins. As noted elsewhere, follicular dynamics in the mare are such that a secondary ovulation, occurring a day or two after normal oestrus and ovulation, is not uncommon; although an oocyte is released, fertilization would not occur because of the endocrine environment.

8.2. Development of Control Measures

8.2.1. Historical

Books dealing with sheep husbandry, dating back to the early 19th century, refer to a method of increasing the lamb output of ewes. This age-old practice of 'flushing' sheep, whereby ewes are provided with better feeding just prior to and during the mating period, has long been the object of research to explain the mechanisms involved. Of more recent vintage has been the use of 'endocrinological flushing', where the aim has been to use exogenous gonadotrophin (pregnant mare's serum gonadotrophin (PMSG)) to increase twin births in sheep. This approach was first used on some scale in the former Soviet Union among Karakul sheep and was reported more than 60 years ago. Elsewhere, serious efforts to examine the possibility of augmenting sheep fertility by gonadotrophins date from the work of Terry Robinson in Cambridge in the late 1940s. He developed a technique using PMSG, in which the ewe received an injection of the gonadotrophin in the latter stages of the oestrous cycle. Although ewes varied considerably in their response, such PMSG treatment proved effective in significantly increasing the average ovulation rate and the subsequent twinning rate (Fig. 8.3). Unfortunately, for the farmer, identifying ewes in the late stages of their cycle was no easy task, involving a sterile teaser ram and time-recording pre-injection heats in the flock – perhaps acceptable under collective-farm conditions in the USSR, but not necessarily appealing in other parts of the world. For farmers in Ireland, it was a matter of waiting until the 1960s and beyond when it became possible to combine a mild form of superovulation with accurate oestrus control in sheep.

8.2.2. Physiology and endocrinology of multiple ovulations

The normal process of multiple ovulation in farm mammals depends on the balance between the stimulatory effects of pituitary gonadotrophins on developing antral follicles in the ovaries and the negative-feedback effects of hormones such as oestradiol and inhibin (Fig. 8.4). Knowledge of this was to lead eventually to the development of immunization procedures which sought to weaken the feedback effects of androgens, oestrogens and inhibin; this

Fig. 8.3. Clun Forest ewe with quadruplets after PMSG treatment (1956). One of the members of a flock of 750 ewes treated with PMSG for the induction of multiple ovulations on the Rushbrooke Estate of Victor Rothschild in Suffolk.

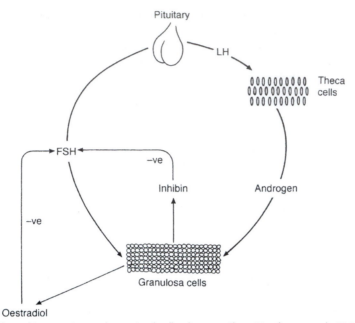

Fig. 8.4. Gonadotrophin secretion and negative feedback agents (from Henderson *et al.*, 1984).

had the effect of lifting the inhibition on gonadotrophin secretion and causing extra ovarian follicles to develop and ovulate.

Sensitivity of granulosa cells

Studies in sheep have sought to examine the endocrine factors involved in ovulation rates, as shown in high- and low-fertility breeds. Those reviewing mechanisms underlying genetic variation in ovulation rate in the ewe have concluded that the ovaries have a more important role in the control of ovulation than previously realized. French studies, for example, compared granulosa cell sensitivity to gonadotrophins in sheep breeds of high (Romanov) and low (Île-de-France) prolificacy; evidence was found suggesting that changes in the sensitivity of granulosa cells during the follicular phase of the cycle probably participate in the regulation of ovulation rate in these sheep. It seems likely that differential dynamics of follicular development underlie differences in poly- and mono-ovulating breeds of sheep. It appears that the factors resulting in high ovulation rates are not exerted through circulatory patterns or

concentrations of follicle-stimulating hormone (FSH) but involve a shorter growth phase and smaller maximal diameter of follicles. In a study of luteal function in breeds with different ovulation rates, it was found that the prolific Finn ewe produced more but smaller corpora lutea and had lower progesterone levels to support the greater number of fetuses carried compared with less prolific breeds; it appeared that breed-specific mechanisms controlling luteal tissue formation may exist.

Twins in cattle

The incidence of twin births in cattle varies among breeds and according to factors such as age and environment. In the UK and Ireland, the incidence recorded in surveys usually falls somewhere between 2 and 3%. On a worldwide basis, twinning rate in different breeds varies from close to zero to almost 10%, usually with a higher rate in dairy than in beef cattle, although considerable variation exists between breeds (Karlsen *et al.*, 2000). One difference between sheep and cattle appears to be in the survival of embryos in relation to the distribution of multiple ovulations in the ovaries. In cattle, there is much less evidence than in sheep of transuterine migration of embryos, a phenomenon that apparently affects embryo survival. Studies in the UK in sheep have shown that the site of ovulation has no effect on embryo survival, with embryos from unilateral ovulations as likely to survive as those from bilateral ovulations. Surveys of sheep reproductive tracts have recorded evidence of transuterine migration of embryos in 12% of single-ovulating ewes and in all ewes pregnant with twins which had double ovulations in one ovary.

Right vs. left ovaries

In cattle, there is ample evidence that the right ovary is significantly more active than the left, the usual balance being 60% ovulations in the right and 40% in the left. The same story holds good for the ewe. In Scotland, for example, a higher incidence of ovulations in the right ovary was observed in Booroola–Texel ewes; multiple ovulations in conjunction with or in the absence of ovulations from the contralateral ovary also occurred more frequently in the right ovary.

8.2.3. Nutritional and environmental factors

Nutrition and flushing in sheep

As noted earlier, the influence of nutrition on the lamb output of sheep has been recognized for the past two centuries and longer. Farmers are well aware of the benefits of 'flushing' their ewes before and during the early part of the breeding season. The way in which nutrition exerts its influence on ovulation has been the subject of many studies up to the present time. In Aberdeen, for example, studies provided evidence that the effects of food intake on ovulation rate may be expressed through differences in the late stages of follicle development, probably through effects on intrafollicular steroid levels. It is known that glucose, when infused into the peripheral circulation of the ewe, can increase ovulation rate

without stimulating FSH secretion. There is also evidence suggesting that ovaries in the ewe take up glucose from the systemic circulation via an insulin-regulatable GLUT4 transporter; the indications are that such glucose transport and follicle size are linked.

Short-term flushing

It is possible to influence ovulation rate in the ewe using very short-term treatments, based on current knowledge. In Argentina, for example, workers have demonstrated the potential for modifying ovulation rate through short-term glucogenic treatment in ewes during the rapid follicle development phase after treatment with intravaginal progestogen; a single dose of a glucogenic formulation given at sponge withdrawal, with ewes exposed to rams subsequently, increased ovulation rate. In Ireland, glucose infused into ewes for the last 5 days of a progestogen-controlled oestrous cycle produced some evidence of a positive effect on ovulation rate. In Australia, it was found that insulin-induced hypoglycaemia inhibited luteinizing hormone (LH) secretion in cyclic ewes, implicating glucose as a mediator of normal hypothalamic–pituitary function.

Prolific ewe characteristics

In sheep, ovulation rate is known to vary not only with breed but according to the stage of the breeding season (see Chapter 7). It appears to be a characteristic of highly prolific sheep breeds (e.g. Finn, Romanov, Chios) that they have long breeding seasons. In Greece, for example, work with Chios sheep showed that the percentage of ewes showing oestrous behaviour was 92% and 100% in spring and autumn, respectively; a significantly lower ovulation rate was observed in spring-mated than in autumn-mated ewes (2.85 vs. 3.33).

Seasonal effects on ovulation rate

In Scotland, workers sought to determine the relative importance of seasonal changes in ovulation rate, fertilization rate and embryo survival as the cause of reduced lambing rates in ewes mated in February (late season) compared with those mated in November (mid-season); it was concluded that a seasonal decline in ovulation rate was the primary cause of the reduced lambing rates observed in February.

Ovulation rate in pigs

Workers have examined the effect of various nutritional factors on litter size in pigs. In Poland, studies have shown that the addition of β-carotene to the diet of sows before and after mating increased ovulation rate; folic acid supplementation from the time of mating decreased the incidence of embryo mortality. Attempts to understand ovulation rate control in pigs led workers in the USA to examine hormone levels in gilts with high and low ovulation rates; their results indicated that increased concentrations of both FSH and LH were associated with increased ovulation rate during the ovulatory phase of the cycle but only FSH increased during much of the luteal phase in such pigs (Knox *et al.*, 2003).

Ovarian follicle population

Studies in Germany have shown that pig breeds exhibiting different fertility parameters also differ in their ovarian follicular populations. The mean number of primordial and primary follicles in Duroc gilts was seven times higher than in Mini pigs; fertility data showed 10.11 piglets born per litter for Duroc and 8.33 for the Mini pig.

Teat numbers and litter size

Workers in the USA investigated factors affecting the number of teats in pigs and found a significant relationship between the number of teats and litter size and the proportion of males in the litter; a greater number of teats on the dam and a lower proportion of males in the litter resulted in a greater number of teats on the gilt. It has also to be recorded that there is some evidence of identical twins in the litters

of some pigs; it appears that monozygotic twinning may be more prevalent in Meishan than in European pig breeds.

8.2.4. Genetic factors

Prolific sheep

The success of a lowland sheep enterprise depends on many factors, among which the most important is the number of lambs sold per ewe per annum. In many countries, efforts to maximize this figure have concentrated on the selection of more prolific breeds of sheep and improved nutrition of the ewe; improvements in understanding the nature of neonatal lamb losses and in controlling disease in both ewes and lambs have helped greatly. Some of those dealing with the mechanisms underlying genetic variation in ovulation rate in sheep have classified the breeds showing high ovulation rates into two groups: (i) a group that includes the Romanov, Finnish Landrace and D'man, in which the high ovulation rate reflects polygenic differences; and (ii) a group that reflects the action of a single major gene, or a closely related group of genes, such as the Booroola Merino, Icelandic, Javanese, Fat-tail (Indonesian Fat-tailed), Inverdale, Belclare, Cambridge and Olkuska. In France, as noted earlier, studies have suggested that the higher ovulation rate in Romanovs than in the non-prolific Île-de-France breed is likely to be the result of greater responsiveness to gonadotrophins during the early follicular phase rather than to any increase in gonadotrophin levels themselves.

In Poland, attention has been drawn to the exceptional reproductive potential of Olkuska sheep, which showed an ovulation rate ranging from one to eight with a mean of 3.0; it was estimated that, assuming a single gene to be responsible for the effect, one copy of the putative gene in heterozygous carriers increased ovulation rate by 1.03 and litter size by 0.63. In Ireland, the high prolificacy of Belclare sheep (a composite breed developed from stocks selected for high ovulation rate and litter size) is believed to be under the control of two major genes.

Elsewhere, a single gene of major effect has been identified that affects fecundity; the hyperprolificacy of Booroola ewes is now known to be due to a single amino acid substitution in the type-IB receptor for bone morphogenetic proteins (BMPR-IB). In Scotland studies revealed no qualitative or quantitative differences in the pattern of secretion of pituitary gonadotrophins or ovarian hormones between ewes carrying the *Booroola* gene and controls; it was concluded that the gene exerted its action at the level of the ovary, as revealed by ovulatory follicles and corpora lutea which were significantly smaller than those in control ewes. A report from New Zealand suggested that the higher ovulation rate in carriers of the Booroola gene is attributable to effects at the level of ovarian follicular development as well as at the level of pituitary FSH release.

It is now known that, in the ovary, the presence of the Booroola mutation leads to an advanced maturation of follicles due to more precocious differentiation of granulosa cells. Studies in the UK, reported by Campbell *et al.* (2003), support the view that the *Fec-B* gene acts at the level of the ovary to enhance ovarian sensitivity to gonadotrophic stimulation. Elsewhere, a gene with a similar effect has been reported in Icelandic sheep (the *Thoka* gene) and in Javanese sheep. Workers in Scotland reported on the introduction of the *Thoca* gene on lamb production in Cheviot ewes; introduction of the gene to such breeds increased fecundity and the potential for more efficient lamb production. It should be noted, however, that, in prolific breeds such as the Finnish Landrace and the Romanov, there is no evidence of a major gene for litter size; the indications are that at least two genetic control mechanisms for high prolificacy operate in sheep.

Selective breeding

Selective breeding among ewes of relatively low prolificacy can result in dramatic increases in litter size; certainly, lowland sheep farmers can ill afford to keep ewes with a tendency for single lambs (Fig. 8.5). Genetic studies of prolificacy in New Zealand sheep breeds have been reported by workers using data recovered over an 8-year period on 2180 élite (highly prolific) ewes maintained at a research station; ovulation rates for descendants of the foundation ewes averaged 2.15, 2.43, 2.15 and 2.96 for Romneys, Coopworths, Perendales and Inverdales, respectively.

Fig. 8.5. Only a single lamb: not good enough.

Endocrine tests

Some workers have examined the possibility of using early endocrine tests to identify ewes of high fertility. There have been studies showing that high concentrations of plasma FSH found in prepubertal ewe lambs of prolific sheep breeds are not evident in breeds or strains of sheep with a limited ovulation rate. In Greece, evidence was found that the plasma FSH concentrations from 6 to 12 weeks of age of ewe lambs could be used for early selection for litter size in Chios sheep. In Morocco, similar evidence was reported showing that the high prolificacy of D'man ewes (mean ovulation rate 2.9) is characterized early in life by elevated basal plasma FSH concentrations.

Multiples and milk yield in goats

In goats, there is a well-recognized relationship between multiple births and milk yields in intensively managed dairy goats. In Canada, for example, studies showed that Anglo-Nubian does that produced multiple kids gave significantly higher milk yields than goats with singles (807.6 kg vs. 664.6 kg). In Italy, workers dealing with the performance of Alpine goats in Lombardy suggested that litter size should be considered when identifying correction factors for comparing milk yields among herds.

FSH levels and litter size in pigs

Research reported by Cassady *et al.* (2000) in boars and gilts sought to determine whether plasma concentrations of FSH were genetically correlated with ovulation rate; they concluded, because selection for FSH could be practised in both sexes, that this seemed to be a practical method for increasing ovulation rate in pig breeding programmes without the need to examine ovaries by laparoscopy. Data presented by Christenson and Leymaster (2003) indicated that ovulation rate and uterine capacity in pigs are genetically independent; the implication of such data is that simultaneous selection emphasis on both ovulation rate and uterine capacity is necessary to produce increases in litter size.

Marker-assisted selection

Litter size is of considerable economic value in pig production and marker-assisted selection (MAS) by pig breeding companies for favourable alleles is likely to become increasingly important; the B allele, discovered initially in Chinese pigs, has been shown to be significantly associated with a higher litter size. Although the B allele is associated with larger litter size, this does not appear to be due to differences in ovulation rate or embryonic survival but to differences in fetal survival. In The Netherlands, Van Rens *et al.* (2000) demonstrated that fetal survival may be related to differences in placental size, which results in placental insufficiency in pigs with oestrogen receptor (ESR) genotype AA but not in those with ESR genotype BB.

Beef cows

Attempts to increase the incidence of multiple ovulations in the beef cow have been many and varied. For those who have taken the genetic selection approach, it has been a lengthy process. A twinning population was established at the US Meat Animal Research Center a quarter-century ago as part of efforts to increase the economic efficiency of beef production; by the end of the 20th century, the frequency of natural twin births in the experimental herd was reported to exceed 35%.

8.2.5. Use of gonadotrophins

Early work in sheep

There are many sheep farming conditions in which a simple technique for increasing the twinning percentage would be of value. Selective breeding, feeding and management and the use of highly prolific sheep breeds all have to play a part in this; there are, however, instances in which the hormonal induction of multiples may be considered, especially in sheep flocks that show a low twinning percentage and are not following any systematic selective breeding policy to increase litter size. In fact, the hormonal induction of multiple births in sheep, especially Karakul sheep, received much of its early attention in the former USSR. As mentioned earlier, the technique used at that time was effective but tedious, involving a single dose of PMSG in the follicular phase of the ewe's oestrous cycle; the need to use teaser rams to identify ewes in their pre-injection heat did not make for easy application.

Progestogen–PMSG combination

With the advent of the progestogen sponge in the 1960s, it became possible both to control oestrus and to influence the ewe's ovulation rate. It was a simple matter to combine a dose of PMSG with the intravaginal treatment, the gonadotrophin being administered as intravaginal devices were withdrawn. Whether the technique was of farmer interest would be a matter of cost and labour relative to what could be gained by the extra lambs. In Ireland, the PMSG treatment was a standard part of the oestrus control protocol in the non-breeding season and it was a matter of adjusting the gonadotrophin dose to suit the type of ewe involved. In the sheep breeding season, when oestrus control did not require gonadotrophin in addition to progestogen, a judicious dose of PMSG could be the means of increasing lamb output.

In the past 30 years, studies in many countries have combined oestrus control with treatment for augmented fertility in sheep. In Turkey, for example, in more recent times, workers used 40 mg fluoro-gestone acetate (FGA) sponges and 500 iu PMSG to increase the incidence of multiple births in cyclic Kivircik sheep (48 vs. 21%). Occasionally, melatonin treatment is also employed, on the basis that this can influence the incidence of multiple ovulations. In France, melatonin pretreatment (3 weeks) was employed prior to FGA–PMSG and workers reported an increased number of large follicles and a significantly higher ovulation rate when compared with controls (2.6 ± 0.4 vs. 1.45 ± 0.35).

Cattle twins

The feasibility of inducing twin births in beef cattle by gonadotrophin treatment (PMSG or pituitary FSH preparations) has been examined by many workers; their efforts, however, made it all too apparent that a major drawback to the gonadotrophin approach was the marked individual variability that occurred. While many cows produced the usual single calf, the treatment occasionally resulted in the birth of triplets, quadruplets and even quintuplets; it quickly became apparent to researchers and farmers alike that litters in excess of twins had no merit. Based on recently acquired knowledge of follicular dynamics in the cow, some studies have involved giving low FSH doses during day 3 of the oestrous cycle (during the follicle selection phase) and then ovulating the follicles by inducing luteolysis on day 5; data from such studies showed that a 50% twin ovulation rate could be achieved when FSH treatment was initiated during the first wave of follicular growth. However, as already mentioned, twin ovulations are not necessarily followed by twin calves.

Deer

Attempts have been made to increase the twinning rate in deer. In Spain, Garcia et al. (2001) used progesterone-releasing internal devices (PRID) in combination with prosta-glandin (PG) and PMSG in Iberian deer (*Cervus elaphus hispanicus*); although the twinning incidence was low (13%), the authors recorded less mortality in twins than in singles, probably due to their reduced birth weights and easier births.

8.2.6. Immunization approach

Androstenedione vaccine

Ovarian physiology is a complex process involving both extragonadal (e.g. FSH and LH) and intragonadal regulators (steroids and peptides, such as inhibin). As a means of increasing ovulation rate, it has been recognized for some years that ovarian activity in sheep can be enhanced by immunizing the ewe against certain of the ovarian androgens/oestrogens or peptides (inhibin) that are involved in the negative-feedback control of gonadotrophin release. Although little was known in the 1970s of the physiological function of the androgen androstenedione, ewes immunized against this particular ovarian steroid showed significant increases in ovulation. Androstenedione is believed to be an active regulator of ovarian activity by way of its feedback action on the hypothalamic–pituitary axis. However, it is also believed that immunization against androstenedione increases ovulation rate in sheep by decreasing the incidence of atresia in preovulatory follicles; the mechanism responsible for this phenomenon is not fully understood. One possibility is that immunization decreases the effect of locally produced atretic factors.

In the early 1980s, an anti-androstenedione vaccine was developed in Australia and released commercially under the trade-name of Fecundin. The treatment called for two injections, given at 8- and 4-week intervals, before the start of matings; in subsequent years, a single booster injection was all that was required, administered a month before introducing rams. Apart from the cost of the product, the long intervals between initiating ewe treatment and the time of expected response raised difficulties in the commercial acceptability of the technique. In New Zealand in the mid-1990s came a report noting the development of Androvax by the state research organization AgResearch. The product was said to increase the lambing percentage by 24% and was given 8 weeks before mating; there were claims that results were more consistent with Androvax than with Fecundin.

Inhibin vaccine

Although it was widely accepted that the negative-feedback effects of ovarian steroids influenced FSH secretion and ovulation rate, the early 1980s saw compelling evidence that the ovary also produces non-steroidal compounds, inhibins, which are involved in regulating FSH secretion. It is believed that inhibins provide the chemical signal indicating the number of growing follicles in the ovary to the pituitary gland to reduce the secretion of FSH to the level which maintains the species-specific number of ovulations. There are actually several forms of inhibin; workers in the USA concluded that eight different dimeric forms of inhibin may be involved in regulating basal FSH and gonadotrophin-releasing hormone (GnRH)-induced LH secretion by the pituitary.

The sequence of events illustrated earlier (see Fig. 8.4) is believed to suggest the way in which inhibin produces its effect. After the initiation of luteal regression, progesterone concentrations decrease; this releases the pituitary from the negative-feedback effect of progesterone and there is a rise of LH concentration. Together with elevated FSH concentrations, the LH promotes the growth and maturation of those follicles that will eventually ovulate. Follicular inhibin production is stimulated by androgens, produced by LH action on the theca interna, stimulating granulosa cell inhibin production. While LH and oestradiol concentrations continue to rise during the follicular phase, FSH level falls during the mid-follicular phase due to the inhibitory effects of oestradiol and inhibin on the pituitary. When the ewe is actively immunized against inhibin, it is believed that this reduces the suppression of FSH during the follicular phase, thereby enabling more follicles to mature to preovulatory status.

Studies in the UK have shown that active immunization of ewes against inhibin (the bovine alpha 1–29 peptide, conjugate form) is capable of increasing FSH levels, ovulation rate and lambing rate; it was also demonstrated that sheep can respond to an inhibin vaccine with a sustained (at least 3 years) antibody response and a recurrent increase in litter size. It has been suggested that an inhibin vaccine could form the basis of a practical, low-input system for promoting a recurrent increase in multiple births in the less fecund breeds of sheep.

Beef cows

In Australia, much research in the early 1990s was directed towards developing inhibin immunization treatments for commercial use in both cattle and sheep. Although a prototype vaccine against inhibin was developed and tested in that country, it was found that ovulation rate in the cow was much more difficult to manipulate by inhibin vaccination than in the ewe. Despite encouraging responses in small experimental groups, the vaccine apparently failed to provide the targeted level of twin births required for its commercialization. In Ireland, research effort was also devoted to developing an immunization approach to cattle twinning, mainly based on the immuno-neutralization of inhibin.

Goats

The effect of active immunization against inhibin in goats has been examined by various workers. One group in China actively immunized two strains of goat, white and black, with inhibin (alpha subunit, 1.32 fragment), recording a 12.5% higher ovulation rate in white goats and a 9% higher rate in black.

Pigs

In the USA, Kreider *et al.* (2000) evaluated the effects of immunization against ovarian steroids on ovulation rate and litter size in gilts; they reported evidence indicating that litter size could be increased by immunization against 17-α-hydroxyprogesterone. In a study reported by Christensen *et al.* (2000), primiparous sows were actively vaccinated against follistatin (a protein isolated from follicular fluid) in an attempt to modify litter size. Pigs were vaccinated four times against a recombinant form of porcine follistatin and allowed to reach sexual maturity prior to breeding; from the results, the authors concluded that vaccination of gilts against this agent increased litter size in those pigs which achieved a high antibody titre to follistatin.

8.2.7. Embryo transfer

Cattle twinning

Research over a period of three decades in Ireland was directed towards a cattle twinning technique based on embryo transfer (ET) (Fig. 8.6). ET was used to introduce a second (donated) embryo into the

Fig. 8.6.   Cattle twins by embryo transfer.

contralateral horn of the cow's uterus a few days after she was bred in the normal way (AI or natural service). The cow was given the means of carrying twins in the form of her own calf and a calf that originated from the donated embryo; the additional embryo could be produced by *in vitro* techniques or by conventional superovulation technology.

Irish studies of the late 1980s involved a series of farm trials, using *in vivo*-produced embryos, which showed that a combined AI and ET technique was capable of consistently producing a twin calving rate of 40–50% in those cattle that became pregnant. The largest field trial in which *in vivo*-produced cattle embryos were employed was in the Czech Republic. In this, embryos were obtained from cull cows superovulated prior to slaughter; one-embryo transfers were made to the contralateral horn of 7500 cattle (Table 8.2).

In Ireland, the opportunity to employ low-cost beef embryos derived from *in vitro* production technology arose in the early 1990s. High-quality beef *in vitro*-produced

(IVP) embryos were produced using oocytes from Limousin, Simmental and Charolais-cross heifers; the oocytes were fertilized using sperm from appropriate bulls (those with proven records for easy calving and good growth rates). Suitable recipient animals were selected on suitable farms (those already achieving good calving results with AI); the belief was that producers on such farms might be more likely to benefit from the twinning application. The recommended procedure was to select cows that would normally be identified as suitable for breeding to a Continental bull, taking into account body size and condition of the recipient cow. A field trial to examine cow reproductive efficiency, twinning rate and overall calf output following the non-surgical transfer of IVP embryos was then carried out. The technology involved a simple in-straw freezing and thawing procedure and direct transfer on the farm; data shown in Table 8.3 deal with 469 cows that received an additional embryo in large-scale field trials.

Similar applications have been attempted in several other countries, notably Japan; such applications have revealed problems not encountered in the Irish work. In Japan, for example, Sakaguchi *et al.* (2002) recorded results suggesting that one cause of a high rate of abortions and stillbirths in twin-bearing dams was the difference in the mean gestation length between native fetuses and those derived from transferred IVP embryos.

8.2.8. Litters and neonatal mortality

Increasing the number of young is of no avail if the extra births are lost by greater neonatal mortality. Keeping young alive

Table 8.2. Large-scale twinning trial in Czech Republic (from Riha and Petelikova, 1990).

Mean no. of eggs recovered per superovulated donor	12.08
Average no. of embryos recovered per donor	9.32
Embryos as % of all eggs recovered	70.4
Average no. of transferable embryos per donor	4.81
No. of transfers carried out	7185
No. pregnant to first service	4124
Conception rate in recipients (%)	57.4
No. of calved recipients	3077
Twins in total	1487 (48.3%)
Single calves	1590
Total calf production	4571
Per parturition	1.49

Table 8.3. Twins after IVP embryo transfer (from Bourke *et al.*, 1995).

	Embryo transfer	Controls	
Number of cows	469	858	
Calves born per cow	1.35 ± 0.02	1.02 ± 0.01	$P < 0.001$
Calves alive at 48 h per cow	1.25 ± 0.02	0.99 ± 0.01	$P < 0.001$
Percentage of multiple births	34%	2%	$P < 0.001$

is essential to reap the full benefits of additional numbers.

Lamb survival

Several authors have pointed to the difficulty in the sheep industry of looking after the needs of the individual (management and disease control) when it is nearly always part of a large group. The survival of lambs may be influenced by the treatment received by their mothers during pregnancy, whether in terms of diet or management. The general level of lamb losses around the time of birth is usually estimated as 10–25%. In The Netherlands, for example, Van Eerdenburg (2003a) gave some recent figures, estimating lamb mortality in that country at 12–14%, with approximately 50% dying within 24 h of birth. Some of the methods employed to keep weakly lambs alive are illustrated in Fig. 8.7.

Shearing and lamb birth weights

Many studies have reported on the effect of shearing during pregnancy on fetal growth and development. In Argentina, significant differences were recorded for birth weights and length of gestation between lambs born from shorn (5 weeks before lambing) and unshorn sheep (4.1 kg and 150.9 days vs. 3.8 kg and 149.5 days, respectively); such differences could not be attributed to feeding levels enjoyed by animals. In New Zealand, selective enhancement of growth in twin fetuses has been reported after the shearing of ewes in early pregnancy; although time of shearing did not affect the birth weight of single-born lambs, the weights of twin lambs rose with increasing interval from shearing to lambing to a maximum of 0.7 kg (per lamb) in those born to ewes shorn on day 70 of pregnancy.

In Australia, Revell *et al.* (2002) investigated the effect of mid-pregnancy shearing (day 70), showing that this increased lamb birth weight without increasing herbage intake or placental weight. In New Zealand, Kenyon *et al.* (2002) showed that mid-pregnancy shearing significantly increased the birth weight of twin-born lambs; the authors concluded that to achieve an increase in survival rate to weaning, these

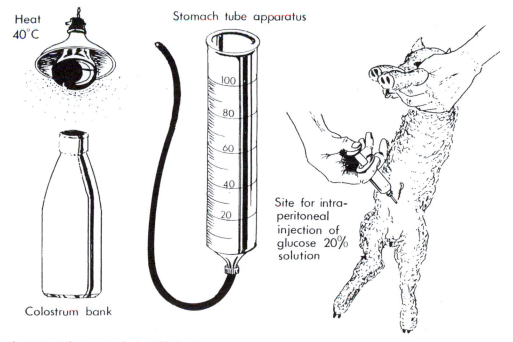

Fig. 8.7. Aids to survival of weakly lambs.

lambs must be born within a birth-weight survival range. Other studies in the same country showed that ewes can be fed maintenance during mid-pregnancy after shearing, which could enable good-quality feed to be provided during the final trimester of pregnancy (Jopson *et al.*, 2002).

Gender

It is well known that many physiological factors may affect maternal behaviour at parturition and the development of bonding between ewes and their lambs at birth and during lactation. Numerous reports have dealt with lamb losses around the time of birth and in the weeks following. Factors associated with some losses are more surprising than others. In Poland, for example, workers dealt with lamb losses on 6557 lambs born over several years on four farms; mortality of ram lambs was significantly higher than that of ewe lambs (13% vs. 9.6%); the same study revealed significantly greater mortality in single-born than in twin-born lambs (9.0% vs. 14.1%). In such situations, presumably it is a matter of lambs being of optimal size to suit the birth canal rather than a question of gender or multiple births.

Temperature during pregnancy

As well as evidence showing that shearing sheep during pregnancy can influence lamb birth weights, it is also known that temperature can exert a very definite effect on pregnant sheep. In fact, birth weight may reflect the ewe's ability to regulate its body temperature in a hot environment. Ways and means of increasing the birth weights of multiple-born lambs may increase survival rates to weaning, and are clearly of much practical interest.

Nutrition

In pregnant sheep, particularly those with twins and multiples, it is well known that requirements for energy and protein increase markedly during the final 2 months of pregnancy. The same is true for cattle carrying twins rather than the usual singleton. It is known that maternal undernutrition in pregnancy may result in low birth weights and impaired postnatal survival in sheep. Studies in Scotland by Dwyer *et al.* (2003) led them to conclude that even a moderate level of undernutrition (35% reduction in nutritional intake) can impair the bond between ewes and lambs by affecting the maternal behaviour expressed at birth. The same authors also found evidence that levels of nutrition resulting in a decrease in birth weight are likely to affect neonatal lamb behavioural progress. In dealing with pregnant ewes, changes in live-weight and body condition as a guide to the adequacy of nutrition during late pregnancy are not necessarily of great practical value, since such measures only provide information retrospectively. For those with access to laboratory facilities, a more immediate assessment of the adequacy of contemporary nutrition may be provided by the quantitative measurement of certain blood metabolites, such as plasma β-hydroxybutyrate.

Pulse oximetry

It is known that hypoxaemia during and immediately after delivery is a major cause of perinatal mortality in lambs; some studies found it to be the cause of 35% of perinatal deaths. Recent work, however, has shown that percentage arterial haemoglobin oxygen saturation (SpO_2) can be assessed by pulse oximetry; in human medicine, it is routinely used to monitor perinatal babies. Pulse oximetry is a non-invasive technique for measuring SpO_2 based on the differential absorption of red and infrared light across vascular non-pigmented tissue. The tail of the newborn lamb is the most reliable site for the application of the oximeter probe; those who have employed the technique suggest that the instrument may have a valuable role to play in improving the perinatal care of lambs and improving lamb survival rates by identifying those that need extra care. In the Cotswolds, one veterinary practitioner has been operating an on-farm education policy to educate sheep farmers in methods to minimize lamb losses; among the measures covered is the use of a resuscitation method involving ventilation of the lungs, which is now widely used by farmers in the area.

Maternal behaviour in sheep

Much research into various aspects of animal behaviour has focused on the care given to young at time of birth, and the factors that may affect this behaviour. It is known, for example, that ewes learn to recognize the odours of lambs within 2 h of giving birth; workers at Cambridge proposed that such olfactory memory formation induced at birth involves several changes in the functional circuitry of the olfactory bulb. Factors controlling maternal behaviour in ewes and their implications for sheep production systems have been the subject of various studies; a greater understanding of physiological factors affecting maternal behaviour at parturition and the development of bonding between ewes and their lambs at birth and during lactation could be of practical benefit. In Canada, ewe and lamb behaviour at parturition was the subject of studies in prolific and non-prolific sheep; several differences were found between such sheep in behavioural traits at birth. In Scotland, it was found that suckling behaviour of the lamb was modified by ewe behaviour and this could affect the strength of the bond between the two.

Maternal behaviour in pigs

In a study of sow behaviour reported from the UK by Marchant et al. (2001), it was found, in common with many other studies, that piglets are most vulnerable to crushing during the first 24 h of life, when they are spending much of their time near the udder and have relatively poor mobility; it is clear that coordination of behaviour between the sow and her litter is important to reduce the risk of crushing. It is also important to ensure that the design of open farrowing systems incorporate knowledge about how crushing deaths occur. Many studies have been reported on the effect of various diets and ingredients in the diets of pregnant pigs

on neonatal mortality. In the UK, Cordoba et al. (2000) reported that feeding salmon oil to pregnant and lactating sows reduced deaths caused by crushing, probably by increasing gestation length; it also seemed possible that reduced deaths reflected improved neonatal vigour.

Pigs at risk

In polytocous species such as pigs, there is strong competition among siblings immediately after birth for access to milk and other resources; it is known that almost 80% of pre-weaning mortality occurs during the perinatal period, that is, during farrowing and the first 3 days after birth, a figure that appears to remain constant, despite the many developments in pig production and management. In Switzerland, workers studied the effects of management on sow piglet-rearing ability; of sows with litters of > 12 piglets, almost 100% lost one or more piglets before weaning vs. 15–50% of sows with smaller litters. The same workers observed 537 cases of pre-weaning mortality, of which 60.1, 23.6 and 16.2%, respectively, occurred on the day of birth, 2–7 days after birth and > 7 days after birth.

Although pig mortality has traditionally been regarded as a matter of economic importance, there is growing sensitivity to the welfare of farm animals, giving added impetus to research in this area. A study by Tuchscherer et al. (2000) provided information on determining traits that might be of prognostic value for identifying piglets at a high risk of death; they concluded that reduced physiological maturity of piglets at birth is mainly responsible for a high risk of mortality during the first days of life. The same workers recommend supervision during the time of farrowing and the postnatal period, provision of an appropriate thermal environment and a guaranteed early colostrum intake as procedures to prevent pig death in the early postnatal period.

Piglet mortality and birth weights

The effect of birth weight on piglet mortality to 21 days of age was the subject of studies in Brazil; mortality rate was found to be significantly higher for piglets with a low birth weight (< 1.1 kg for females and < 1.4 kg for males) than for piglets with a birth weight of 1.5–2.3 kg. An analysis of the number of stillbirths in 7817 litters from sows on farms in The Netherlands, Belgium and France was made by workers in the late 1990s; a significant positive relationship was found between the number of stillborn piglets per litter and the number of live-born piglets that died before weaning. The occurrence of sows lying on their young, or piglet 'crushing', is a significant cause of piglet mortality in current production systems. Although mortality rates of piglets in farrowing crates are lower than in pens, loss due to crushing is still estimated to be between 5 and 18%. During the first few days after farrowing, piglets are highly attracted to the odour of their dam's udder. With this in mind, some workers have used a simulated udder to move the piglets away from the sow by competing with the sow's udder; this strategy was found to work better than heat lamps alone.

Fig. 8.8.   Testing for the freemartin.

Freemartins

Relevant to a discussion of multiple births are problems that arise in births of freemartins, which are not found with singletons. Intersexuality is a problem that has been reported in all domestic species; intersexuality includes conditions of true hermaphroditism and pseudohermaphroditism, abnormalities of accessory genital organs, gonadal dysgenesis and freemartinism. Any discussion of cattle twinning inevitably raises the question of the freemartin, the sterile heifer born co-twin to a bull; they are also known to occur in buffaloes and rarely in other farm species. Reports in the literature put the incidence of freemartins in naturally occurring cattle twins of unlike sex at 92% (Fig. 8.8); among hormonally induced twins, there has been a suspicion that the incidence of freemartins may be greater in unilateral than in bilateral twins. However, in the context of an induced twinning programme in beef animals, freemartins should present no difficulty;

they are destined for slaughter rather than breeding and grow much the same as normal heifers.

The practical importance of being able to identify the freemartin among animals destined for breeding requires no emphasis; there are many examples cited in the past of dairy heifers sold in open market for breeding turning out to be freemartins. Many diagnostic techniques have been developed for early fertility assessment in heterosexual twin female cattle, including tests based on differences in the reproductive organs, cytogenetic tests, sex chromatin tests, PCR analysis and endocrine tests.

Effect of freemartin on co-twin bull

The reproductive normality of bulls born co-twin to freemartins was accepted until evidence surfaced in the early 1960s showing the presence of X–X germ-cells in the testes of newborn bull calves. There was also evidence from other reports showing that germ cells entered the circulation of the 25–34-day-old cattle embryo at a time when chorionic vascular anastomoses are already established between partners. Apart from its biological interest, an understanding of germ-cell chimerism is of practical importance in view of evidence clearly showing that certain bulls born co-twin to freemartins may be either sterile or below average in semen quality and fertility. Statistics show that more than 50% of chimeric bulls are likely to be culled for poor fertility in the first 10 years of life in comparison with a figure of 5% for single-born bulls. Occasionally, there has been debate as to whether germ cells that migrate from the heifer into the bull testes are eventually capable of giving rise to sperm, with some observers reporting individual bulls born co-twin to freemartins producing excess female offspring.

Freemartins in sheep

Although the incidence of freemartinism is well documented for cattle, there are few reports of the condition in sheep; those who have reported freemartinism have recorded a figure of 1–5%, while others have been as low as 0.033%. It is all a matter of how the fetal membranes develop (Fig. 8.9). Most lowland breeds of sheep in the UK and Ireland give birth to twins and occasionally triplets in the normal course of events and it is to be expected that there might be some interactions between neighbouring fetuses during pregnancy, particularly when fetuses of opposite sex are side by side in the uterus. Although freemartins are very uncommon in sheep, some studies have revealed that the extent of embryonic mortality may be influenced by the position of the ewe lamb in intrauterine life relative to male fetuses.

8.2.9. Animal welfare considerations

Animal welfare concerns often centre on protecting the weak and innocent and, for that reason, the welfare of the newborn farm animal, whether singleton or one of multiples, deserves considerable scrutiny. Taking sheep as a case in point, perinatal lamb mortality, usually defined as losses occurring within the first 3 days of life, is a major cause of lamb deaths – estimates ranging from 10 to 25% of total lambs born – and represents a major animal welfare concern. It appears that there has been little reduction in the incidence of perinatal lamb deaths over the past several decades, which makes concern all the greater. Many physiological factors are believed to contribute to lamb losses, during the pre-partum, intra-partum and post-partum periods; such

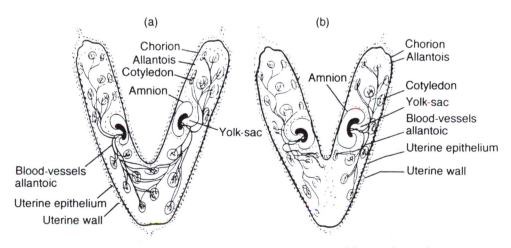

Fig. 8.9. Diagrammatic representation of fetal membranes in (a) cow and (b) ewe (from Robinson, 1957).

factors include inadequate dam nutrition during late gestation, dystocia and the mis-mothering, hypothermia and starvation of lambs. Losses occurring during the three periods can be effectively controlled by good husbandry, adequate dam nutrition, 24 h flock supervision during the lambing period and ensuring that lambs have access to sufficent good colostrum within the first 6 h of life, with the provision of adequate shelter.

Account should also be taken of the fact that the resistance of sheep to gastrointestinal parasites tends to break down around parturition. There are studies showing that the resistance of ewes bearing and suckling twin lambs may be significantly less than that of ewes with singletons; it is also known that the resistance of ewes, measured in terms of faecal egg concentration and worm burden, may be sensitive to dietary protein intake and, to a lesser extent, to energy intake and body condition. The lesson is that care of the newborn involves care of the mother as well.

9

Pregnancy Testing Technology

9.1. Advantages of Control Measures

The practical importance to the livestock farmer of ensuring that breeding animals become pregnant needs little emphasis. There are clear economic advantages in determining the pregnancy status of livestock at the earliest opportunity after mating; the simplest method, that of observing return to oestrus, may often prove unreliable. Estimates made some time ago, for example, suggested that, of the cattle that do not exhibit a 'repeat' heat 3 weeks after breeding, 15–25% are actually not pregnant. Even with a high oestrus detection rate, there will be some percentage of animals that are not seen to return to service and will incorrectly be assumed to be pregnant; although these may well be detected at a later stage, this may involve a loss of several weeks. In goats, to take another example, the majority of does in which oestrus control is applied out of season would be expected to return to anoestrus if not pregnant; even in the natural breeding season, persistence of the corpus luteum may lead to serious errors in distinguishing does that are pregnant from those that either have a long cycle or are pseudopregnant. In terms of methods used in identifying the pregnant female, what works well in one species may be quite inappropriate in the next (Table 9.1).

9.1.1. Cattle and buffaloes

The availability of a simple on-farm early pregnancy test has much to offer dairy farmers, especially one that can alert them to the fact that the cow is not pregnant before the first opportunity arrives to rebreed her (i.e. 3 weeks after first mating); it should be noted that such a test is not available but remains a desirable objective. Methods are available, however, by which beef cows can be bred at the first opportunity, but this

Table 9.1. Pregnancy tests for the cow.

Stage of gestation	Method of pregnancy diagnosis
18–24 days	Failure to return to oestrus
18–24 days	Persistence of the corpus luteum
22–26 days	Milk or plasma progesterone assay
30 days	Scanning by real-time ultrasonics
30–65 days	Palpation of the amniotic vesicle
35–90 days	Disparity in size of uterine horn
35–90 days	Palpation of the allantochorion (membrane slip)
70 days to term	Palpation of caruncles
90 days to term	Fremitus in middle uterine artery of gravid horn
105 days to term	Oestrone sulphate in milk assay
150 days to term	Fremitus in middle uterine artery of non-gravid horn

involves oestrus control technology (see Chapter 5, Section 5.5.2).

and appropriate nutrition for the different animal categories.

Gender identification by ultrasonics

In cattle, the use of real-time ultrasonics now makes it possible for researchers to follow the course of pregnancy with considerable accuracy through the early months of gestation; this has added a new dimension to research dealing with early pregnancy by providing precise information on the progress of the embryo and fetus. From the farmer's viewpoint, such technology can be employed to identify the sex of the calf from about day 70 or even earlier. There may be a demand in commercial practice to know the sex of the unborn calf because this may influence the monetary value of the pregnancy. For example, cows pregnant after embryo transfer may be exported to customers far afield with greater confidence if they are found to be carrying the calf of the desired sex, usually a heifer; on other occasions, cows carrying bull calves sired by certain bulls may need to be considered for induced calving to avoid a difficult birth.

Dairy goats

Early pregnancy diagnosis is an important aspect of reproductive management of dairy goats. It is essential in optimizing milk yields and feeding regimens and permits the early detection of fertility problems. The use of milk for the determination of hormones or other compounds associated with pregnancy in the goat has obvious attractions; sampling avoids the possible stressful effects of venepuncture and collection is relatively simple to carry out. Ultrasonography is a useful research tool for determining the extent of embryonic mortality in goats, based on the presence or absence of heartbeats from day 21 post-mating. Measurement of the crown–rump length enables the gestational age of the embryo to be accurately estimated up to day 40, using a linear regression equation.

9.1.3. Pigs and horses

Pregnancy diagnosis is of considerable importance in pig production; sows that either do not become or do not remain pregnant after insemination and are not detected non-pregnant at an early stage are the cause of considerable economic loss. Several techniques for pregnancy diagnosis are available, each with its own merits and disadvantages. Widespread application of ultrasonics in pigs followed the development of this technology in human medicine, dating back to the 1960s. Ultrasonography is a sensitive method of detecting pregnant animals but its detection of non-pregnancy is often variable and sometimes low. Ultrasonography is performed sow-side, but the apparatus can be expensive and the degree of experience of the operator may greatly influence the results; there is also the question of minimizing disease risks in the application of the technology to several farms. In Holland, for example, where pig artificial insemination (AI) centres launched a routine on-farm

9.1.2. Sheep and goats

In most commercial flocks of sheep and herds of goats, the dates of natural service are usually unobserved or unrecorded, making it difficult to ascertain the stage of gestation in individual animals. Accurate pregnancy diagnosis could provide much useful information for the management of sheep and goats, whether in the culling of non-pregnant females or determining the number of fetuses in those that are pregnant. Current pregnancy diagnosis techniques for sheep include progesterone tests in blood and milk, ultrasonic diagnosis, vaginal biopsy, laparoscopy, radiography, palpation, rectal probes and udder examination. Early pregnancy diagnosis can be a particularly useful management tool in the sheep industry; separating the flock into pregnant and non-pregnant animals or into those carrying single or multiples allows better control of management pregnancy testing service, concern has been expressed about the disease risks posed by such a practice.

Pregnancy testing can also be based on detecting hormonal changes in sows. The embryonic origin of oestrogens makes these steroids of especial interest; many studies have shown that measurement of plasma oestrone sulphate concentrations shows both high test sensitivity and test specificity, but blood sampling may pose problems in applying the test on the farm. In measuring the levels of steroids in pigs, blood is usually collected from the cranial vena or caudal-auricular vein by retention of the nose, which involves considerable physical effort and the possibility of accidents; the procedure is also evidently stressful for the pig. For such reasons, oestrogen measurement in faeces, although aesthetically less appealing, may be a possible alternative, as sample collection is easy and non-invasive.

Importance in mares

Efforts are constantly being made to improve reproductive management practices in horses. In this context, there is ample scope for methods which can be employed for the very early detection of pregnancy in the mare; a technique which can be employed to detect non-pregnancy early enough for the mare to be rebred at the first possible heat period is likely to be particularly useful. In horses, perhaps more than in any of the other farm animals, it is often important to know as soon as possible whether a valuable mare has conceived.

9.1.4. Deer

Consideration of deer brings certain unique aspects of pregnancy to light in certain of the species. The roe deer (*Capreolus capreolus*), for example, is a ruminant that exhibits embryonic diapause, embryos remaining static in their growth in the uterus for several months before embryonic development resumes. Several authors have examined the possible factors involved in this phenomenon; clearly it is of interest to

those concerned with storing embryos of related species temporarily (days and months) at ambient temperatures rather than embarking on long-term cryopreservation. The duration of pregnancy in red deer is 227–234 days and in fallow deer 229–234 days. In roe deer, in which delayed implantation may occur, the period of gestation may extend to 280 days. In the UK, workers have recorded a gestation length of 283–284 days in Père David's deer and concluded that this may imply a period of embryonic diapause or slow fetal growth in this species.

9.2. Factors Affecting Establishment of Pregnancy

In the days before sensitive hormone assays and ultrasonics, the usual way of determining whether a farm animal was pregnant was by observing the occurrence or absence of oestrus. If the sheep or goat was pregnant, she would normally not show heat one cycle interval after mating (Fig. 9.1).

The problem with the non-appearance of oestrus as an indication of pregnancy lay in failure to detect the oestrus or the loss of an embryo at an early stage of pregnancy. In goats, for example, a Norwegian study examined pregnancy diagnosis in dairy goats using progesterone assay (radioimmunoassay (RIA)) and oestrus detection; the accuracy of diagnosis by RIA and heat detection was 93 and 86%, respectively. Initially, progesterone assays in the study were based on plasma samples, but this was quickly superseded by the simpler and more acceptable milk test.

9.2.1. Physiology and endocrinology of early pregnancy

Prolongation of luteal function

Pregnancy in the farm mammal starts at the instant an oocyte is fertilized and continues through until the fetus, fluids and membranes are expelled at the time of parturition. Prenatal life can be divided into two

Fig. 9.1. Ready to catch the 'repeat' ewes.

periods: one covering the development of the embryo and the other dealing with the fetus. Although the longest period is that of the fetus, the first period is usually the most critical, for it is then that the greatest mortality can be expected. An essential feature of the establishment of pregnancy in the cow, as well as in the other farm mammals, is the prolongation of luteal function beyond the approximately 2-week duration that is found in the non-pregnant animal. This is now known to be due to mechanisms that suppress the release of luteolytic quantities of prostaglandin $F_{2\alpha}$ ($PGF_{2\alpha}$) into the circulation, which in the cyclic animal brings regression of the corpus luteum and the end of progesterone production.

Process of attachment

The process whereby the embryos of farm mammals become attached to the uterus involves the proliferation, differentiation and migration of various cell types, both in the embryo and in the uterus. Quite remarkable growth in the dimensions of the embryo can take place in a matter of 1 or 2 days (Fig. 9.2).

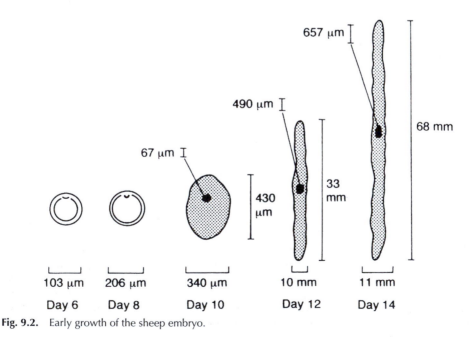

Fig. 9.2. Early growth of the sheep embryo.

The term implantation is commonly used in texts rather than attachment, although implantation has its origins in species where the embryo passes through the inner lining of the uterus (endometrium). It should be noted that in farm animals placental formation is markedly different from that found in rabbits and mice; prior to attachment to the endometrium, ruminant and pig embryos float free for several days in the lumen of the uterus as the trophectoderm elongates dramatically. In horses, the embryo, within its capsule, also passes through a period of mobility before it finally becomes fixed to the surface of the mare's uterus.

The ovarian steroid hormones, oestrogens and progesterone, play a major role in the period around attachment; it is also clear that several growth factors are intimately involved in embryonic development and attachment; in Spain, workers have detected the presence of epidermal growth factor (EGF), transforming growth factor-alpha (TGF-α) and transforming growth factor-beta (TGF-β) around the time of attachment in the goat. In ruminants, the early weeks of pregnancy are a time of extensive growth and remodelling of the blastocyst, which continues to expand after hatching until it reaches a length of 15 cm or more, 3 or 4 days later, resembling a long filament composed mainly of trophoblastic cells; much the same story holds for pigs as well.

The process of attachment starts relatively late in development (by day 15 after mating in sheep) in the vicinity of the embryo and spreads towards the extremity of the elongated conceptus. The period of embryo attachment extends over a long period, during which the trophoblast is composed of attached and non-attached areas; the extensive growth of the blastocyst in early pregnancy occurs in a sequential order, suggesting that specific genes are expressed according to a finely adjusted schedule. Various studies have shown that the ruminant conceptus is free-living in the uterine lumen until day 16 (ewe) or 20 (cow) of pregnancy.

Embryo attachment during pregnancy in pigs is a non-invasive process that results in simple attachment between the trophoblast and the uterine epithelium; this type of placentation is termed epitheliochorial to distinguish it from other more invasive types of placentation, such as the haemochorial placentation that is found in humans. Unlike most laboratory species, pig blastocysts undergo a dramatic change in morphology prior to implantation. Around day 11 of pregnancy, blastocyst morphology changes from small spheres to threadlike structures of considerable size; at the same time, maternal recognition of pregnancy takes place, with the blastocysts synthesizing oestrogen. On days 11 to 12 of gestation, the pig conceptus elongates from a 10 mm sphere to a long filamentous form in the course of some 12 to 24 h, largely through changes in trophectoderm morphology; there is currently no clear understanding of the genetic pathways involved in this rapid and dynamic morphogenesis.

Attachment in the sow

In the sow, it is known that the early embryo synthesizes oestradiol, which plays an important role in the attachment process. In France, workers showed that, at the same gestational age, gilts with small embryos and/or few embryos may not always present enough uterine oestradiol to support the attachment process. In Poland, steroid hormone concentrations were examined during early pregnancy in pigs on three different occasions; day 12 (maternal recognition of pregnancy), day 18 (formation of the corpora lutea (CL) of pregnancy) and day 35 (elongation of the placenta). It was found, during the maternal recognition and embryo attachment period, that the oviduct and uterus are supplied with locally greater concentrations of such hormones as progesterone, oestrone, oestradiol, androstenedione and testosterone. It appears that an adaptation of the local blood circulation provides for the needs of the oviduct and uterus without elevating hormone concentrations in systemic blood to levels that would affect nervous system functions and influence animal behaviour.

Placentation

The placenta is an important determinant of fetal growth rate and there is ample evidence suggesting that the insulin-like growth factor (IGF) system plays a pivotal role in regulating placental development. Many components of the IGF system have been identified in the placenta, including IGF-binding protein 5 (IGFBP-5). Studies have shown IGFBP-5 to be expressed in ovine uteroplacental tissues throughout gestation and the binding protein is thought to be an important modulator of IGF-mediated placental growth. Another member of the IGF-binding protein family (IGFBP-1) is thought to play a role in regulating the transfer of IGFs between the endometrium and the uterine lumen; it is also believed that the conceptus may enhance IGFBP-1 expression during early pregnancy.

Binucleate cells

Embryo development in the pig, including blastula formation, hatching and migration of blastocysts, gastrulation, embryo attachment, placental formation, sex differentiation and embryo position in relation to the embryonic membranes, has been described by various authors. Placentation as it occurs in ruminants, both large and small, can be regarded as superficial and relatively non-invasive. In this type of placentation, the chorionic epithelium contains both mono-nucleate and binucleate cells. Binucleate cells in the ruminant placenta are differentiated from fetal mononucleate trophoblast cells and are known to secrete glyco-proteins. As attachment is initiated, the trophoblast binucleate cells migrate from the trophoblast to fuse with maternal endometrial epithelial cells. It is at this stage that the contents of binucleate cell secretory granules are released directly into maternal tissues; among these secretory products are many of the pregnancy-associated glycoproteins (PAGs).

Placentation in camels

Placentation during the first 2 months of pregnancy in the dromedary camel has been the subject of several studies. By day 14, the majority of the trophoblast has become closely apposed to the uterine luminal epithelium; in some places, the start of micro-villar interdigitation is evident and by day 25 a well-developed microvillar junction had formed between the fetal and maternal tissues. Between 1 and 2 months of gestation, large multinuclear trophoblast cells of a syncytial nature have developed at frequent, but irregular, intervals along the trophoblast.

Unique features in the mare

There are several unique physiological and endocrinological features of pregnancy in the mare, the first occurring at the start of pregnancy with the selective transport of embryos through the oviduct. While the embryo is transported safely down the oviduct into the uterus, the unfertilized oocyte finds itself retained, near the ampullary–isthmic junction; here, it slowly degenerates over a period of several months. It appears that the embryo can bypass degenerate, unfertilized oocytes from previous ovulations without dislodging them. Although the mechanisms responsible for the tubal transport of horse embryos remained unclear for some time, it was eventually established that it involves the secretion of PGE_2, a known relaxant of oviductal musculature. The time-scale of important events during the first 50 days of pregnancy is set out in Fig. 9.3.

The mare is also unique among the farm animals in the way in which the embryo shows mobility during the early period of its life in the uterus. In the USA, studies with pregnant pony mares showed that fixation of the embryonic vesicle occurred about day 15 of pregnancy, the conceptus diameter, about 22 mm, being similar at that time to the endometrial diameter of the caudal segment of the caudal uterine horn. It is known that the early equine conceptus is extremely mobile, travelling to all parts of the mare's uterus 10 to 20 times a day, especially on days 12 to 14. Studies have also suggested that PGE_2 is the likely stimulator of both uterine contractions and uterine

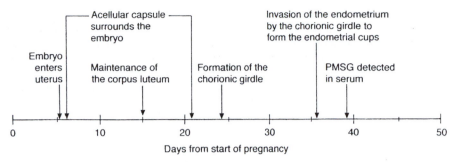

Fig. 9.3. The first 50 days of pregnancy in the mare.

tone during the time of embryo mobility in mares.

Hormonal events

In the mare, the hormonal events in pregnancy include several unusual features. One feature is that equine chorionic gonadotrophin (eCG, pregnant mare's serum gonadotrophin (PMSG)) regulates luteal steroidogenesis in pregnant mares. In the USA, studies have shown that PMSG stimulates luteal androgen and oestrogen production during pregnancy in mares. A prerequisite for a successful pregnancy is the maintenance of a relatively quiescent myometrium. In species such as sheep and goats, myometrial activity is inhibited by progesterone; however, the factors regulating myometrial activity in the mare are still poorly understood. It is known that progestogen requirements of equine pregnancy are met by the fetoplacental unit, commencing at 70–80 days of gestation. However, the concentration of progesterone in maternal plasma falls to very low levels by mid-gestation and remains low throughout the second half of pregnancy.

It is evident that the mare possesses unique steroid hormone metabolic activity during pregnancy in that peripheral progesterone is undetectable by 220 days of gestation. Various studies have investigated steroid transformations in pregnant mares, seeking to explain the absence of progesterone and the presence of other progestogen metabolites in maternal circulation during mid- and late pregnancy. It is believed that progesterone is reduced to simpler compounds (5-α-pregnanes and pregnenes) within the placenta and possibly within the fetus and it is believed that some of these (5-α-pregnanes) take over the role of progesterone in maintaining myometrial quiescence in the mare.

Immunological aspects of pregnancy

Much useful information has been gained about the immunological aspects of pregnancy in mammals, using the sheep as the domestic animal model. Pregnancy involves major adjustments in the way that the animal body normally rejects foreign tissue; clearly, adjustments must occur in the reproductive tract to prevent immunological destruction of sperm and the developing conceptus. Pregnancy usually goes ahead despite the immunological threat to the conceptus because expression of major histocompatibility (MHC) antigens on the trophoblast is altered and there are changes in the normal function of maternal endometrial lymphocytes due to molecules produced by the trophoblast (e.g. interferon-tau (IFN-τ)).

9.2.2. Nutritional and environmental influences

Sheep

Nutrition and pregnancy in sheep have been the subject of numerous studies. In the UK, some research groups have focused on the production, regulation and function of the IGF system in the reproductive tract and fetal membranes of sheep and cattle; this system is believed to be important in altering placental development according to the nutritional status of the dam. In New Zealand, workers found that low nutrition in early to mid-pregnancy could affect placental development in high-fertility ewes; this, in turn, could affect fetal development, lamb birth weight and lamb survival. These workers concluded that ewes with twins could not afford to lose more than 4 kg from mating to mid-pregnancy if the birth weight of twins was not to be adversely affected. Studies in goats have also examined the effect of diet and other factors during pregnancy. Work in Norway concluded that plasma progesterone level was affected by nutrition during pregnancy; higher progesterone levels were found in goats with > 1 fetus than in those carrying singles.

Role of amino acids

There are studies showing that amino acids play a vital role in the development of the conceptus (embryo/fetus and associated placental membranes). In the USA, workers recently reported finding an unusual abundance of traditionally classified non-essential amino acids in allantoic fluid and suggest that this may raise important questions regarding their roles in ovine conceptus development. The developing fetus, surrounded by the amniotic fluid compartment and connected with the allantoic sac via the urachus and placental vasculature, receives nutrients mainly via the umbilical vein. The amniotic fluid provides the fetus with a unique aqueous environment in which it can develop symmetrically; swallowed by the fetus, amniotic fluid is also recognized as being a significant source of nutrients, whereas allantoic fluid is traditionally regarded as a reservoir for fetal wastes.

The studies of Kwon *et al.* (2003), however, have shown that remarkable changes occur in the concentrations of amino acids in ovine fetal allantoic fluid between days 30 and 140 of gestation, findings that raise important questions regarding placental and fetal metabolism and may have important

implications for both intrauterine growth retardation and the fetal origin of diseases shown by humans in later life (e.g. diabetes, hypertension and coronary heart disease). In Australian studies, high concentrations of inhibin, activin and follistatin have been found in amniotic fluid surrounding the ruminant fetus; it is believed that the presence of such hormones indicates a role in embryonic/fetal development.

9.2.3. Maternal recognition of pregnancy

It is now well accepted that normal pregnancy in the farm animal depends upon the early embryo signalling its presence to the maternal system, a process termed maternal recognition of pregnancy (Fig. 9.4); it is well recognized that maternal recognition involves the immune, vascular and endocrine systems.

The establishment of pregnancy in the cow is highly dependent on the precise timing of a functional relationship between mother and embryo. The timing of the progesterone rise is of critical importance; an inadequately developed embryo is less able to inhibit the development of the luteolytic mechanism, thus compromising the normal progress of pregnancy. Important areas of research have involved IFN-τ, the pregnancy recognition signal in ruminants, and metabolic factors affecting pregnancy

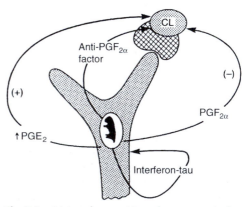

Fig. 9.4. Maternal recognition of pregnancy in the ewe (from Niswender *et al.*, 1994).

recognition; it is well accepted that for the establishment of a successful pregnancy there is a crucial need for certain conceptus–endometrial and epithelial–stromal interactions that maintain the endometrium under progesterone dominance.

Interferons

In ruminants, it is known that interferon produced by the trophectoderm (IFN-τ) is recognized as the embryonic signal responsible for maternal recognition of pregnancy. IFN-τ is believed to act by down-regulating oestrogen receptors, thus preventing the appearance of oxytocin receptors responsible for the release of $PGF_{2\alpha}$ by the endometrium. At the time of their discovery in 1957, the term 'interferon' referred to certain molecules produced by cells in response to a viral infection. It is clear that the original concept of one IFN with one biological function was too simplistic; IFNs are now known to be members of a large family of regulatory proteins, some of which are involved in the regulation of pregnancy.

Goats and deer

Studies in France led workers to conclude that the goat conceptus secretes IFN-τ during the period of maternal recognition of pregnancy but its rapid decrease suggests that other factors are required by day 18 to take over its role in the maintenance of luteal function. It is also clear that the embryonic pregnancy recognition signal in red deer is an IFN; recent studies suggest that secretion of oxytocin and $PGF_{2\alpha}$ are suppressed in pregnant hinds by IFN as they are in sheep.

Pigs

Japanese workers have reported on events in the maternal recognition of pregnancy in the sow; it appears that four embryos are necessary for the establishment of pregnancy after normal mating in the pig, although two embryos may be sufficient to establish a pregnancy after embryo transfer.

During early pregnancy, embryos are required to synthesize signal substances, both oestrogens and IFNs, above the threshold level for maternal recognition of pregnancy. As the conceptus starts to elongate during week 2 of pregnancy, concomitant oestradiol-17β, synthesized and secreted from the trophoblast, is the first maternal recognition signal; beyond 2 weeks, IFNs are known to play a role, although pig trophoblastic IFNs, unlike those of ruminants, do not apparently exert an anti-luteolytic effect in the sow. French studies have suggested that, in early pregnancy, trophoblastic cells of pig embryos secrete two distinct IFNs, one of which is IFN-γ and the other IFN-δ.

French researchers examined the involvement of porcine trophoblast IFNs in the maternal recognition of pregnancy; results confirmed that pig IFNs, unlike those of ruminants, do not have an anti-luteolytic effect. Concepts of luteolysis and inhibition of luteolysis in pigs have been reviewed by Ziecik (2002). Results reported by Joyce *et al.* (2003) in the USA indicated that, during pregnancy in the sow, induction of IFN-stimulated genes (ISGs) is not exclusively a result of high levels of oestrogen secreted by elongating conceptuses. It is known that ISGs are regulated by IFNs and that porcine conceptuses secrete both IFN-γ and IFN-δ on days 12–18.

Horses

In the mare, the growing equine conceptus must suppress the cyclical release of $PGF_{2\alpha}$ from the endometrium to effect maternal recognition of its presence in the uterus. In the mare, as in ruminants, oxytocin plays an important role in generating the endometrial $PGF_{2\alpha}$ releases, which bring about luteolysis in cyclic mares during days 12–16 after ovulation; it is this luteolysis which must be suppressed by the young developing equine conceptus during early pregnancy. However, the equine conceptus itself secretes physiologically effective quantities of $PGF_{2\alpha}$ and PGE_2 to stimulate the uterine contractions that drive the equine capsule to migrate throughout the uterine lumen to distribute its anti-luteolytic signal. Studies reported by Stout and Allen (2002) led them to conclude that the equine conceptus effects maternal recognition of pregnancy by inhibiting the ability of the endometrium to release $PGF_{2\alpha}$ during days 12 to 16 after ovulation; the mechanisms whereby luteolysis is prevented in the period 18–32 days after ovulation remain to be determined.

Mobility of capsule

Recent studies in the USA have indicated that an embryo-derived factor is the most likely cause of the suppression of endometrial-produced $PGF_{2\alpha}$ and interruption of the oxytocin–$PGF_{2\alpha}$ interaction in mares during early pregnancy. Although PGE_2 and embryonic oestrogens have been shown to be involved in maternal recognition of pregnancy in some species, there is no evidence that either has anti-luteolytic properties in the mare. The nature of this embryonic signal remains unknown in the mare but, during the period when it must be transmitted, the spherical conceptus is enveloped by a unique acellular glycoprotein capsule and it remains mobile so as to distribute the signal throughout the uterine lumen. It is believed that the prostaglandins and the oestrogens produced by the equine conceptus play important roles in local fetomaternal dialogue, especially the former, by stimulating myometrial contractions to induce conceptus mobility

within the uterus. Although the functions of the equine blastocyst capsule have not yet been fully characterized, it is believed to be essential for embryonic survival and both its development and its dissolution appear to be stage-specific.

9.3. Pregnancy Testing Methods

9.3.1. Palpation *per rectum*

Cattle

In large ruminants (cow, buffalo) and in camelids and the horse, pregnancy diagnosis can be readily performed by palpation *per rectum*; the small body size of the sheep and goat generally rule out this approach. Rectal palpation has been used for pregnancy diagnosis in cattle for many years and has remained one of the most simple and valuable methods (Fig. 9.5); it has the obvious advantage, in comparison with some other methods, of allowing an opinion to be given immediately. However, this is likely to be a veterinary opinion, which obviously comes at a cost. In some countries, the stockperson may carry out this examination; this is a question of scale, with the large beef herds of the USA providing ample opportunity for stockpersons to maintain skills in a way not open to them in many other countries. With the veterinarian, on the other hand, if the cow is not pregnant, then some indication of the cause

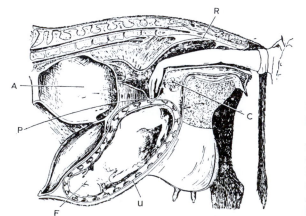

Fig. 9.5. Rectal palpation to determine pregnancy in cattle. The hand is inserted into the rectum (R); the cervix (C) can be felt, but the uterus (U) and ovaries have dropped into the body cavity and only the lower end of the uterus and fetus (F) and some of the caruncles (P) can be touched. A is the cavity of the rumen. (From Hammond *et al.*, 1983.)

may be evident and steps taken to correct problems. It is possible to apply rectal palpation from about 35 days after breeding, which should enable a careful watch to be kept for oestrous symptoms in non-pregnant animals around the 6-week mark.

Horses

For many years, the most commonly used method of diagnosing pregnancy in the mare was rectal palpation, which can be carried out at an early stage of gestation and yields an immediate answer. Numerous reports have appeared in the past 50 years showing that a satisfactory diagnosis could be made between 20 and 30 days after service by a person with sufficient experience of the technique. It is also known that it is possible to estimate the stage of pregnancy within a week or so by the rectal palpation method (Fig. 9.6).

Sheep and goats

Although a variety of techniques have been reported, there is no simple clinical method for detecting early pregnancy in small ruminants; those that have been used include external palpation, palpation of the caudal uterine artery through the vaginal wall, abdominal ballottement and Hulet's recto-abdominal palpation technique. Then there are diagnostic methods based on radiography, ultrasonography and several hormonal assays. None of the clinical methods provides a reliable diagnosis earlier than about 3 months of gestation, and techniques such as ultrasonics may find little application, on the basis of expense, in developing countries. In India, work recently reported suggested a manual palpation approach may provide a simple, inexpensive and reliable clinical method for detecting pregnancy in the goat as early as 28–30 days after breeding; it is suggested that the technique may be useful in the reproductive management of small ruminants.

9.3.2. Progesterone and oestrogen assays

Cattle

With the advent of sensitive RIA and enzyme immunoassay (EIA) techniques in

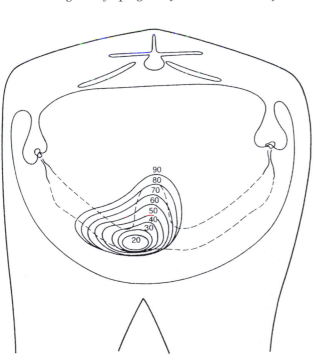

Fig. 9.6. Relative size of conceptus in the mare at different stages of pregnancy (from Day, 1940).

the 1970s came the development of methods whereby several of the hormones of pregnancy could be readily detected, not only in blood plasma and tissue fluids, but also in milk, faeces and even saliva. Non-pregnancy in the cow can be routinely detected with almost 100% accuracy by way of the progesterone assay; the high degree of certainty associated with the diagnosis of non-pregnancy is regarded as the most valuable feature of the test. The herdsman can confidently take appropriate action towards dairy cattle that have clearly not conceived and at a much earlier stage than previously possible.

RIA and EIA techniques

Until the development of enzyme-linked immunosorbent assays (ELISAs), progesterone determination was based on RIA and for that reason confined to suitably equipped laboratories; ELISA test kits removed many of the constraints imposed by RIA. Kits for on-farm pregnancy and oestrus confirmation came on the market in the mid-1980s; evaluation of results was usually based on either a colour or an agglutination reaction, which was compared with a known standard.

Progesterone and embryo survival

The progesterone-based pregnancy test can only be applied at times when a non-pregnant animal can be expected to have a low progesterone level; where cows are bred by natural service without observing heats, the test is of little avail. Progesterone in the cow's blood and milk is derived from the cells of the corpus luteum and not from the products of conception; for that reason, luteal secretion in the absence of a viable embryo can lead, on occasions, to a false diagnosis of pregnancy. The cow's CL can continue secreting for some time beyond the day at which embryo death occurs. Studies carried out in the UK have shown that low post-ovulatory systemic progesterone or a delay in the normal increase in progesterone concentration during the early luteal phase is associated with reduced embryo survival rate in dairy herds. In Ireland, Stronge et al. (2003) have also demonstrated an association between milk progesterone concentration on day 5 after breeding and embryo survival rate; low concentrations of progesterone on day 5 were associated with a low probability of embryo survival.

Buffaloes

Progesterone concentrations measured by RIA in buffaloes have given an average progesterone concentration at oestrus of 0.8 ng/ml, increasing to 8.5 ng/ml 24 days later in pregnant animals; it is also reported that after day 16 the concentration decreases in non-pregnant animals but in pregnant buffaloes it continued to increase until day 22. Elsewhere, other studies with progesterone levels in buffaloes have indicated that a proportion of animals submitted for AI may have a progesterone concentration too high for conception to occur; such evidence supports the view about difficulties in accurately detecting oestrus by observation in this species.

Pigs

Many reports have appeared on pregnancy testing in sows using an on-farm blood progesterone assay. In Taiwan, workers have developed a test in which results can be read within 30 min by spectrophotometry or the naked eye, using heparinized fresh blood samples collected from the ear vein of sows 17–22 days after breeding. The accuracy of diagnosis with this enzyme immunoassay was 93.1% for a positive diagnosis and 83.3% for non-pregnancy. In Japan, workers compared the accuracy of early pregnancy diagnosis in the sow by saliva and plasma progesterone measurements; they concluded that the accuracy of saliva progesterone measurements was little different from that of the blood test.

Sheep

Numerous studies on early pregnancy diagnosis in sheep by RIA for plasma progesterone have been reported. In Turkey, the accuracy of the test was given as 82.6% for pregnant ewes and 100% for non-pregnant sheep. Some studies suggest that plasma progesterone levels can be used to predict the number of fetuses in Awassi sheep during the second half of pregnancy; in the USA, workers were able to estimate the number of fetuses with 88% accuracy in ewe lambs and 74% accuracy in ewes, using a single blood progesterone measurement at 100 days of gestation. In Spain, reports showed that from weeks 12 to 20 of gestation, ewes carrying twins had significantly higher progesterone levels than those carrying a single lamb.

Goats

In goats, serum progesterone levels reported in Indonesia were significantly higher in twin- than in single-bearing goats in animals examined in the final 2 months of pregnancy; similar evidence was reported for serum oestradiol. A high plasma progesterone level was found during pregnancy in goats in Zimbabwe; the level reaching a peak prior to parturition and reducing to basal levels at kidding. In Brazil, working with dairy goats, Azevedo *et al.* (2002) concluded that pregnancy diagnosis based on serum oestrone sulphate concentrations can be successfully made starting on the 15th day after breeding (Table 9.2).

Oestrogens in cattle and buffaloes

It was realized at an early stage that the detection of substances produced by the conceptus would have a particular value for the early positive confirmation of pregnancy; studies from the 1970s onwards demonstrated that oestrone sulphate in blood or milk can provide such confirmation. However, it is not until beyond day 100 of gestation that the oestrone sulphate test can be reliably used to diagnose pregnancy in the cow. In the buffalo, increasing concentrations of oestrone sulphate, beginning at the fourth month of pregnancy, have been recorded; such evidence provided the basis for pregnancy confirmation by milk oestrone sulphate after 110 days of gestation in this species.

Camelids

In a study of hormonal indicators of pregnancy in llamas and alpacas, it was found that oestrone sulphate concentrations peaked 21 days after breeding and again during the last month of pregnancy; the first increase in oestrone sulphate concentrations over basal values was thought to indicate early interaction between mother and embryo, whereas the second increase was thought to reflect fetal viability. It was concluded that the use of oestrone sulphate concentrations to diagnose pregnancy in llamas and alpacas is highly dependent on time of sampling. Other studies on steroids in late pregnancy in llamas and alpacas showed plasma concentrations of oestrone sulphate starting to increase 80 days before parturition, reaching peak concentrations just before parturition in both species. Serum progesterone and oestradiol-17β levels during pregnancy in the Bactrian camel were measured by RIA by Chinese workers; serum progesterone concentrations increased from 15 days after breeding and remained elevated throughout most of the 13-month gestation period. The same study showed oestradiol increasing

Table 9.2. Pregnancy diagnosis in goats by oestrone sulphate assay (from Azevedo *et al.*, 2002).

Goat breed and no. of animals	Days after breeding by AI		
	15	20	25
Alpine Brown (11)	0.90 ng/ml	0.39 ng/ml	0.43 ng/ml
Toggenburg (13)	0.007 ng/ml	0.07 ng/ml	0.07 ng/ml
Half-bred (16)	0.03 ng/ml	0.04 ng/ml	0.04 ng/ml

significantly from the 11th month of pregnancy and peaking at 11.5 months.

9.3.3. Faecal testing

Oestrogen and progesterone metabolites

Faecal oestrogen and progesterone metabolite evaluations are now well-established procedures for monitoring reproductive function in several mammalian species, including farm animals. The route of excretion of steroid hormone metabolites varies considerably among species and also between steroids within the same species. Steroid concentrations in faeces exhibit a similar patteren to those in plasma, but have a lag time varying from 12 h to > 2 days, depending on the species. Faecal steroid metabolites in mammals are mainly unconjugated compounds; faecal oestrogens consist predominantly of oestrone and/or oestradiol-17α or -17β. Faecal oestrogen evaluations have been used as reliable indicators of pregnancy in several farm animals. There may be occasions on which it is desirable to know the pregnancy status of free-ranging sheep, such as Rocky Mountain Bighorns (*Ovis canadensis*) in North America. Workers in the USA developed an EIA for an immunoreactive progesterone metabolite (pregnanediol-3-glucuronide (iPdG)) in faeces to monitor the reproductive status of these animals; on the basis of this work, it was concluded that this EIA has potential to monitor pregnancy in these animals.

Progesterone assay for pig faecal samples

In Japan, workers used a cattle milk progesterone qualitative test on pig faecal samples; the accuracy of pregnancy at 21–25 days after last mating was 97.6% for positive cases and 100% for negative ones. Further work in the same laboratory to determine the best period for pregnancy diagnosis concluded that 20 to 25 days after mating was the optimum time for pregnancy diagnosis by this technique. The Japanese workers noted that applying the

commercial cow-milk testing kit took no more than 10 min and was easy to carry out.

Faecal oestrogen levels in pigs

It has been known for some time that the pig blastocyst synthesizes substantial amounts of oestrogens, mainly oestrone and oestradiol-17β, which pass through the uterine wall, where they are sulphoconjugated. These oestrogen metabolites in maternal blood and urine or the deconjugated oestrogens in the faeces between days 23 and 30 after mating are good indicators of the presence of live embryos. The estimation of oestrogens is therefore a useful method for early pregnancy diagnosis in sows; there is also the practical point that faeces are more easily collected than blood or urine. In The Netherlands, workers reported an ELISA test for faecal oestrone in pigs using monoclonal antibodies against 6-keto-oestrone-carboxymethyloxime conjugated to bovine serum albumin; results showed that faecal oestrone levels were significantly higher during days 23–35 of pregnancy compared with the levels during the oestrous cycle.

Sow-side test

Hungarian workers evaluated the accuracy of ultrasonographic pregnancy diagnosis in pigs by measuring the concentration of unconjugated oestrogens in faeces; practically all sows that had a false-negative diagnosis by ultrasonography on day 26 of gestation were correctly diagnosed pregnant by unconjugated oestrogen measurement (> 11.7 ng/g faeces indicative of pregnancy). In Holland highly significant differences were recorded between faecal oestrone concentrations in pregnant and non-pregnant sows; the distribution of these concentrations around the cut-off value appeared to offer good prospects for the development of a simple, sow-side pregnancy test. It was concluded that the ELISA faecal oestrone test had a sensitivity (correct diagnosis of pregnancy) of 96.5% and a specificity (correct diagnosis of non-pregnancy) of 93.6%, using 3.65 ng oestrone/g faeces as a cut-off value; although an increase in faecal concentrations was observed for increasing litter sizes, it was concluded that these levels could not reliably predict litter size.

Horses

In New Zealand, faecal samples from pregnant and non-pregnant Thoroughbred mares were assayed by workers using an antigen-coated EIA; non-pregnant mares had an oestrone sulphate concentration of 23 ± 09 ng/g, whereas pregnant animals showed levels rising to 122 ± 9 ng/g in samples collected 161–180 days postmating. After foaling or fetal death, faecal oestrone sulphate concentrations returned to non-pregnant levels within a few days.

Deer

In Japan, sika deer (*Cervus nippon*) are sometimes regarded as pests, damaging planted forests and crops; in order to make plans for appropriate management of this species, it is important to collect basic information about the animal. Measurement of steroid hormone levels in the blood has been a popular technique for estimating reproductive status in many species, but blood sampling can obviously be difficult in wild species. In such circumstances, a non-invasive method for monitoring reproductive function may have particular attraction. Studies have been reported in which faecal steroid analysis has been used to gather information on the annual reproductive profiles of sika deer; it was concluded that the application of this method to sika deer made it possible to accumulate the information necessary for wildlife management.

9.3.4. Predicting litter size

Cattle and sheep

It has been shown, in the UK, Ireland and elsewhere, that the measurement of oestrone sulphate by EIA confirmed the progressive rise in oestrogen previously found using RIA (Fig. 9.7) and that such a test could be used to identify cows pregnant with twins after 120 days. Although such a hormonal test looked promising at one time and has its uses in research, it failed to measure up to commercial needs due to the overlap between single- and twin-bearing cows in their oestrone sulphate levels. Attempts to explain this include the suggestion that mammary gland tissue actively synthesizes oestrone sulphate and this may be a factor in masking genuine differences between twin- and single-carrying cows. Although studies have shown that oestrone sulphate levels in sheep vary according to the number of live fetuses, as in the cow, the considerable variation and overlap between individual ewes rule out its use in commercial practice.

Placental progesterone

Placental production of progesterone is known to increase markedly between days 70 and 100 of gestation in the ewe after the placenta has reached its maximum size; in the ewe, unlike the goat, because steroid level is related to litter size, it is possible to make some estimate of fetal number by

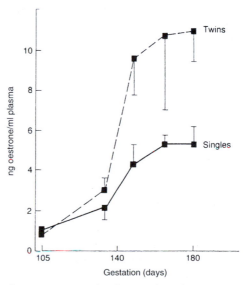

Fig. 9.7. Oestrone levels in single and twin-bearing cattle (from MacDonnell *et al.*, 1993).

quantitative progesterone determinations. However, attempts to classify litter size in prolific sheep in the UK, by measuring progesterone concentration between days 91 and 105 of gestation, only achieved a 65% success rate, too low to have any practical relevance.

Pigs

Several studies have dealt with the relationship of certain hormonal levels with litter size in pigs and, to a lesser extent, in identifying multiples in sheep and goats. In Taiwan, workers showed that the number of viable fetuses was positively correlated with plasma oestrone sulphate levels on days 22–24; the accuracies for estimating viable litter size (> 8 or < 8) based on oestrone sulphate levels on days 22 and 24 were 86.4 and 89.8%, respectively.

9.3.5. Use of ultrasonics

The advent of modern real-time scanning techniques permitted the visualization of the internal organs of the various farm mammals, including the ovaries and the uterus. In the mare, French workers were among those reporting encouraging results; visualization of the conceptus was possible from day 14 of pregnancy and a very accurate (> 92%) early pregnancy diagnosis shown to be possible. Comparisons between real-time ultrasonics and progesterone assays showed that diagnosis could be made earlier (day 14 vs. day 18) and there was no error arising from a persistent CL; the scanning technique was capable of being learned quickly and had several points in its favour compared with other methods. Danish workers also showed that the technique could be employed to determine the gender of the equine fetus, from day 64 of gestation.

Real-time ultrasound scanning for pregnancy diagnosis in large animals involves the use of a rectal probe, which transmits harmless ultrasound waves through the body tissues; waves are reflected to the transducer when they reach the fetus and are converted to produce an image on a display screen. Although real-time ultrasonics was first used in farm animals to detect early pregnancy in the horse, use of the method soon spread to other species, including the cow. Applications of this technology in sheep involves the transducer being applied externally to the ventral abdomen and the uterus scanned transcutaneously. Manufacturers initially marketed equipment with transducers of low frequency (i.e. 3.5 MHz); the lower the frequency, the greater the penetration of sounds and the greater the depth at which images could be obtained. However, this also meant poor resolution of detail in the image. High-frequency transducers were therefore developed (i.e. 7.5 MHz) to permit high resolution of the image obtained by their use *per rectum*. For research purposes, for example, the bovine conceptus may be visualized as early as 9–10 days and with some accuracy by day 12.

Need for rechecks

Some operators, while accepting that early pregnancy diagnosis by scanning is valuable, emphasize that a recheck at 60–70 days is essential to ensure that the conceptus has not been lost since the initial positive diagnosis. Scanning operators are usually capable of dealing with 60 cows an hour on a pregnancy detection basis; if an estimate of calving date (i.e. fetal age) is required, the rate would be about 40. A standard cattle crush is sufficient in most instances; with dairy cattle, the test can be carried out in the milking-parlour.

Twin-bearing cows

Of the several methods, hormonal and otherwise, tested for the detection of twin-bearing cows, the most useful has proved to be real-time ultrasonics applied between 50 and 60 days of gestation. For several practical reasons, it is highly desirable that farmers should be aware of cows carrying twins, particularly in any commercial exploitation of twinning technology in beef production systems. The sooner this can be carried out in pregnancy, the more useful it will be; in any event, it should always be prior to the cow entering the final 2 months of the gestation period. It is well established that the energy requirements of cows carrying twins are higher than for those with singles. In New Zealand, workers scanned twin-carrying cows by real-time ultrasonics at 50–70 days of pregnancy and examined them rectally at 111–156 days; prediction accuracy was highest when the scanning equipment provided the highest degree of visual detail and the operator was experienced in distinguishing fetal abnormalities; the large size of the uterus after about 80 days of gestation can make it difficult to determine the number of young by transrectal ultrasonics.

Buffaloes

The use of linear array, real-time, B-mode ultrasonics in the detection of early pregnancy in the buffalo has been described by several workers. Monitoring the growth and anatomical features of the early buffalo conceptus showed growth of the embryo to be slower than in the cow. It is believed that this is due to the length of gestation in buffaloes, which is substantially longer (305–320 days) than in taurine cattle. In view of the fact that early detection of pregnancy is important in buffaloes bred by AI and because of the difficulty in detecting returns to service by observation, a non-invasive method of early pregnancy diagnosis by scanning could be very useful. Unfortunately, the cost factor in operating such a service may prove prohibitive.

Pigs

Real-time ultrasonics has gained much attention in the pig industry for its ability to diagnose early pregnancy. Ultrasonics can also be employed as a non-invasive method of assessing the reproductive status of the reproductive tract (Mejia Silva *et al.*, 2001). In many different production settings, external transabdominal real-time ultrasonics is commonly employed to determine the pregnancy status of pigs between the third and fourth weeks after breeding. An example of the accuracy of the method is shown in Table 9.3, taken from French studies in the mid-1980s; they showed that diagnosis could be made from 20 days after breeding if the sow was tied up during examination and 2 days later if the animal was restrained in a retention box. Real-time ultrasonics allows examination of ovaries and reproductive tract in a way not possible using either A mode ultrasonics or Doppler equipment. Results of work in pregnancy testing in pigs by ultrasonics reported from Germany dealt with scanning of the uterus in 4086 gilts and 12,506 sows 20 to 110 days after mating or insemination; pregnancy was correctly identified in 99.5% of pigs at 20–23 days of gestation, rising to 100% at 26–32 days. Absence of pregnancy was confirmed in 97% of non-pregnant pigs at 20–28 days.

Transrectal ultrasonics in the pig can permit earlier and more accurate pregnancy detection (Clark and Knox, 2000); the earliest pregnancy detection methods are based on the fact that during early pregnancy, between days 16 and 18 after mating, fluid

Table 9.3. Ultrasonics in pregnancy diagnosis in pigs (values are percentages with the no. of examinations in parentheses) (from Martinat-Botte *et al.*, 1985).

	Day of echography after insemination					
	18–19	20–21	22–23	24–25	> 26	Total
Closed building						
Tethered	90[b] (113)	93[b] (239)	96[a] (265)	95[a] (134)	96[a] (633)	95[a] (1384)
Retention box	80[c] (90)	87[c] (158)	94[a] (113)	92[c] (91)	93[c] (294)	90[c] (746)
Semi-outdoor building						
Retention box	67[c] (9)	80[c] (40)	94[a] (18)	100[a] (13)	96[a] (47)	89[c] (127)

a vs. b, *P* < 0.05; a vs. c, *P* < 0.001.

begins to accumulate inside the embryonic membranes within the uterus. This fluid from the early embryonic vesicle can be detected inside the well-defined uterine horns when using a 7.5 MHz transrectal probe; for pregnant animals, this is in marked contrast to the day 16–18 non-pregnant uterus, in which fluid is typically absent. Such early pregnancy diagnosis can be useful in breeding management. If females are found to be non-pregnant between days 18 and 21 after mating, they can be checked more intensely for oestrus and quickly rebred when they come into heat or culled if they fail to express oestrus and show no ovarian activity. In Thailand, Kaeoket (2003) described the uses and benefits of real-time B-mode ultrasonics for pregnancy diagnosis; the reliability of diagnosis was dependent on the experience of the operator and the type of ultrasound apparatus employed; the author concluded that the availability of an inexpensive, lightweight, portable ultrasound machine will prove valuable in making management decisions based on accurate pregnancy diagnosis of gilts and sows.

Sheep

B-mode ultrasonography is an accurate, rapid and safe technique for diagnosing pregnancy in small ruminants; trans-abdominal or transrectal approaches can be used with considerable accuracy. One important advantage of the transrectal approach in sheep and goats is that it does not require a special transducer; a regular linear transducer can be used, just as for examination of larger animals. In Germany, workers used real-time ultrasonics on a single occasion between days 13 and 69 in 1159 merino ewes in a standing position using a 5.0 MHz probe; the highest predic-tion accuracy (89.1%) was in ewes scanned between days 35 and 46; in a second group of ewes scanned in dorsal recumbency with a 7.5 MHz probe between days 15 and 64, the prediction accuracy was at least 80% from day 29. The German work-ers also recorded that prediction accuracy decreased as litter size increased and that

dead fetuses were detectable by the absence of a heartbeat. The effectiveness of trans-rectal ultrasonics in sheep was also exam-ined in Spain; workers concluded that pregnancy diagnosis could be reliably made from day 23 of pregnancy, with an accuracy of ~98%.

Goats

In goats, accurate information on the stage of gestation would be useful in drying off lactating does at the appropriate time and in monitoring goats near term. In Canada, workers were able to diagnose gestational age in 96% of animals with a margin of ±14 days on the basis of measuring placentomes by ultrasonics. In Argentina, workers exam-ined goats using real-time B-mode ultra-sonography from day 13 to day 40; they recorded the average age at which they first detected heartbeats in at least one embryo as 21 days (range 19–23). It has been evident from the early 1980s that real-time ultrasonic scanning is a very useful method for use with goats and is one of the means of identifying the false pregnancies that can be troublesome in this species. Spanish workers in the 1990s, using a transrectal 7.5 MHz ultrasonic probe, recorded 100% accuracy in detecting pregnancy at day 24 and 100% accuracy in detecting multiples at day 30.

Deer

Traditional methods of pregnancy diagno-sis, such as rectal palpation or progesterone testing, are generally not options for use in deer. New Zealand workers have presented data on the ultrasonography of the non-pregnant reproductive tract and of the very early stages of pregnancy.

Camelids

The use of ultrasound in early pregnancy diagnosis in New World camelids was studied by workers in Chile, who showed that diagnosis could be carried out as early as 9 and 7 days after mating in alpacas and llamas, respectively.

Gender determination

Ultrasound has been employed in horses to determine the gender of equine fetuses. A paper in Denmark in the late 1980s reported that, from the 64th day of gestation, it was possible to determine the sex of the fetus; the ultrasonic technique was described as suitable for routine use. A 3.5 MHz linear-array transducer was employed in ultrasonic scanning studies aimed at sex diagnosis reported by workers in the USA. It was concluded that gender diagnosis was readily accomplished and most accurate (90%) when carried out during the fifth and sixth months of pregnancy; after that time, the areas that could be used for diagnosis were usually inaccessible. In Germany, in the late 1990s, workers examined Thoroughbred mares between 50 and 70 days of pregnancy; diagnosis of sex was possible in 51% of mares and was correct in 85% of these. Male fetuses were found to be easier to diagnose due to a larger genital tubercle.

Fetal sex determination has been achieved in sheep, using ultrasonics to locate the genital tubercle in ewes; using a real-time diagnostic scanner equipped with a linear-array 5 Mhz transducer between days 60 and 69, workers reported 100% accuracy with male and 76% with female fetuses. According to Lamb and Dahlen (2002), fetal sexing is fast becoming a common management tool in North American beef cattle enterprises with an accuracy of sex determination exceeding 97%; knowing whether the beef cow carries a bull or heifer calf obviously enables the farmer to manage its feed and management accordingly.

9.3.6. Early dipstick tests

In view of the practical problems associated with progesterone assays (relatively low accuracy of early positive diagnosis), researchers have sought a specific method for detecting pregnancy by assaying proteins secreted by placental membranes and detectable in the maternal circulation, similar to that available for many years in humans in the form of the human chorionic gonadotrophin (hCG) assay. Most of the embryonic substances found in the maternal circulation of the cow in early pregnancy are produced by binucleate trophoblast cells; these binucleate cells constitute about 10% of the trophoblast cells on days 18 and 19 of gestation and subsequently about 20% until close to calving. The cells first appear in the uterine epithelium at about day 19, where they apparently fuse with uterine epithelial cells; it is believed that the function of these binucleate cells is to facilitate the transfer of complex molecules across the embryo–maternal junction into the maternal circulation.

Pregnancy-specific protein B (PSPB)

Pregnancy-specific protein B (PSPB), a family of five glycoproteins secreted by trophoblast cells, has been characterized. A pregnancy test based on PSPB has been suggested as suitable for application 30 days after breeding; however, the slow disappearance of PSPB from the circulation of cows after calving is a limitation to this method. A further pregnancy serum protein (PSP60) has been tested as a possible diagnostic molecule in cattle; it was concluded that, for the farmer, the PSP60 test was reliable, easier to use than rectal palpation and less expensive than ultrasonics, but involved a few days' delay before results came back from the laboratory. Early pregnancy factor (EPF) is one of the pregnancy-associated proteins and has been detected in the serum of many pregnant animals shortly after fertilization. It is believed that the bovine uterus becomes more immunotolerant because of the action of EPF. Based on the presence of EPF in the serum of pregnant ewes, pregnancy can be detected as early as 24 h after fertilization by a rosette inhibition test; unfortunately, as in cattle, the test is too complex to be applied in the field.

Pregnancy-associated glycoproteins

The purification and characterization of a bovine pregnancy-associated glycoprotein (bPAG) from fetal cotyledons led to an assay which could be applied from 42 days of gestation onwards in cattle. Pregnancy-associated glyoproteins (PAGs) and PSPB both belong to the aspartic proteinase family and are secreted by the trophoblastic binucleate cells. These proteins are detectable in the maternal blood around the time of attachment of the fetal placenta, when the trophoblastic binucleate cells start to migrate and fuse to the endometrial cells, forming the fetomaternal syncytium; the two glycoproteins are considered to be good indicators of pregnancy and fetoplacental well-being. Using heterologous RIAs, sheep PAG and sheep PSPB can be detected in the blood of pregnant ewes around day 20 after mating. Levels of PAG were determined by RIA in a study in Brazil. Concentrations of this glycoprotein were significantly affected by week of pregnancy and number of fetuses; after parturition, PAG levels decreased rapidly. It is known that, throughout pregnancy, sheep PAG concentration varies according to the breed of the ewe and the sex and number of fetuses; after lambing, PAG and PSPB concentrations decrease rapidly, reaching their basal level 2–4 weeks after parturition.

A report by workers in the late 1990s dealt with an improved RIA for the detection of PAG in the blood of the goat, which can discriminate between pregnant and non-pregnant does as early as day 21 after breeding; the value of the test lay not only in its ability to diagnose early pregnancy but also in its ability to discern between goats with extended inter-oestrus intervals and those that are pregnant, which cannot be determined either by oestrus behaviour or progesterone assay. In a comparison of various diagnostic techniques in dairy goats, Rodriguez et al. (2003) reported that transrectal ultrasonic scanning and the determination of PAG concentrations provide a very accurate pregnancy diagnosis at 26 and 24 days after breeding, respectively; progesterone assays performed on day 22 after breeding, on the other hand, while accurate in detecting pregnant animals, were not accurate for detecting non-pregnant animals.

Working with sheep in Hungary, Karen et al. (2003) evaluated the accuracy of a PAG–RIA test for early pregnancy and showed that it was a reliable method for using from day 22 after breeding and that one advantage of PAG assay over a progesterone assay was in its ability to differentiate between pregnancy and prolonged inter-oestrus intervals.

Some workers have suggested that PAG levels could be an early indicator of trophoblast distress and prove useful in detecting individuals that are at high risk of pregnancy failure; sequential measurement of PAG can detect the onset of disturbance of trophoblastic activity associated with the death of a fetus. In one such study, PAG concentrations were determined weekly by workers in Belgium from 20 days of gestation; it was found that PAG fell under the positive pregnancy threshold when the trophoblast died. In other studies, in goats carrying two or three fetuses, PAG profiles have shown a marked fall in concentration, which could indicate placental distress.

9.3.7. Other approaches

Radiography

Although radiography was used to determine fetal numbers in sheep many years ago, this was usually only in small-scale research programmes. In the 1980s, New Zealand workers reported multiple pregnancy diagnosis in 4700 sheep in 21 flocks using a real-time ultrasonic body scanner or X-rays (videofluoroscopy) and checked the accuracy of the two methods at lambing time; the percentage of ewes accurately diagnosed for lamb numbers was 96.1–100% (19 flocks) for ultrasonics and 94.3–99.6% (13 flocks) for X-rays. However, the cost of the X-ray system was about 12 times that of ultrasonics; clearly, although technically on a par, the ultrasonic scanner was by far the winner. For sheep farmers looking for an accurate answer on single- and multiple-bearing ewes, the scanner was to prove very successful; greatest accuracy was achieved in sheep tested on days 30–60 of gestation.

Detecting PMSG (eCG)

There may be occasions, with mares that are either too fractious or too small for pregnancy diagnosis by rectal palpation or ultrasonics, when a blood test may be useful. One way in which a positive diagnosis is possible is by detecting PMSG in the mare's blood; this test does entail, however, the mare reaching at least 40 days after conception. The gonadotrophin PMSG is secreted by the endometrial cups and is present in the blood between days 40 and 130 of pregnancy; although early methods to detect PMSG involved the use of biological assays, these were replaced some time ago by extremely accurate immunological tests. Several groups have reported on pregnancy diagnosis using a PMSG–latex test in mares. In Turkey, workers reported using Rapi Tex, a commercial latex agglutination test for PMSG, which gave results within 5 min; its accuracy was 85, 94 and 100% at 40–45, 46–55 and 60–65 days of pregnancy, respectively and 100% in non-pregnant mares. In Germany, other workers used a monoclonal antibody assay to distinguish between concentrations of PMSG in mares; pregnancy could be diagnosed from day 37 of pregnancy, using the assay on a single blood sample.

10

Controlling Parturition

10.1. Advantages of Control Measures

There are many reasons why researchers and farmers alike are interested in the control of parturition. It may be a matter of providing more care to the newborn by having them born at a predictable time when the care and attention necessary for their survival are at hand. On occasion, there may be the need to advance or delay the onset of parturition. Advancing parturition in a cow which otherwise might have a difficult calving could be a means of reducing stress in both mother and calf. Postponing calving for a few hours might enable the cow to be shifted safely to a more suitable location. In some situations, a valuable cow may find itself bred to the wrong bull and there may be need for recourse to a pregnancy termination treatment. In following any or all of these possibilities, the treatment must be based on a reasonable understanding of the physiology and endocrinology of pregnancy and parturition. In that context, one name that immediately springs to mind is that of Mont Liggins, a fetal physiologist in Auckland, New Zealand, who did so much in the 1960s to elucidate factors that play a part in triggering birth in sheep and other animals.

10.1.1. Pigs

The induction of farrowing within a reproductive management framework already plays a considerable routine role in the husbandry of the sow herd in many countries. The application of methods designed to synchronize the commencement of the birth process and the management of farrowing can be traced back to the 1970s. With the availability of prostaglandins (PGs) at that time, various treatment protocols for synchronized farrowings were developed. Controlled farrowing would enable all piglets to be born during the normal working day or at other times, including the night, which might be considered most appropriate to ensure the survival of as many young as possible; ample evidence exists to show that the majority of stillbirths in pigs could be prevented if a skilled stockperson could be in attendance at time of the farrowing (Fig. 10.1).

10.1.2. Cattle and sheep

Calving is a major event in the life of the cow and any attempt to control or manipulate the process artificially should only be

Fig. 10.1. Early moments in a piglet's life. Piglet survival rate depends on sound management systems and attention to detail. It is essential for the newborn piglet to start sucking the mother at the earliest opportunity – many of the deaths that occur in the very young arise because of a failure to suck in the early hours of life.

undertaken after careful consideration of the possible consequences, both short- and long-term. From the farmer's viewpoint, there are two main reasons why he might consider the need to control calving. One is as a management tool in preventing what would otherwise be late calvings in a herd where he is attempting to synchronize the grazing season with the main bulk of calvings and the onset of milk production. The other reason is to help to ease calvings and ensure the viability of the calf, either because of the particular value of the newborn pedigree animal or as a means of minimizing the general level of perinatal mortality. In this context, induced calving techniques are seen as a means of avoiding dystocia by shortening the duration of gestation and thereby reducing the size and weight of the calf. One problem with this may be that of deciding the exact pregnancy stage, for accurate conception dates may not always be available for beef cows and normal gestation periods can be variable. In Ireland, interest in controlled calving has usually been directed towards avoiding difficult calvings by developing a protocol suitable for beef heifers bred to Continental bulls (e.g. Limousin, Charolais and Simmental) as part of the beef production systems which sought to exploit the growth rate and lean meat advantages of the Continental breeds.

Ultra-compact lambings

In the context of making the most efficient use of labour on the farm, ultra-compact synchronized lambings may be of value in some conditions; it could mean that oestrus-synchronized ewes, which would normally lamb down within the space of 7–10 days, could be further aligned to give birth within a 48 h period, with most lambs being born during daytime. There might be some unusual applications, such as in Karakul sheep, where high-quality pelts are a primary consideration; in such flocks, induction of parturition has been used as an alternative to Caesarean section.

10.1.3. Horses

The control of parturition in the mare is primarily one of ensuring that sufficient care and attention are at hand when the foal is born. According to reports in the literature, induction of parturition is a common procedure with valuable or problem mares and has become an established, accepted practice on many breeding farms. Because most mares give birth at night, the elective induction of parturition during the day can help to ensure that professional help is present or available.

10.2. Development of Parturition Control Measures

Methods now in use to control parturition have been developed as a result of probing the mysteries surrounding birth, some of it in humans and much of it using the sheep as the model. The onset of parturition appears to be very much a matter of the activity of the fetal hypothalamus and the events that flow from such activity.

10.2.1. Physiology and endocrinology of late pregnancy and parturition

Although it may appear that there is considerable diversity in the way that the endocrine systems of the various farm mammals operate during pregnancy, closer inspection reveals several fundamental characteristics that have been retained by most species. One important characteristic common to all is the requirement to maintain an elevated concentration of progesterone throughout much of pregnancy; the production of oestrogen during late pregnancy also appears to be a feature common to all. In the various farm mammals, with the possible exception of the horse, there is the need for the production of progesterone to be maintained at or above the level normally found during the luteal phase of the oestrous cycle throughout pregnancy. Progesterone acts to prevent resumption of cyclicity, prepare the uterus for attachment of the embryo and maintain myometrial quiescence.

Indonesian workers examined correlations between lamb birth weights and the concentrations of various hormones in the maternal circulation during pregnancy. Progesterone and oestradiol concentrations in the maternal circulation during week 8 of pregnancy in sheep were found to have the highest correlation with lamb birth weight; the higher the progesterone level, the heavier the birth weight. Studies of oestradiol profiles in sheep in Brazil during pregnancy showed concentrations of this hormone to be low at the start of gestation but markedly increased in the days running up to parturition; the same study found that oestradiol secretion showed 24 h (circadian) rhythms. It is believed that placental factors contribute to the inhibition of follicular growth during late pregnancy in the ewe.

Differing sites of progesterone production

In most farm species, including ruminants and pigs, plasma concentrations of progesterone rise during early pregnancy, are maintained at elevated levels, and then fall prior to the onset of labour; an exception is the mare, which is able to maintain pregnancy with very low levels of progesterone. There is a major difference in the site of progesterone production between those farm animals (e.g. cow, goat and pig) in which the main source of progesterone is the corpus luteum (CL) and those species (e.g. sheep) in which the placenta is the major source of this steroid. As well as progesterone, several peptides/factors are believed to contribute to the maintenance of myometrial quiescence, including relaxin (a hormone also important in the preparation of the birth canal for the passage of the fetus), nitric oxide, prostacyclin and catecholamines.

The cow's CL is believed to be the only significant source of progesterone during pregnancy, and it usually stops secreting progesterone some 30–40 h before parturition. The uterine musculature and birth canal are prepared for calving by the rising concentration of oestrogen, which is believed to increase uterine contractility and enhance responsiveness to oxytocin and PGs. There is also an increased release of β-endorphin into the cow's peripheral circulation.

Increased oestrogen levels

During late pregnancy in the cow (Fig. 10.2), several hormones are involved in maintaining and ensuring a successful live birth (see Kindahl *et al.*, 2002). These hormones include progesterone, which is secreted in high concentrations during the whole pregnancy period by the CL and, to some extent, by the maternal adrenals and the placenta. A second steroid, oestrone sulphate, shows elevated levels from about mid-pregnancy until the third stage of parturition (expulsion of fetal membranes). For the onset of normal parturition and the calving process itself, a change from progesterone to oestrone synthesis is crucial. Observations on hormonal events during the second half of pregnancy by Hoffmann and Schuler (2002) led them to suggest that placental oestrogens and progesterone are important factors controlling caruncular growth, differentiation and function.

Changing hormone balance

In most farm mammals, the onset of labour is associated with a change from a state of progesterone domination to one in which oestrogen exerts the strongest influence. This is achieved either by conversion of progesterone to oestrogen (as in cattle, sheep, goats and pigs) or conversion of progesterone to inactive metabolites (the horse). The resultant oestrogen : progesterone ratio plays an important role in the increased synthesis and/or release of uterotonins, activation of the myometrium and ripening of the cervix. Among the stimulatory uterotonins, PGs, particularly PGF, play a central role in the stimulation of myometrial contractions during labour in most species. These PGs may act to induce luteolysis (in species such as pigs, cattle and goats, which are CL-dependent) or directly stimulate myometrial contractility, or both. In recent years, it has been shown

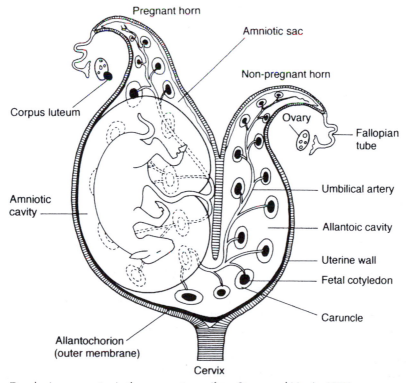

Fig. 10.2. Developing conceptus in the pregnant cow (from Steven and Morris, 1975).

that the increase in PG synthesis during labour is primarily due to an increase in the functional expression of the inducible form of cyclo-oxygenase (the *COX-II* gene product) in uterine and intrauterine tissues. The role of oxytocin has also been clarified, with the recognition that expression of this peptide is increased in intrauterine tissues during late pregnancy, providing a local source of this hormone.

Role of oxytocin

Oxytocin is the most potent endogenous uterine-stimulating hormone; this peptide, or a close analogue of it, is present in all mammals that have been studied so far. Release of oxytocin during parturition has been demonstrated in cattle and several other mammals, but diverging opinions remain about the importance of oxytocin in the mechanism of parturition. Oxytocin receptors have been found in the myometrium and endometrium of all species studied; in most mammals, uterine oxytocin receptors increase markedly at term, supporting the view that oxytocin plays a physiological role in the process of normal parturition.

In cattle, receptor concentrations in the endometrium increase markedly at term, whereas those of the myometrium greatly increase as early as mid-term and change little thereafter. It is known from various studies that, during pregnancy in the cow, oxytocin is secreted intermittently, the amplitude and frequency of oxytocin spurts remaining low until about 2 weeks before parturition. The rate of oxytocin secretion then increases stepwise rather than gradually, suggesting that ovarian hormones may regulate its release. There may also be a neural reflex arising from the pressure of the fetal head against the cervix.

Opioids and oxytocin release

The increase in oxytocin secretion 2 weeks before calving also suggests that, in addition to its excitatory effects on contractions at labour, it participates in the preparation of the tract for calving. Although the posterior pituitary is the major source of oxytocin

during parturition, a minor part of the total hormone derives from a utero-ovarian source, presumably the CL. Endogenous opioids may also be involved in regulating the release of oxytocin at the onset of labour. Oxytocin-producing cells in the hypothalamus and the noradrenergic inputs to oxytocin neurones are known to be extremely sensitive to inhibition by endogenous and exogenous opioid peptides; withdrawal of exogenous opioid inhibition may be necessary for neural stimuli from the reproductive tract (e.g. head against cervix) to act shortly before the onset of labour. It is well known that there is a period of characteristic restlessness in cows shortly prior to labour, which may well result from such withdrawal of inhibitory opioids.

Onset of lactation

One of the important events accompanying parturition is the onset of lactation. In pigs, for example, the periparturient decline in plasma progesterone concentration provides the trigger for lactogenesis in that species; it is believed that progesterone blocks lactogenesis in part by suppressing up-regulation of prolactin receptors. There are also hormonally induced changes in behavioural patterns to provide the care necessary for the welfare of the newborn piglets. Periparturient sows allowed freedom of movement in their pens will show intense physical activity connected with nest building occurring about 12 h prior to the start of farrowing (Klocek *et al.*, 2000); the average duration of parturition was 219 min. Sows were observed to lie down very carefully to nurse the piglets; this behaviour favoured the protection and safety of piglets and resulted in a low mortality rate (7.2%).

10.2.2. Nutritional influences

Nutrition is an important factor affecting growth of the fetus during pregnancy and many reports have dealt with it. Workers in Scotland, particularly John Robinson, have reviewed research into the nutritional effects on fetal growth, its implications on developmental biology and the impact on animal production; much of the work covered relates to sheep, but studies on cattle and pigs are also included.

10.2.3. Factors influencing gestation length

If the process of parturition is to be controlled, then it is necessary to have accurate information on factors that may influence the length of gestation in the different farm species. Some of these factors in cattle and other farm animals are outlined in Fig. 10.3. The normal duration of gestation in the cow is 9 months; the existence of statistically significant differences between cattle breeds in the duration of gestation is well documented. In general terms, it is possible to group the gestation periods of dairy cows into those considered normal and those regarded as abnormal. New Zealand workers, for example, designated Holstein–Friesian cows showing gestation periods of less than 251 days as possible abortions, 251–271 days as possible premature births, 272–293 as normal gestation periods and 294 days and beyond as possible prolonged gestations.

Reducing gestation length

There are those who advocate a reduction in the gestation length of cattle through selective breeding. In the USA, for example, it was calculated that mean gestation lengths could be decreased by 10 days in three generations of dairy cattle if 5% of males and 50% of the heifer calves resulting from the shorter gestation periods were retained as breeding stock. The attraction would be that such cattle could be kept on an annual breeding cycle more easily in view of a longer time span between calving and breeding.

Effect of male

Although many reports on the duration of pregnancy in cattle refer to cows bred to bulls of the same breed, it is well recognized that a bull can confer a characteristic gestation length on the cattle with which he is mated. There is ample evidence to show that Holstein–Friesian cows in calf to a Charolais or Simmental bull can be expected to show an average gestation period several days longer than if pregnant to a Hereford or Aberdeen–Angus bull.

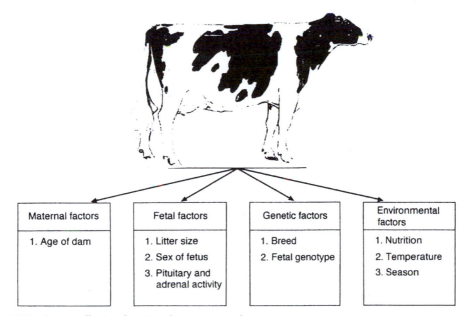

Fig. 10.3. Factors affecting duration of pregnancy in the cow.

Although less noticeable, the same principle would apply in sheep and goats. As well as breed of sire and dam, factors such as sex of the calf play a minor but significant role in influencing gestation length, bull calves being carried a day or more, on average, beyond that of heifer calves. One factor which can influence pregnancy duration in cattle is the birth of twin calves, which can be expected to arrive a week earlier than a singleton.

Gestation length in the mare

The duration of pregnancy in mares has been recorded by many authors. The literature reveals large variations in the gestation periods of horses, with differences usually much greater than in cattle, sheep and pigs; ten factors that influence gestation length are listed in Table 10.1. The average length of pregnancy has been quoted as 335–343 days. In Ireland, workers reported an average of 340.7 days in Thoroughbred mares; in the USA, studies have shown gestation periods varying from 327 to 357 with an average duration of 341 days. In the mare, it

is generally accepted that 'full term' begins at 320 days and foals born before this time would usually be designated as premature; foals delivered by Caesarean section before 310–315 days may be difficult to maintain, even in intensive care.

Gender and season

A paper by Karadjov (2001) found sex of progeny and month of parturition to be significant factors influencing gestation length in mares; pregnancy was longer in mares carrying male progeny and in spring parturitions. Data from 433 Thoroughbred pregnancies in the UK over a 10-year period were provided by Morel et al. (2002), the information being obtained by sequential ultrasonic scanning to enable the true gestation length (fertilization–parturition) to be ascertained. The overall gestation length was 344.1 ± 0.49 days (range 315–388 days), with colts being carried significantly longer than fillies (346.2 vs. 342.4 days). Month of birth had a significant effect, with foals born in January having the shortest and those born in April having the longest gestation lengths; mares carrying colts and foaling in April are likely to have the longest gestation lengths.

Litter size in sheep

The gestation period in sheep is known to be affected by many factors. There is a clear decrease in gestation length with increasing litter size (Table 10.2).

Mutton breeds of sheep have a gestation period which is usually several days shorter than that found with fine-wool breeds, such as the merino; even within the merino, some strains may have periods that are significantly different from others. Studies in India

Table 10.1. Ten factors influencing pregnancy duration in the mare (from Vandeplassche, 1986).

1. Breed: ± 1.5 days
2. Sex of foals: colts 2 days longer than fillies
3. Age of mare: no influence
4. Stallion
5. Dam
6. Month of conception: spring longer than autumn
7. Daylight: over 80% of mares foal by night (between 6 p.m. and 6 a.m.)
8. Severe undernutrition: in second half of gestation, adds 4–10 days
9. Overfeeding in second half of gestation: adds a few days
10. Stress: delays foaling by 1–2 days

Table 10.2. Litter size and gestation length in the ewe.

	Single	Twin	Triplet	Quadruplet
No. of ewes	178	164	31	10
Gestation length (days)				
Range	138–153	142–152	142–152	140–150
Mean	147.5	146.8	146.7	144.8

recorded the gestation period in three breeds and eight crosses of fine-wool sheep; values averaged 148.6 days in Stavropol sheep, 149.1 in Grozny and 150.1 days in Nilagiri, the period in Stavropol being significantly shorter than in other breed types. The same work recorded significantly shorter gestation periods in twin- than in single-lamb pregnancies (149.0 vs. 149.9).

The gestation period is usually shorter in twin- than in single-lamb pregnancies and shorter still in triplet and quadruplet pregnancies. There are many studies which support the concept that the length of gestation, in the sheep, is determined by the fetus; as well as that, it is also clear that the sheep fetus can influence both the length and ease of parturition. As well as litter size influencing gestation length, it also has a marked effect on the birth weight of lambs (Fig. 10.4).

Embryo transfer techniques were used in Scotland to show that the increased gestation length and birth difficulty of many Suffolk lambs contributed to the lower activity levels and teat-seeking of these lambs immediately after birth in comparison with Scottish Blackface lambs.

Goats

The duration of gestation in the goat is similar to that in the sheep. One study, covering a large series of goats in the 1960s, gave a mean of 150.8 days, with 86% of gestation periods falling between 147 and 155 days; as in sheep, there is an effect of litter size on gestation length.

Sows

Data for Landrace, Yorkshire and Duroc sows bred by artificial insemination (AI) and reported by workers in Taiwan showed an average gestation period of 114.8 ± 1.9 days, with 80% of sows showing gestation periods of 112–117 days. The effect of various factors on the duration of pregnancy of pigs was also reported in a paper from Russia; this dealt with 18,082 pregnancies in a pig production complex, where gestation period averaged 114.94 days. The averages of individual parities ranged from 114.80 days for seventh farrowings to 115.05 days for third farrowings; average gestation period also ranged from 114.6 days in winter to 115.0 days in spring.

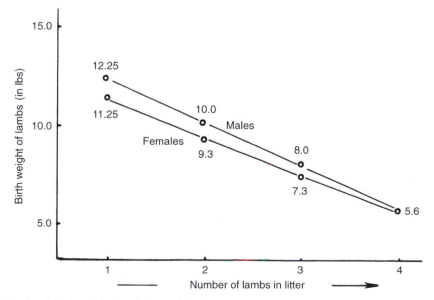

Fig. 10.4. Lamb birth weights in relation to litter size.

Camelids

Seasonal effects on gestation length in alpacas were measured in studies reported from New Zealand; a comparison of pregnancies from spring and autumn matings showed that gestation length was significantly longer (by 12.5 days) in spring (348.9 ± 1.4 days vs. 336.4 ± 1.2 days); it was concluded that it would be difficult to maintain spring-mated females in a 12-month breeding cycle.

10.2.4. Predicting the onset of parturition

Hormonal profiles and udder scoring

There have been attempts to predict calving dates on the basis of hormonal changes in the cow. Some attempts have been on the basis of the decline in progesterone levels; in one study, more than 95% of cows calved within 24 h when their plasma progesterone concentrations declined to less than 1.3 ng/ml. A decline in total oestrogen levels could form the basis of similar predictions. Predicting the onset of parturition in cows in late pregnancy was studied by Davis and Macmillan (2002) in New Zealand, using udder scoring and hormonal profiling from day 272; plasma oestradiol concentration on day 274 was a useful predictor of whether an individual cow would calve within 8 days. The same workers used plasma oestradiol levels to build models to predict whether the cow could calve within the next 5 days from day 272 to 275; progesterone, PG metabolites and cortisol were found to be unreliable indicators. Studies in Saudi Arabia suggested that, together with other parameters, oestradiol-17β, progesterone and cortisol may be used as indicators to predict the time of parturition in camels.

Mammary secretions

In horses, American studies investigated different changes in the concentrations of sodium and the other mineral elements in mammary secretions; it was thought that this might provide a method for predicting parturition and assessing the readiness of the fetus for birth. In New Zealand, on the other hand, workers measured concentrations of potassium, calcium, citrate, lactose and sodium in the 2 weeks prior to the expected date of foaling but found variation too great between mares to be practically useful (Douglas *et al.*, 2002); it was concluded that the use of a model combining mammary secretion electrolytes and physical and behavioural factors may better predict foaling than those based on mammary secretions alone. In France, work showed that the size of the mammary glands and the calcium concentration of mammary secretions (determined using test strips) gave an approximate indication of imminent foaling; 90% of mares studied had a high calcium concentration at foaling and 20% had a high concentration in the week prior to foaling.

Fetal well-being

Apart from attempting to predict the time of foaling, others have sought to use real-time ultrasonics to evaluate fetal well-being and the intrauterine environment (Tullio and Kozicki, 2001); by scanning transabdominally and transrectally from day 300 to parturition, they were able to establish a biophysical profile which was informative in terms of fetal well-being and perinatal morbidity. Workers in Scotland have presented results suggesting that sequential measurement of pregnancy-specific protein B (PSPB) may provide a reliable indicator of fetal distress and adverse pregnancy outcome in singleton-bearing sheep; the same workers also suggest that PSPB and progesterone analysis may have a prognostic value as a biochemical marker of suboptimal placental growth and function in sheep.

10.2.5. Physiology and endocrinology of parturition

The late 1960s and early 1970s witnessed a marked increase in knowledge of factors involved in the hormonal regulation of parturition in farm mammals; much of this was due to studies conducted with the sheep as the experimental model (Fig. 10.5). The classic experiments of Mont Liggins and associates in New Zealand did much to establish the role of the fetal pituitary–adrenal axis in the initiation of parturition in the ewe.

Liggins – the pioneer

Thanks to the work of Liggins in New Zealand and various associates, it is now well recognized that the fetus plays a major role in determining the time of parturition. Liggins used the sheep as his model and proposed that, a few weeks before parturition, the fetal hypothalamus becomes active, stimulating the anterior pituitary to secrete adrenocorticotrophic hormone (ACTH). The pituitary hormone acts on the fetal adrenals, triggering the production of cortisol. The cortisol, in turn, acts to induce placental enzyme activity, which enables the placenta to metabolize progesterone to oestrogen, bringing about an increase in the oestrogen : progesterone ratio. These changes result in the production of PGs, which in due course initiate the involuntary muscle contractions of labour. It is believed that a similar chain of events occurs in the cow and goat. In the mare, currently there is insufficient evidence of all the factors involved in the initiation of parturition.

Fetal hypothalamus

Much interest has centred on the possible factors that activate the fetal hypothalamus. There is evidence suggesting that it may be a matter of fetal demand for glucose or other metabolic essentials; the neuroendocrine control of ACTH by the fetal brain may be strongly influenced by changes in the sensitivity of the hypothalamus to a variety of stimuli (Fig. 10.6).

Among the brain chemicals known to be produced by the hypothalamus is neuropeptide Y (NPY), which is usually known as a potent stimulant of appetite and is

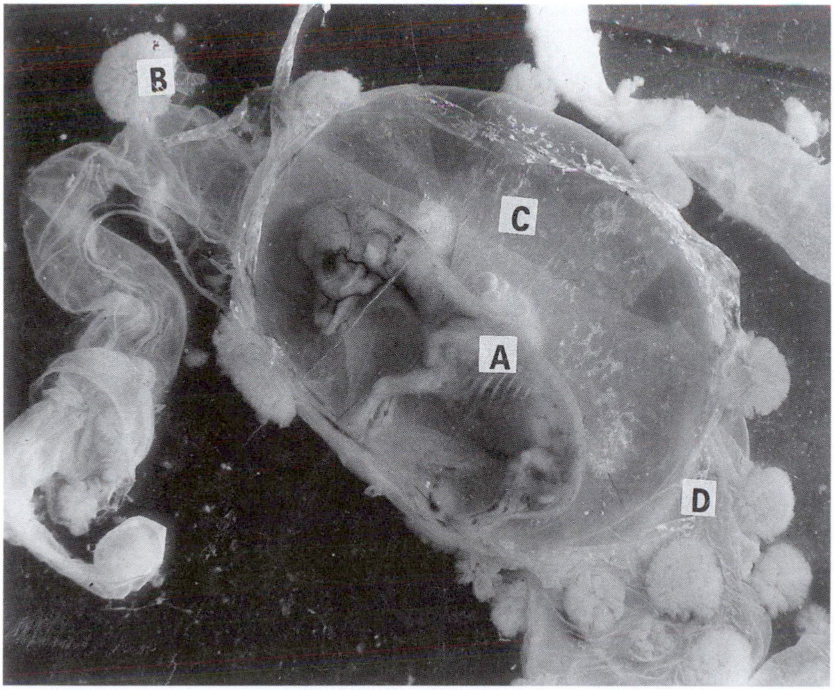

Fig. 10.5. Lamb fetus and its membranes. (A) Lamb fetus; (B) caruncle; (C) amniotic sac; (D) allantochorion membrane.

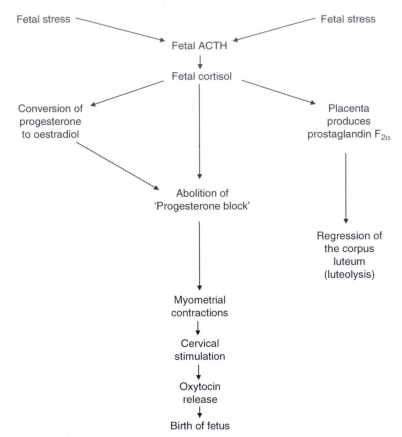

Fig. 10.6. Sequence of events leading to parturition in the ewe.

regulated by leptin. It is known that the levels of NPY in the fetal hypothalamus may increase fourfold in the final 50 days of the sheep's pregnancy and it seems possible that this compound may be an important factor leading to the secretion of ACTH. There are studies showing that the pulsatile output of $PGF_{2\alpha}$ does not increase around the time of luteolysis in the pregnant goat; for this reason, it is concluded that $PGF_{2\alpha}$ is not the principal luteolysin in goats during late pregnancy. An earlier report from Australian workers concluded that the fetal signal for parturition in the doe precedes luteolysis by some 5 days.

Contractures

In the cow, the fetal endocrine signals that initiate the calving process result in pre-partum luteolysis; the consequent

withdrawal of progesterone is a prerequisite for a normal calving. The rather abrupt declining influence of progesterone is followed by a cascade of physiological processes in the myometrium and cervix. A review by Taverne *et al.* (2002) noted that the myometrium in cows is not completely inactive during pregnancy; so-called 'contractures' were recorded during the final weeks of pregnancy, showing low frequency (average, 13.6 daily) and long duration (average, 12.1 min).

Timing of parturition

It is well known that mares give birth when secluded and undisturbed and predominantly during the hours of darkness; some authors have reported 80–85% of foaling occurring at night, with most of these within 2 h of midnight. Workers have concluded that circadian rhythm was responsible for the great majority of foalings occurring at night, but also believed that this could be influenced by husbandry and other factors (e.g. disturbance). Workers in Australia considered that decreased light intensity was the most significant factor triggering foaling. In a recent UK report, data showed a marked preference of mares to foal between 18.00 and 06.00 h; seasonal day length and onset of darkness had no effect on the time of foaling. The same report noted that mares probably prefer to foal in a quiet but not necessarily dark environment and that continued disturbance during the night may adversely affect the onset of normal foaling. There is plenty of anecdotal evidence showing that mares subjected to disturbances at night will postpone foaling until quieter times, even though this may be in daylight hours.

Environmental influences

There is ample evidence in farm ruminants of environmental factors such as feeding time influencing the time of parturition. In the USA, although workers found that the proportion of lambs born at night was not changed by lighting management, they did conclude that sheep producers could spend less time lambing at night if a morning feeding time (hay and grain fed between 7 and 8 a.m.) was included as part of the management system. Data on kiddings in Nubian goats in Uruguay were examined in a study to determine the influence of factors such as kidding season (winter/spring vs. summer), parity, number of kids and sires; maximum kidding occurred at midday and minimum kidding at midnight, with 78% of parturitions occurring during daylight hours. Work elsewhere has shown evidence of significant differences between sheep and goats in birth patterns, and it has been suggested that this may be due to differences in behaviour in the two species.

Final trimester of pregnancy

The pregnant ewe's requirements for energy and protein increase markedly during the final trimester of pregnancy, particularly in the last few weeks. Although body condition and live-weight provide a general indication of the adequacy of feeding, these parameters do not provide an immediate assessment of the animal's needs; these needs may be assessed on the basis of circulating concentrations of certain blood metabolites, such as β-hydroxybutyrate and glucose. It is well recognized in sheep that the resistance of ewes to gastrointestinal parasites tends to break down around parturition, the sheep being sensitive to dietary protein intake and, to a lesser extent, to energy intake and body condition.

Temperature changes

Changes in body temperature have been recorded around the time of parturition in some studies. In The Netherlands, work showed that the average temperature in the cow increased from 38.2°C before parturition to 39.1°C the day after parturition and remained at an elevated level of 38.8°C during lactation; after weaning, the temperature decreased and there was no effect when oestrus was shown. In Scotland, studies examined the effect of piglet expulsion in the sow on plasma cortisol, ACTH and β-endorphin levels, observing that all three concentrations increased as the number of piglets born increased during farrowing.

10.2.6. Periparturitional events

Behaviour

The behaviour of farm animals at the time of giving birth has been reported by many authors; the behaviour of the mother and the newly born is likely to differ markedly according to the species. A summary of information for the ewe and lamb is in Table 10.3.

In goats, Indian workers have made observations of behaviour around the time of birth and immediately afterwards. Movement to an isolated location and intolerance of companions were observed in 80% of does 4–6 h before kidding and intense

Table 10.3.   Ewe and lamb behaviour at
parturition.

Pre-partum behaviour
  Restless
  Seeks isolation
  Interested in lambs and birth fluids of other
    ewes
Duration of birth
  Approximately 1 h
Time for dam to stand after birth
  Usually less than 1 min
Grooming
  Intense for first hour or two
Abnormal maternal behaviour
  Desertion of young
  Moves from suckling attempts
  Butting young
Progress of young after birth
  Time to stand approximately 15–30 min
  Time to suck approximately 1–2 h
Frequency of suckling
  Approximately hourly initially
  In mid-lactation approximately every 2 h

restlessness occurred in 79% of does 1–4 h
before kidding. Primiparous and multiparous does behaved similarly, except that
restlessness was more intense among
primiparous does. Delivery in does while
sitting was more frequent than while standing. More than 90% of kids were born head
first and with one foreleg preceding the
other while passing through the birth canal.
The duration of parturition (from appearance of the water bag to complete expulsion
of the fetus) was 20 and 6 min in first and
second parities, respectively; the interval
between individuals in a twin birth was
6 min.

Expulsion of placental membranes

Behavioural studies in Spain recorded
multiparous goats standing during most
of the first hour post-partum; expulsion of
the placental membranes was completed
after an average interval of 2.5 h. The cow's
placental membranes are usually expelled
between 30 min and 8 h after the calf has
been delivered. The difficulties posed by
retention of fetal membranes (RFM) in dairy

farming have been the subject of many
reports; retention is generally accepted to
be when the membranes remain attached
more than 8–12 h after calving. The incidence of RFM is usually quoted as about
7%, with a range of 3–12% in herds free
from disease; the problem is often found
after abnormal deliveries (premature births,
difficult calvings, birth of twins). The problem is much less apparent in beef cows,
presumably because it either occurs less
frequently or does not have the same
undesirable sequelae reported for dairy
cattle.

Orientation of young at delivery

Studies of the farrowing process show that
about 80% of piglets are delivered head
first; it is believed that tail-first deliveries
are due to a fetus passing from one uterine
horn to the other before being delivered in
its reversed orientation. A summary of the
main features of the farrowing process is
provided in Table 10.4.

Management for live births

Calf mortality is a worldwide problem
and methods of minimizing it, particularly
where the calf is to be reared for beef or
as a herd replacement, justify thought and
attention. Figures in Ireland have shown
an incidence of calf mortality of 12%, this
figure covering 1.5–2.0% abortions, 3–5%
perinatal deaths and 6–8% mortality
between birth and the age of 3 months. The
duration of parturition, as this affects the
birth of live young, differs among the various farm species. In horses, there is little
room for delay in the delivery of the foal if
it is to be born alive; dystocia in the mare
can be regarded as an emergency. Results
reported by Byron *et al.* (2003) showed that
dystocia duration has a significant effect on
foal survival and methods should be chosen
to minimize this time; the difference
between mean dystocia duration for foals
that lived and those that did not was
13.6 min. Such information is important for
horse care personnel and veterinarians if
they are to manage cases effectively.

Table 10.4. The farrowing process in the sow (from Signoret *et al.*, 1975).

Feature	Value
Vulvar swelling	1–7 days before farrowing
Milky secretions on udder	6–48 h before farrowing
Sow settles, lying on one side	10–90 min before farrowing
Vulvar discharge	1–20 min before delivery of first piglet
Time to give birth to entire litter	3.5 h (0.5–16 h)
Additional time to expel fetal membranes	4.5 h (0–12 h)
Interval time between piglets	16 min (1 min to several hours)
Piglets born tail-end first	20%
Piglets in head-first presentation	80%
Piglets with umbilical cord intact	60–70%
Perinatal mortality	6%
Time for piglets to suck on teats	10–35 min

10.2.7. Induction agents

Pigs

The history of controlled farrowing using PGF$_{2\alpha}$ and its analogues goes back to the mid-1970s. The earliest time of treatment with PGF$_{2\alpha}$ is generally taken as day 113 of gestation, but the agent is usually administered on the morning of day 114. In large field trials in Germany, it was shown that approximately 95% of sows farrowed within 36 h of treatment with 87.5–175 µg of the PG analogue cloprostenol. Most farrowings are likely to occur in the afternoon of day 114 and the remainder in the morning of day 115. Additional methods have been developed to enable even tighter control of parturition timing using a depot oxytocin analogue (carbetocin, Depotocin) 22–24 h after cloprostenol. A further report from Germany gave the earliest advisable date for PGF$_{2\alpha}$ injection as day 114 of pregnancy in sows and 115 in gilts; synchronized farrowing times ensured both a minimum suckling period and a virtually equal suckling period for all litters in the group. Higher and uniform live-weights of piglets at weaning were recorded; the numbers of stillborn piglets and piglet losses during the suckling period were also reduced.

Further reports from Germany on the use of PGs in controlled farrowings showed that the proportion of stillbirths was slightly less after induced parturition than after

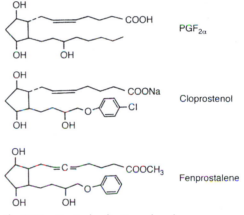

Fig. 10.7. Prostaglandin F$_{2\alpha}$ and analogues.

natural farrowing in a pig breeding facility; parturition was induced (day 114 or 115) with cloprostenol (Fig. 10.7) or cloprostenol in combination with a long-acting oxytocin preparation. The average duration of spontaneous parturition was 164.1 min, which was significantly shorter than sows treated with cloprostenol on day 114 (194.4 min) or day 115 (200.7 min); the addition of the oxytocin preparation reduced the average duration to around 130 min.

According to Wahner and Huhn (2003), the injection to induce farrowing should not be given prior to the 114th day of gestation, in order to ensure minimal risk of affecting the final and important growth surge of the fetuses. The combination of PGF with a long-acting oxytocin preparation allows for a further reduction in the variation of the onset of farrowing among sows and reduces the duration of farrowing for each parturient sow. Induced farrowing eases the observation, recording and husbandry tasks required for the sow and neonatal piglets, facilitates an effective cross-fostering system and can reduce the incidence of dystocia and metritis, mastitis and agalactia (MMA) syndrome type disorders.

Sheep

In sheep, there are two ways in which the timing of lambing may be influenced by exogenous hormones; it is either a matter of prolonging gestation or shortening it. With the shortening option, there are known limits as to how far this may be taken. Workers in Oxford in the late 1950s concluded that lambs of gestational age less than 95% of normal were unlikely to be viable; in practical terms, this suggested that, if lambs were to survive, an induction treatment should not be applied earlier than about 1 week before the average date for lambing in the sheep in question. With oestrus-synchronized ewes, it is possible to use induction treatments to achieve a remarkable concentration of lambings, which may be of practical appeal in certain commerical and research conditions (Fig. 10.8).

Induction of parturition has been induced by highly potent corticosteroid analogues such as dexamethasone, a glucocorticosteroid with about 25 times greater potency than cortisol; alternative steroids are betamethasone and flumethasone. The usual dose of dexamethasone employed as an induction agent in sheep is 15–20 mg; the more potent flumethasone is used at a dose level of 2 mg. The actual day of gestation when the agent is given can be expected to vary with the breed in question; generally, it would be 4–5 days ahead of the mean gestation length for the breed. Dexamethasone injection leads to a lambing peak at 36 h, with births virtually completed by 72 h.

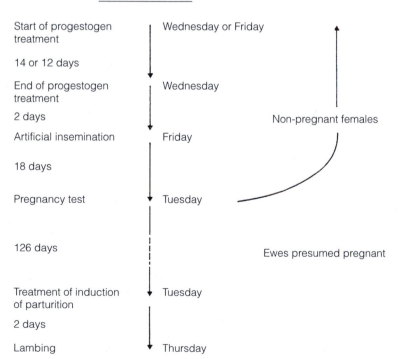

Fig. 10.8. Avoiding weekend lambings by induction treatment.

There are also studies showing that oestradiol benzoate (ODB) at the 15–20 mg dose level, administered in the last week of gestation, may be effective in the induction of lambings.

In the early 1970s, New Zealand workers produced evidence implicating PGF in the normal parturition process in sheep. PGs, however, have proved relatively ineffective in initiating parturition earlier than about a week from full term. In one comparison between a corticosteroid (2 mg flumethasone) and a normal luteolytic dose of PGF (15 mg) administered on day 141 of gestation, 89% and 33% of ewes, respectively, delivered lambs within 72 h. Other work, this time in the USA, showed that a luteolytic dose of the PG analogue, cloprostenol, was without effect as an induction agent. Studies have also been reported using epostane, which prevents progesterone synthesis, and RU486 (mifepristone), an anti-progestogen, in the induction of parturition in sheep.

Dairy sheep

The induction of parturition in dairy ewes, using 10 mg dexamethasone on day 143, was reported from Argentina; the treatment had a significant effect on gestation period (145 days vs. 147.6 days for controls) and on lamb mortality to 1 month of age (4% vs. 15%). In Canada, other workers reported the induction of parturition in ewes with dexamethasone or dexamathasone and cloprostenol; 29 of 33 ewes treated with dexamethasone and/or cloprostenol lambed within 72 h, with all lambs viable and no ewe with retained placenta.

Goats

Various reports in goats have dealt with the use of dexamethasone and $PGF_{2\alpha}$ in varying doses, administered on day 141 of gestation. The induction of parturition in goats by means of the PG analogue cloprostenol was reported by workers in Brazil; the PG was given as an intravulvar injection on day 145; the average time between treatment and parturition was about 30 h; significant differences in kidding date between goats giving birth to singles or twins were not observed. In Turkish goats, it was concluded by Alan and Tasal (2002) that, when compared with an intramuscular injection of $PGF_{2\alpha}$, oral or cervical applications of a PG analogue (misoprostol) was less effective at inducing parturition in goats; applying misoprostol cervically may be an alternative method for cervical priming and induction of parturition or abortion.

Cattle

The first report on the use of corticosteroids in the induction of calving came towards the end of the 1960s. A considerable volume of literature became available on the response of cows to these agents as a result of research in Europe and North America; it was in New Zealand, however, that the procedure was to become a feature of dairy farming for some years. In that country, induced calvings were to rise from 2000 in 1970 to 400,000 by 1978, all in an effort to align calving dates with pasture growth.

Many different corticosteroid formulations and treatment schedules have been assessed; parturition can be induced reliably after day 255 and less reliably as early as day 235. When the aim is to induce calving within the last 2 or 3 weeks of gestation, single injections of a number of short-acting preparations of dexamethasone, betamethasone and flumethasone are known to produce reliable and predictable results. Earlier in pregnancy such short-acting corticosteroids are less effective, and long-acting formulations are required. RFM was to be a consistent feature in cows receiving the short-acting preparations, but the calf was usually viable; with long-acting formulations, there was a high perinatal mortality rate. In Ireland, researchers used a combination of long-acting and short-acting corticosteroids and showed this to be effective in about 90% of cattle within 80 h of the second injection; the combination resulted in satisfactory maternal preparation for delivery and acceptable udder development. The effectiveness of the corticosteroid is believed to be dependent on the permeability of the ruminant placenta to the molecule. In sheep, evidence suggests that the placenta remains relatively impermeable to the agent until about 1 week before the due lambing date; in the cow, on the other hand, the corticosteroid can apparently be active for more than a month before full term.

Prostaglandins

Reports dating back to the early 1970s have dealt with the use of $PGF_{2\alpha}$ in the induction of calving; births were recorded 2–3 days after administration, with acceptable results being achieved in cows within 2 weeks of term, using the normal luteolytic dose of the PG analogue cloprostenol; in dealing with less advanced pregnancies, the protocol was amended to include a priming dose of long-acting corticosteroid prior to the PG. $PGF_{2\alpha}$ and its analogues have also been successfully employed to control parturition in goats, again dating back to the early 1970s. Reports have described injecting does at 144 days of gestation in the early morning and expecting kids to be born during the afternoon of the following day. There have also been reports in which workers used dexamethasone as the induction agent.

Oxytocin in horses

The induction of parturition has been used in mares for the management of high-risk pregnancies as well as for research and for convenience. Several agents and methods have been used, including glucocorticoids, PGs and oxytocin. The latter product, oxytocin, is generally considered to be the best induction agent, since it has a rapid effect and delivery usually occurs within 15 to 90 min (Table 10.5). In addition, the pattern of events during induced parturition with oxytocin is consistent and the hormone appears to have minimal negative effects on full-term foals. Several methods of oxytocin induction have been described, including bolus intramuscular or intravenous injection of 60–120 iu of oxytocin, intramuscular or subcutaneous injection of 2.5–20 iu of oxytocin at 15 min intervals and intravenous administration of 60–120 iu of hormone in 1 l of 0.9% NaCl solution as a drip feed (1 iu/min).

While some workers have found that inductions may proceed successfully in mares with a tightly closed cervix, others found that mares with a detectably dilated cervix (assessed by digital examination *per vagina*) prior to induction delivered foals more rapidly with improved vigour (i.e. foals were better oxygenated at birth, stood and nursed sooner and were less likely to be maladjusted) than mares with a closed cervix. In the USA, workers found that cervical ripening prior to induction favoured a shorter delivery period with greater foal vigour. It was concluded that intracervical administration of PGE_2 could prove useful in induction treatments. In the USA, workers used a combination of oxytocin and $PGF_{2\alpha}$ and showed it to be a safe, effective

Table 10.5. Foaling after oxytocin induction (from Carleton and Threlfall, 1986).

Stage	Time (minutes)	Remarks
I	5–10	Restlessness, anorexia, colicky pains (looking at its flanks), minor tail movement
	15–20	Stall walking, frequent defecation, getting up and down, elevation of the tail, repeated stretching, sweat on shoulders and neck behind the elbows
II	20–25	Restlessness, tail switching, sweat on ribs and flanks
	25–30	Acceleration of respiratory and pulse rates, with rupture of the allantochorion membrane and the start of abdominal pressing
III	35–40	Amniotic membrane at vulvar lips
	45–50	Stage II completed with passage of foal
	30	Initial phase completed with passage of placental membranes

and predictable method for induction of foaling when used in conjunction with pre-determined criteria indicating readiness for induction; such criteria included a milk calcium concentration of 200 ppm or greater, a gestation length exceeding 330 days, correct presentation of the foal and cervical dilation greater than 2 cm.

10.2.8. Delaying parturition

In farm animals, the sympathetic innervation of uterine smooth muscle contains few alpha receptors but many beta receptors. Two distinct types of beta receptors have been identified, the beta-1 and beta-2 receptors; stimulation of such receptors can result in relaxation of the uterus. In the 1970s, a beta-2 mimetic agent (clenbuterol) became available for use in farm animals (Fig. 10.9). In cattle, it was shown to be capable of delaying calving by up to 10 h by inhibiting uterine contractions; the pharmacological profile of this agent is such that it is highly selective for beta-2 receptors. Clenbuterol may prove useful in certain practical situations, such as in delaying night-time calvings until the next day or to enable the cow to be moved to a more suitable location. Cows with dystocia, caused by fetal malpresentation or for other reasons, may be dealt with more easily using clenbuterol; with the relaxation of the myometrium, corrections can be made with less trauma to the animal.

Although the synchronization of farrowing in pigs may be readily achieved by slightly shortening the normal length of the gestation period using PGs, it may also be achieved, with much less success, by prolonging gestation a few days longer than usual. On the basis of evidence made available some time ago by various studies, control of parturition by extending gestation beyond day 115 or shortening it to less than 112 days is unlikely to be of practical value because of unacceptable piglet mortality rates. On the other hand, if talking of a temporary delay in farrowing, this can be achieved, as in the cow, using a tocolytic agent, such as clenbuterol, to inhibit myometrial activity.

10.2.9. Terminating pregnancy

Cattle

There are some practical situations in which it may be desirable to terminate an established pregnancy in a cow. It may be a matter of misalliance in the breeding of a valuable animal (wrong bull used in AI). It may be to avoid having pregnant beef heifers in a feedlot, where pregnancy can interfere with growth rate and carcass quality. In using PG to induce regression of the cow's CL and precipitate abortion, it is now recognized that the efficacy of the agent decreases as the stage of gestation approaches 150 days; up to that time, PG-induced terminations are rapid and uncomplicated and the fetal membranes are delivered along with the fetus. Beyond the 150-day mark, however, the efficacy of PGs decreases rapidly. Cows that are induced with a PG before 80 days of gestation will generally exhibit a standing oestrus after expulsion of the fetus and, if mated, become pregnant at that time; cows pregnant for more than 100 days usually retain the fetal membranes and show no heat period. A report by Fernandes *et al.* (2002) in Brazil showed that terminating pregnancy during the first trimester of gestation did not affect heifer fertility.

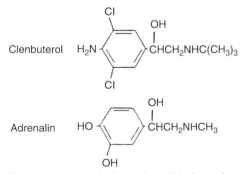

Fig. 10.9. Structural relationship of clenbuterol to adrenalin.

Horses

Among the methods available for terminating pregnancy in the mare, whether for
reason of unwanted twins or otherwise, is
the use of PGs, either the natural $PGF_{2\alpha}$
product or one of the analogues. Some
reports showed that a single dose of 250 µg
of the fluprostenol analogue given on day
35 terminated pregnancies, whereas the
same treatment given at 70–77 days proved
ineffective; however, multiple doses of the
analogue resulted in all pregnancies being
terminated. When fluprostenol was given
on day 35, oestrus and ovulation were
exhibited within a day of abortion; at day
70, oestrus and ovulation were delayed for
some 40–50 days.

Goats

There may be occasions when it might be
necessary to terminate pregnancy in goats
that have been inadvertently bred by an undesirable buck. $PGF_{2\alpha}$ has been found effective; one report dealt with goats estimated
to be about 3 months pregnant in which PG
induced termination very effectively.

Camelids

In camelids, workers terminated an unwanted pregnancy in a llama with cloprostenol
(150 µg) given intramuscularly. In this
instance, the fetus was expelled 108 h after
injection and the animal was successfully
remated 20 days after abortion.

11

Controlling Onset of Puberty

Some amount of research has been directed towards the use of various management strategies and hormones (progestogens, prostaglandins, oestrogens and gonadotrophins) in the induction of puberty in cattle and other farm species. There have been many studies dealing with the role of genetic selection, the effect of exposure to males and the influence of nutrition in achieving an earlier onset of puberty. Within reason, the sooner the breeding female ends the non-productive phase of its reproductive life the better it suits the farmer. In the case of seasonal breeders such as sheep, missing the boat in their first breeding season may mean an extra year before they are bred. A relatively simple treatment may mean they can become pregnant in their first year. In pigs, inducing puberty a week or two earlier may mean a useful saving in cost.

11.1. Practical Implications of Early Puberty

11.1.1. Cattle

Reducing the non-productive phase

Factors affecting puberty in cattle have not received a great deal of attention in the past, relative to many other aspects of bovine reproduction, although clearly there is much to be said, in both dairy and beef

cattle, in favour of reducing the non-productive phase of the animal's life, i.e. the period which ends at the time of its first breeding. In his classic work on cattle reproduction, John Hammond in Cambridge in the 1920s estimated the average age of puberty in heifers of dairy breeds, maintained under normal conditions of feeding and management, as being about 9 months, with a range from about 3 to 15 months. In parts of the world, such as the USA, where first calving of beef cattle normally occurs at 2 years of age, animals are mainly early-maturing and feed resources are relatively inexpensive and not severely limiting.

Puberty in bulls

The age of puberty in bulls varies according to breed and the plane of nutrition; as a general rule puberty is usually attained when the animals attain about two-thirds of their mature body weight and when they are still under a year of age. The rate of sperm production by the bull increases gradually with age and is directly related to the size of the testes (Table 11.1).

In terms of hormone activity, sexual maturation in bulls can be divided into three phases: the infantile, prepubertal and pubertal. The infantile period is characterized by low episodic discharge of luteinizing hormone (LH); during this period, the pulse frequency of gonadotrophin-releasing hormone

Table 11.1.   Sperm production in the dairy bull (from Hammond *et al.*, 1983).

Age	Number of bulls	Gross weight of testes	Daily sperm production (in millions)	
			Total	Per gram testis
0–4 months	25	20	0	0
5–7 months	15	97	104	1
8–10 months	20	284	1750	7
11–12 months	15	370	3300	10
17 months	13	480	4480	10
3 years	10	586	6040	11
4–5 years	11	647	6530	11
> 7 years	11	806	8000	11

(GnRH) release is low. From 6 to 20 weeks of age, an early transient increase in LH secretion has been observed in what is known as the prepubertal phase. It it also known that GnRH pulse frequency increases twofold during the transition from the infantile to the prepubertal period and the sensitivity of the pituitary also increases at this time. The pubertal period is one of active reproductive development from 20 weeks of age until puberty (defined as 50 million sperm per ml ejaculate with a minimum of 10% motility), which occurs at about 1 year of age.

Testicular size (e.g. a scrotal circumference of about 28 cm) and blood levels of LH and testosterone are other important traits included in studies on male sexual maturation. The development of the reproductive tract in bulls has been studied by measuring scrotal circumference; the weight of testes and vesicular glands increase slowly during the first half of the sexual development in bulls and then more rapidly to the time of puberty. Spermatogenesis is initiated around 3–4 months of age, spermatocytes appear by 6 months of age and elongated spermatids and sperm by 8 months of age.

Dairy heifers

In dairy animals, the possibility of improving the economic efficiency of milk production by the early mating of heifers is clearly a valid objective, but problems associated with dystocia and with possible reduced lifetime milk yield might outweigh possible advantages. The general rule observed in the UK is that Friesian heifers should not be bred earlier than 60 weeks of age. In Brazil, a study of Holstein heifers in 48 herds showed age at first calving to be about 30 months; it was noted that age of puberty was significantly affected by herd, sire, year of calving and season of calving. In New Zealand, one recent study dealt with improvements in the rearing of dairy heifers under the grazing conditions of that country; it was suggested that dairy heifers should reach 60% mature weight by 15 months of age for optimum fertility at first mating and 90% mature weight by 22 months for optimum milk solids production in their first lactation.

Producing high-quality replacements for the dairy herd at minimum cost is one of the many challenges facing milk producers in the opening years of the 21st century. Various studies have sought to determine how dairy replacement rearing systems affect the lifetime productivity and profitability of the dairy cow. In some conditions, it might be useful to breed dairy heifers so that they calve at 20–24 months of age, but heifers reared much faster than usual during their first year of life tend to give lower than usual milk yields regardless of their age at calving, not only in their first lactation but in later ones as well. In a Wisconsin study, for example, it was recorded that accelerated postpubertal growth and early calving reduced performance during the first lactation of Holstein heifers.

Beef heifers

Age at puberty in beef cattle is a major determinant of their lifetime reproductive performance. The effect of season on age at puberty of beef heifers can be attributed to a range of factors, including photoperiod, temperature, management and other variables. One major factor influencing the pattern of gonadotrophin secretion and the the time of puberty onset is nutrition; many studies show that heifers fed a higher-energy diet are likely to be younger at puberty than those fed low-energy diets. When puberty is not reached before the start of the breeding season, production costs increase. From various reports, it is known that, using standard industry practices in the USA, significant percentages of heifers in beef herds may remain pre-pubertal until after the initiation of the breeding season. Treatments to induce puberty in heifers may help to improve their reproductive efficiency by allowing them to complete their subfertile first oestrus before the start of the breeding season (Fig. 11.1); such heifers can then be bred early in the season, enabling them to calve early and for their calves to reach greater weaning weights. Although calving beef heifers at 2 years of age has become the widely accepted practice among North American beef producers, breeding animals to calve at that age involves other considerations; it is well recognized that losses at

calving are usually much higher in maiden heifers than in older animals. Bearing in mind that calf mortality in beef herds is estimated to range from 5 to 20%, it is a matter of ensuring that appropriate care and attention are directed towards the offspring of the young animal.

11.1.2. Sheep

Puberty in ewe lambs

Age of puberty in sheep can have important practical implications. The efficiency of sheep production can be increased by eliminating or reducing the unproductive periods in a breeding ewe's life; one of these periods is from weaning until first breeding. Under lowland flock conditions, where animals are kept under good feeding and management, it should usually be possible for ewe lambs to be bred to lamb at 1 year of age. The advantages of breeding 7–8-month-old ewe lambs to give birth at about a year include reduced maintenance costs before the start of reproduction, a shortened generation interval, which results in more rapid genetic gains from selection, and increased lifetime production. Induced mating could be of value in ewe-lamb breeding as

Fig. 11.1. Puberty and conception in the beef heifer.

foreknowledge of the lambing date permits close supervision of lambing, which may be necessary to ensure that lamb losses are kept to the minimum. It would also be true to say that proper feeding in late pregnancy may be even more critical in the ewe lamb than in the adult sheep.

Puberty in ram lambs

Advancing puberty in ram lambs may be useful in sheep farming by enabling younger males to be used in the breeding flock; younger males could help to reduce production costs, accelerate the benefits of genetic selection and permit earlier progeny and libido testing. It is also possible that the advantages of early puberty in ram lambs may be accentuated in year-round breeding programmes. It should be noted that the age at which a ram reaches puberty is not necessarily the time when he is fully effective in mating. The ram passes through a period when semen quality is poor and his use may be followed by many repeats to service. A January-born Suffolk ram lamb may well be less fertile when used in July than in the following October, even though he apparently performs satisfactorily in his mating behaviour. Although females, whether sheep, cattle, pigs or other farm species, also have a phase after reaching puberty when their fertility is lower than at later oestrous periods, this phase in the male can extend over a longer period.

The adolescent ewe as a model

Quite apart from farming considerations, there may be occasions in the research laboratory when the induction of oestrus and ovulation in young sheep is employed in studies relevant to important aspects of human pregnancy. In humans, for example, epidemiological data are now available suggesting a strong association between fetal growth derangement and the development of a number of health disorders in adult life. Current understanding of the origins and consequences of prenatal growth retardation are largely derived from studies of pregnant sheep, in which restricted fetal

growth may be achieved in various ways, varying from chronic undernutrition to uterine artery ligation. In Aberdeen, workers have evaluated the growing adolescent ewe (6 months old) as a model for investigating the causes and consequences of prenatal growth restriction; in this work, ewe lambs were induced to ovulate by fluorogestone acetate (FGA) sponge and pregnant mare's serum gonadotrophin (PMSG) treatment at 21 weeks of age and were used as recipients of sheep embryos.

In one Aberdeen study using such adolescent ewes, it was shown that in rapidly growing animals on high dietary intakes the anabolic drive to maternal tissue synthesis was maintained at the expense of the gradually evolving nutrient requirements of the gravid uterus; this resulted in a major restriction in placental growth, leading to a highly significant decrease in birth weight. It is possible that the rapidly growing adolescent sheep could provide a new non-invasive model with which to study the causes and consequences of prenatal growth restriction. It is known that human adolescent mothers have a high risk of delivering low-birth-weight and premature infants, who often die within the first year of life; the sheep model may provide valuable information relevant to such human pregnancies.

11.1.3. Pigs

Many of the piglets produced each year are farrowed by gilts; it follows that the age at which these females attain puberty is of considerable economic interest. In the UK, it has been estimated that between 15 and 25% of litters produced in pig units come from gilts; techniques for advancing the onset of puberty, even by a relatively short margin, are likely to be of practical interest. When considering the question of earlier breeding in pigs, it is clearly desirable to have information on factors likely to influence the attainment of puberty in the gilt; it is also important to examine lifetime reproductive performance in relation to age at first mating.

Accelerating genetic progress

Embryo transfer in the pig is still regarded as a research tool rather than as a practical aid to commercial pig farmers; however, reduced generation intervals and accelerated genetic improvement could be achieved if younger gilts could be stimulated into an early puberty and used as donor animals. Although gilts can be stimulated into heat as young as 100 days by way of a single-dose injection of human chorionic gonadotrophin (hCG) and PMSG (pig gonadotrophin 600 (PG600)), responses are known to be better at 160 days or older.

Lean pigs

Consumer demands for leaner pork have led to genetic selection for increased lean-tissue growth rate and reduced body fat; this has resulted in a delay in the age at puberty and a decrease in energy stores for subsequent growth, pregnancy and lactation. There are workers who suggest that future pig management strategies both to enhance reproductive performance and to satisfy consumer demands need to consider hypothalamic–pituitary maturation in association with body weight and body composition parameters (Evans and O'Doherty, 2001).

Once-bred gilt system

Accelerated puberty is clearly relevant to the feasibility of the once-bred gilt system, where the idea is to take a litter from the pig before dispatching her to the heavy hog market. For such a system, it is necessary to have a predictable throughput of relatively young gilts that are mated on at least their second observed heat period.

11.1.4. Horses

Much information has been published about the age of puberty in horses and ponies. In Iceland, one study investigated reproduction in Icelandic ponies kept under free-range conditions. Puberty was attained by 70% of females by 2 years of age and 90% by 3 years of age; colts had usually reached puberty by 2 years of age. The results of a study reported by Gomes *et al.* (2003) confirmed previous reports that there is an effect of the season of birth on the onset of puberty in fillies, showing that spring-born fillies ovulated younger than summer-born fillies (26 vs. 30 months). Where animal performance in sports horses is improved for competitive purposes, the influence of dietary constituents is a factor to keep in mind. Some reports draw attention to the need to ensure that steroids should not be given to horses of either sex before puberty, for the effects on their fertility may be irreversible; it is also apparent that anabolic steroids are not very effective in building up muscle mass or promoting growth in colts and it has been suggested that protein supplements with a high lysine content should be given instead.

11.2. Development of Puberty Control Measures

11.2.1. Physiology and endocrinology of puberty in farm mammals

Gonadostat theory

It is believed that components of the hypothalamic–pituitary–ovarian axis of the various farm animals are functional for some time before the normal onset of puberty, although little is known of the mechanisms within the brain which bring these various components into an appropriate temporal relationship to initiate puberty. The 'gonadostat' hypothesis is one of the most widely accepted theories dealing with neuroendocrine mechanisms controlling the onset of puberty; according to this theory, there is a decline in sensitivity to oestrogen negative feedback on pulsatile LH secretion.

The gonadostat theory proposed by workers in the early 1960s sought to explain the onset of puberty in rats; subsequently, it has also been accepted as relevant to cattle and other farm animals. In this, there is a change in frequency (increase) of pulsatile LH secretion in response to increasing oestradiol concentrations during development; circulating oestradiol levels that are able to suppress gonadotrophin in pre-pubertal females are less effective in gonadotrophin suppression after puberty. Similar events are believed to occur in cattle. It is also held that puberty in the ewe lamb occurs because of a marked decrease in the response of the hypothalamic–pituitary axis to the negative-feedback action of oestradiol on tonic LH secretion. It is thought that the pig conforms to the gonadostat theory because onset of puberty is immediately preceded by a significant increase in pulsatile LH secretion concomitant with increasing oestrogen secretion.

The similarity of puberty in ewe lambs and the onset of the breeding season in adult sheep, in which there is evidence of marked reduction in response to the inhibitory feedback effect of oestradiol on tonic LH, has led to the suggestion that hypersensitivity to oestradiol feedback on LH secretion may be the final common mechanism at work in both the prepubertal ewe lamb and the anoestrous ewe. According to the gonadostat theory of puberty, the ewe lamb becomes less inhibited by oestradiol as it grows older, this eventually enabling LH pulse frequency to occur at a level sufficient to cause follicle development, oestrogen production and the preovulatory LH surge which leads to the first ovulation and the initiation of puberty. A similar chain of events occurs in the anoestrous ewe as it emerges from the non-breeding season in the autumn.

Stages in brain development

The transition from a sexually immature state to full reproductive activity is timed in the farm animals by signals that act in the brain to increase activity of the hypothalamic–pituitary–gonadal axis. It is clear that profound sex differences exist in the timing of puberty in many species, including farm animals. Those who have written about the development of the gonadotrophic and somatotrophic axes of sheep note that both these axes develop in the fetus from about mid-gestation. Activation of the testes and ovaries is determined by the brain mechanisms controlling GnRH secretion, as exemplified by various studies with sheep. It is known that exposure of sheep fetuses to testicular steroids alters the timing of puberty, principally by reducing responsiveness to the photoperiod; this is shown in the form of an early increase in LH secretion in males or in females exposed experimentally to testosterone before birth. Steroids are also believed to act on non-photoperiodic mechanisms to abolish the preovulatory gonadotrophin surge.

It has become increasingly evident that there are many critical periods of brain development in which the GnRH neurosecretory system is organized; during such periods, the system may be sensitive to different testosterone metabolites. As shown in several studies, the degree of involvement of endogenous opioid peptides (EOPs) in the regulation of LH secretion changes at different reproductive stages and appears to be a function of brain maturational processes. It is believed that EOPs do not play a major role in prepubertal gilts; in postpubertal gilts, however, EOPs influence the GnRH/LH secretory system during the oestrous cycle and their influence appears to be modulated by steroid hormones.

Preliminary ovarian cycles

Although, to the farmer, puberty is indicated by the ewe lamb exhibiting its first oestrus, for the young ewe, in endocrinological terms, this is by no means the first important event occurring around that time. The probability is that two preliminary ovarian cycles may actually have preceded the first heat period; in the preceding cycle there was a well-established normal luteal phase after a 'silent heat' and, before that, another short cycle. The initial short cycle was probably less than half the length of the normal cycle and was apparently initiated by the first LH surge. This sequence of events is not unique to ewe lambs at puberty; it is now well established that short cycles may occur before the commencement of cyclical breeding activity in some proportion of adult ewes after ram introduction in the ewe anoestrus.

Follicular dynamics

Puberty in the heifer is the age at first ovulation and at which regular oestrous cycles begin; it is known, however, that heifer calves show evidence of follicular activity some time before the onset of puberty and the establishment of regular cycles. As early as 2 weeks of age, in fact, ovarian follicles are seen to grow in a wavelike form, similar to those of adult cattle. The onset of puberty in heifers is largely determined by age and weight, although these two factors obviously vary according to breed. If it is a beef heifer and she is to calve at 2 years, she must be cyclic and fertile at 15 months of age.

In contrast to heifer calves and ewe lambs, researchers have usually failed to induce follicular development or ovulation in piglets in their first 2 months of life by exogenous gonadotrophins; this agrees with observations that tertiary antral follicles do not appear in the ovaries of the gilt until it is more than 60 days old. There is also the fact that the pig differs from cattle and sheep in its pattern of oogenesis, which apparently continues after its birth (Fig. 11.2).

It is believed that in the pig, as in the other farm mammals, puberty is brought about by a reduction in the activity of intrinsic neural inhibitory mechanisms and/or a decrease in the negative-feedback effect of ovarian steroids; such events result in the stimulation of pulsatile GnRH release and consequent augmentation of episodic LH secretion, which in turn leads to ovarian activity.

Age and fertility

There is ample evidence showing that the reproductive performance of ewe lambs is often considerably lower than that of adult ewes. This is mainly the result of early embryonic losses, which can exceed 50% in ewe lambs as compared with 20–30% in ewes; recent studies in Wales suggested that such losses may be associated with inadequate luteal function in ewe lambs.

Thyroid activity and puberty

In sheep, it is known that thyroxine plays a role in seasonal reproduction and it is possible to prevent the negative-feedback effect of oestradiol by removing the thyroid gland before the end of the breeding season; this can enable the period of sexual activity

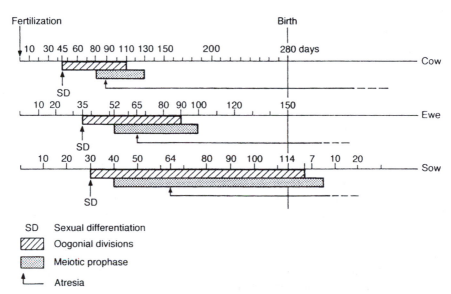

SD   Sexual differentiation

▨   Oogonial divisions

▨   Meiotic prophase

↟⎽   Atresia

Fig. 11.2. Species differences in the pattern of oogenesis (from Thibault and Levasseur, 1974).

to continue longer than normal. It is also known that administration of antithyroidal compounds (e.g. propylthiouracil (PTU)) can also prevent the onset of oestrus. Attempts have been made to examine the effect of such compounds as PTU on puberty in sheep, so far without success (Wells *et al.*, 2003).

Metabolic hormones as modulators of puberty

The onset of puberty in the gilt may be related to attainment of a critical body weight or a minimum percentage of body fat. Two metabolic hormones, insulin-like growth factor I (IGF-I) and leptin, have been implicated as possible modulators of reproductive function; increased IGF-I and leptin concentrations are associated with onset of puberty and follicular development. It is clear that pregnancy rates and litter size after induced ovulation in prepubertal gilts may be much below those in more mature animals. The variable ovulatory response and low pregnancy rate have been attributed to abnormal follicular development and corpus luteum function characterized by increased sensitivity to the uterine luteolysin, low LH receptor numbers and variations in progesterone synthesis. It is believed that variable responses to the induction of ovulation and pregnancy maintenance in prepubertal gilts may be related to maturation of the neuroendocrine–ovarian axis.

Role of leptin

Emerging evidence indicates that leptin is related to reproductive function; this small protein molecule is secreted by adipocytes and is known to play a role in the regulation of body weight and food intake. Since its discovery in 1994, much information has accumulated in the literature on the physiological role of leptin, particularly its effect on the brain (see Williams *et al.*, 2002). The hormone is known to circulate in the blood of humans at levels paralleling those of fat reserves. It is believed that leptin communicates information about nutritional status to regulatory centres in the brain; the hormone circulates in the blood bound to a family of binding proteins.

It is known that the amount of body fat stored in the animal influences fertility, indicating a link between adipose tissue and the reproductive system; it is believed that leptin may act as a peripheral signal indicating the adequacy of nutritional status for reproductive function. It seems that low leptin concentrations indicate a status of inadequate nutritional stores, which could be instrumental in preventing an unwanted pregnancy, with its obvious demands for additional energy. In humans, some believe that puberty in girls may occur when sufficient leptin concentrations are reached; earlier menarche in obese girls than in normal girls may involve leptin action.

Role of melatonin

Melatonin, a hormone known to mediate photoperiodic cues, is secreted from the pineal gland in a circadian rhythm. However, workers who have examined the role of melatonin in the pubertal process in gilts have concluded that nocturnal rises in serum melatonin are not necessary for a gilt to attain puberty. There are also those who have reported the failure of melatonin implants to alter the onset of puberty in gilts. In the USA, one recent study used melatonin implants to elevate blood melatonin concentration five- to tenfold without this affecting the timing of puberty onset.

11.2.2. Genetic and environmental factors

Zebu and taurine cattle

Dairy cattle are known to be more precocious than beef animals; zebu cattle (*Bos indicus*) are known to reach puberty at a later age than taurine breeds of cattle. However, zebu cattle have well-proven advantages over taurine animals in certain regions; Brahman (*B. indicus*) and Brahman-crossbred cattle are popular in the southern states of the USA because of their tropical adaptation. Even within the one breed of cattle, there may be important factors influencing age at puberty; one of these can be the genotype of the animals. In Ireland, certain strains of Holstein–Friesian heifers have been shown to be significantly younger, lighter and smaller than other strains when they show their first oestrus (McGrath *et al.*, 2003). In beef cattle, workers in New Zealand attempted to change the age at puberty in Aberdeen Angus cattle by selecting either for reduced age or increased age; after a 10-year period of such selection, heifers born in the former were 81 days younger at puberty (19% reduction in age) than those born in the latter herd.

Pigs

The onset of puberty in gilts usually occurs between 200 and 220 days of age and is due to the combined effect of genetic and management factors. In pigs, puberty is known to be affected by many factors, including genotype, breed, boar contact, transport, supplementary light, ambient temperature, growth, management and hormone treatment. Crossbred gilts usually attain puberty earlier than pure-breds and the plane of nutrition affects the age at which first oestrus is shown. Various studies have examined gilt management from selection to mating and discussed the many factors relating to puberty. Direct boar contact is regarded to be of prime importance; extraneous factors which can either improve gilt response, such as mixing, relocating and on-farm transportation, or reduce such responses, such as crowding, have also been studied. The effect of age at first farrowing and herd management on long-term productivity of sows was the subject of one recent report; the study involved records from 38,349 sows and 171,178 litters in 976 French herds. Herds were classified according to age at first farrowing as early (337 days), usual (356 days) or late (371 days); it was found that the best reproductive results were in the 'usual' category, with the first farrowing at 356 days. Within each herd, gilts that were oldest at first farrowing had the lowest performance and the youngest gilts had the best performance.

Sheep

The exposure to the long days of spring and summer, followed by the short days of autumn, provides the photoperiodic cues to stimulate the ewe lamb's first oestrus. The cycle of reproductive events in sheep is such that an autumn breeding season occurs, followed by a season of anoestrus. During anoestrus, the pulsatile secretion of LH decreases because of a change in the potency of the oestradiol negative feedback on LH. The effect of lambing season on the age of ewe lambs at first oestrus has been noted by various authors (Fig. 11.3). In Mexico, one study showed a significantly lower percentage of autumn-born lambs reaching puberty by 9–10 months of age than spring-born lambs, thought to be the result of the anoestrous season interfering with the onset of sexual activity.

Nutritional effects

The interrelationships between nutrition, growth rate and age at puberty follow similar lines in both male and female farm mammals. It is well established that ram lambs reared on high as opposed to low planes of nutrition attain puberty earlier in life and that their rate of sexual development is highly dependent on food energy intake and live-weight gain; apparently, their reproductive neuroendocrine system is highly sensitive to changes in nutrition, and photoperiod has much less influence on the timing of puberty. Sexual activity in merino rams in Australia, for example, is not restricted by photoperiod to rigidly timed seasons, but appears to be highly dependent on diet. Males that are reared on low, as opposed to high, planes of nutrition reach puberty at older ages and lighter body weights; in seasonal breeders, such as sheep, goats and deer, undernutrition can delay puberty for a complete year.

Ewe lambs

Restricted growth and development of ewe lambs may have profound effects on subsequent lifetime reproductive

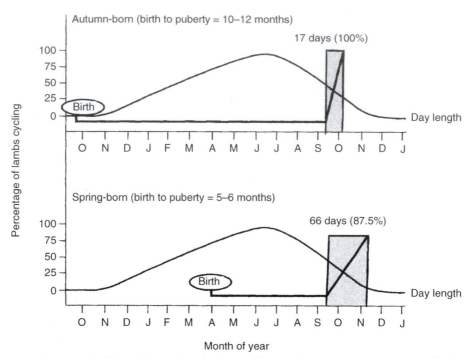

Fig. 11.3. Season as it affects puberty in female sheep. Influence of season (spring or autumn) of birth on the onset of puberty in ewe lambs as shown by the percentage of animals showing oestrous cycles (shaded area). Spring-born lambs reached puberty at a much younger age and less synchronously than those that were autumn-born. (After Senger, 1997.)

performance. In one Scottish study, Cheviot sheep subjected to a restricted level of nutrition before weaning at 4 months of age had a significantly lower lifetime incidence (seven parities) of multiple births than those on a high plane of nutrition; it was also found that birth weights of lambs born to ewes restricted in early life were significantly lower than those born to unrestricted ewes.

Heterosexual experience

A report by workers in the USA has shown how heterosexual experience differentially affects the expression of sexual behavior in 6- and 8-month-old ram lambs. These workers concluded that the sexual responsiveness of ram lambs towards females is sufficiently undeveloped at 6 months (i.e. puberty) for extended exposure to sexually receptive ewes to be needed for many males

to exhibit adult levels of performance; at 8 months, the sexual development of ram lambs has sufficiently matured for relatively brief encounters with oestrous females to release the full expression of adult sexual behaviour.

11.2.3. Defining pubertal status

Cattle

Attempts to define the pubertal status of beef heifers prior to breeding have been reported by various authors. Evaluating heifers prior to breeding would enable animals with poor breeding potential to be eliminated before the costs of breeding have been incurred. Some workers have employed a five-point reproductive tract scoring (RTS) system to estimate pubertal status and subsequent breeding potential of beef heifers via rectal palpation of the uterine horns, ovaries and ovarian structures. Using this system, heifers with an RTS of 1 possess immature tracts, with no uterine tone and with no palpable ovarian structures. Heifers with a score of 2 show a uterine horn diameter of 20–25 mm, no uterine tone and follicles less than 8 mm in size. Heifers with a tract score of 3 show slight uterine tone and follicles of 8–10 mm in size. Heifers with a score of 4 show a uterine horn diameter of 30 mm, tone to the uterus and > 10 mm follicles. Tract score 5 heifers possess a palpable corpus luteum. Animals with a tract score of 1, 2 or 3 are considered prepubertal and those with a tract score of 4–5 are considered to be pubertal. A study by Rosenkrans and Hardin (2003) demonstrated that the RTS system can prove to be a useful screening test for herds, but not a tool to cull individual animals.

Goats

Studies reported from Mexico mention vaginal cell exfoliation as being indicative of cyclicity in goats at the onset of their reproductive activity; the same work found that such exfoliation could be used as a tool to detect responses to hormone treatments in female goats. In male goats, an earlier report from Brazil concluded that penis detachment from the prepuce could be employed as an indicator of sexual maturation; males in which the penis had detached from the prepuce showed all stages of spermatogenesis whereas those in which the penis was not detached showed high levels of spermatids.

Pigs

Some authors have examined factors that may be useful in predicting the early onset of puberty in gilts. In the USA, a study showed that vulva size (1 = infantile; 2 = normal; 3 = large) at 155 days of age is related to the rate of sexual maturation in gilts; more gilts with a vulva score of 2 or 3 achieved puberty by 200 days of age than gilts with a score of 1.

11.2.4. Puberty and conception

Cattle

There are studies showing that heifer cattle have a higher conception rate at their third oestrus than at their pubertal oestrus. It is well known that fertility, or conception rate, in dairy heifers is consistently better than fertility in lactating cows. The dairy heifer is usually bred after several oestrous cycles have occurred. Replacement programmes in Holstein herds are usually designed to have heifers calving at 22–24 months of age; as with cows, reproductive performance in the young heifer is dependent on the efficiency of oestrus detection and conception rate. The reproductive performance of the animals is also influenced by the environment. A study reported from Florida by Donovan et al. (2003) suggests that producers in similar hot environments should avoid breeding heifers during the summer months or restrict breeding to large-frame animals.

Pigs

The usual recommendation for mating gilts after they reach puberty is for them to be bred at the third oestrus rather than earlier; results from a recent study in the USA indicated that differences in embryo viability between gilts inseminated at first vs. third oestrus were related to the number of oestrous cycles and possibly to differential nutrition. Although puberty normally occurs after 150 days of age for most domestic breeds of pig, gilts can be induced to ovulate and pregnancies can be initiated at 90–120 days of age. It is evident that the gilt's uterus is responsive to the hormones of pregnancy well in advance of spontaneous puberty. The ability of such pigs to conceive, however, is marginal and in many gilts their corpora lutea are usually not maintained beyond cyclic lifespan; those that do manage to maintain their pregnancies usually have small litters. North American studies have shown that uterine responsiveness to progesterone develops between 20 and 90 days after birth.

11.3. Applications of Technology

11.3.1. Management guidelines

Management guidelines for an effective strategy to maximize reproductive performance in beef heifers have been the subject of several reviews. Such guidelines include investigation of genetic background, bringing the heifer into early puberty by exposure to bulls to induce calving at 24 months, effective culling strategy, herd size and reproductive tract scores.

The boar effect

The essential steps in applying the 'boar effect' include keeping the gilt out of sight, sound and smell of the boar until she reaches about 160 days of age; she then gets a change of environment and is exposed to a suitably mature sexually active boar. It is believed that the 'boar effect' is mediated by several stimuli acting together; tactile cues originating from the physical contact with the gilt are apparently necessary for olfactory cues to exert pheromonal effects in the gilt. The olfactory stimulus is believed to be due to compounds in boar urine and pheromones present in saliva secreted by the submaxillary glands. Numerous trials have examined the effects of boar exposure in promoting earlier puberty in pigs. In Australia, one study, working with gilts averaging 160 days of age, showed that boar exposure significantly increased the proportion of gilts attaining puberty within 60 days of commencing treatment; gilts receiving boar exposure twice daily attained puberty significantly earlier than gilts exposed once daily.

Stress and housing

In Denmark, workers examined boar-induced puberty in gilts in relation to stress and housing; they found that gilts that reacted less fearfully towards boar stimulation responded more often by attaining puberty within 10 days than gilts that responded more fearfully. Individually housed gilts were more restless during oestrus and more responsive to novel stimuli and their surroundings, compared with pair-housed gilts which showed little increase in restlessness or social activity during the trial. Some work has suggested that exposure of gilts to an oestrous female pig can be effective in reducing the age of puberty. Among studies reporting on factors which may influence the stimulus value of boars in the induction of puberty, mating frequency and the nutritional status were found to have no effect. In Australia, one study reported on the effects of contact frequency and transport on the efficacy of the boar effect on the induction of puberty; a combination of frequent boar contact and transport proved to be significantly more successful in inducing puberty than frequent boar contact alone.

11.3.2. Hormonal induction of puberty

Pigs

As first shown way back in the mid-1930s, after the age of about 3 months it is possible to induce follicle development, oestrus and ovulation in the prepubertal gilt with gonadotrophins such as PMSG; if such animals are bred, fertilization and early embryonic development will usually occur. However, whether the gilt remains pregnant or not may largely be a question of its physiological age at the time of treatment. In this regard, the pig is unlike the prepubertal sheep, in which it is possible to establish pregnancy and see it maintained in the young ewe without undue difficulty. For such reasons, the use of gonadotrophins in prepubertal gilts must be approached with due caution. One preparation commonly used is PG600 (a combination of PMSG/hCG gonadotrophins).

Although oestrus control is relatively well established in cyclic gilts, treatment with exogenous hormones has not been a useful method for the controlled induction of puberty (Fig. 11.4). Problems associated with gonadotrophins that preclude their adoption as a routine procedure include variable oestrous responses, poor

Fig. 11.4.   Pregnancy failure after induced oestrus in gilts. Although oestrus and ovulation can be induced with relative ease in prepubertal gilts, the maintenance of pregnancy may not necessarily follow. This is in contrast to what happens in sheep and cattle, where pregnancy is usually continued much the same as in the postpubertal animal. Various strategies need to be employed with pigs to overcome the problem.

reproductive performance and failure to maintain cyclic activity in non-pregnant gilts. In recent times, a greater understanding of the basic mechanism mediating nutritional effects on reproductive performance has emerged; the metabolic changes mediating such nutritional effects involve change in insulin status, which influences LH secretion and probably increases the sensitivity of the ovaries to exogenous gonadotrophins.

Beef heifers

Although there was a lack of precise knowledge regarding all the endocrine mechanisms involved in puberty, sufficient was known by the mid-1970s to provide a basis for a hormonal treatment capable of inducing oestrus in prepubertal cattle; the regimen at that time involved a 9-day progestogen treatment and adjunct oestrogen/progestogen as in oestrus control measures then available for cattle; the age of the heifer for a given live-weight was found to be important in determining the success of the induction treatment. Such short-term progestogen treatments could result in

approximately 90% of prepubertal beef heifers exhibiting oestrus and more than 50% becoming pregnant to breeding within a week of progestogen withdrawal.

What this stimulus to puberty involved was the application of a standard oestrus control treatment; presumably, this was sufficient to activate the hypothalamic–pituitary–ovarian axis to initiate cyclic breeding activity, as it may also do in many instances when applied to post-partum cows. However, there were those who reported variable results with the induction treatment, with some heifers returning to their previous prepubertal anoestrous condition after one cycle; clearly, age and other factors are likely to influence treatment outcome. For that reason, many papers have dealt with the effect of nutritional management and trace mineral supplementation as well as endocrine treatments on the attainment of puberty in beef heifers.

In Montana, workers used norgestomet implants (Synchromate-B) for a 10-day period in crossbred heifers and recorded a significant increase in the number of heifers that attained puberty (89% vs. 71%). Other work in Montana involved an examination of the effects of age on the induction of puberty in beef heifers by a progestogen (6 mg norgestomet) implanted for 10 days at 9.5, 11.0 and 12.5 months of age; at 12.5 months of age more treated heifers exhibited a pubertal oestrus within 5 days of implant removal than controls (82% vs. 9%), but the progestogen did not induce a response at the earlier ages. It was concluded that progestogens induced puberty by hastening the normal cascade of endocrine and ovarian events associated with spontaneous puberty; age appeared to be a critical factor influencing the efficacy of the hormone treatment. In Kentucky, studies involved the short-term (8-day) treatment of prepubertal Aberdeen Angus heifers with melengestrol acetate (MGA); this was capable of stimulating LH secretion and follicular growth and an earlier onset of puberty.

The effectiveness of treating beef heifers with controlled internal drug release (CIDR) devices and oestrogen to induce puberty has also been the subject of studies in North America; results suggested that a combination of short-term progestogen treatment with oestradiol benzoate adequately mimicked the normal endocrine mechanism for inducing oestrus and normal luteal function in most prepubertal heifers. An additional positive effect was that the induced oestrus occurred within 2–4 days of removing the CIDR devices. In South Dakota, workers examined the effect of progestogen and GnRH on the attainment of puberty in beef heifers; norgestomet-treated heifers were 26.2 days younger and 21.6 kg lighter than controls at puberty; norgestomet plus GnRH-treated heifers were 30.9 days younger and 18.8 kg lighter than controls at puberty.

Sheep

Although there is a considerable literature now available on the control and induction of oestrus in adult ewes, reports dealing with ewe lambs are much fewer. In France, where controlled breeding has been used on some scale in ewe lambs, the FGA-impregnated sponge–PMSG regimen has been employed as the induction method; the general rule is that the animals must be older than 7 months and at least 60–65% of their adult weight. Clearly, there is little merit in inducing oestrus in young sheep that are incapable of rearing lambs because of size or general unsuitability. Companies marketing sponges in France prepared devices specifically for ewe lambs. Care is necessary to ensure that there is no tissue damage as a result of sponge placement, which would otherwise lead to difficulties in the removal of the device.

Immunization in sheep

Among methods employed experimentally in sheep to advance puberty in ewe lambs has been active immunization against inhibin early in life. An Australian report showed that immunization of ewe lambs against one form of inhibin (alpha peptide 1-32) advanced puberty (time of first ovulation); immunization against inhibin also resulted in persistent increases in ovulation rates in subsequent breeding seasons. Although active immunization against inhibin as a method to induce early and enhanced gametogenesis may advance puberty in ewe-lambs, in ram lambs this form of treatment has not been effective (Wheaton and Godfrey, 2003).

Melatonin in deer

New Zealand workers have shown that melatonin treatment can be used to advance the breeding age of red deer hinds by about a month; the majority of 14-month-old hinds treated with melatonin implants for 1 month showed early ovarian activity.

12

Cloning Technology

12.1. Practical Implications of Technology

12.1.1. Historical

Serious research in nuclear transfer (NT) dates back, in amphibians, to the work of Briggs and King in the early 1950s, who reported the first successful transplants in frogs. Despite the long history of work in amphibians, successful cloning in farm animals has taken many years to perfect, partially because of the many technical difficulties in working with the much smaller mammalian oocyte, 1000 times smaller in volume than the frog oocyte. Any historical account of cloning must take account of the part played by Britain in developments in this field; the birth of the first mammalian clone and early work on embryonic stem cells took place in that country.

Splitting the embryo

A limited form of cloning in farm animals has been available for almost two decades by splitting an early embryo or transferring embryonic cell nuclei into enucleated oocytes. Since 1997, cloning has been possible with adult somatic cells. As noted by Wilmut (2003), there have been two periods in the history of cloning in farm animals. The first began with the pioneering efforts of Willadsen at the Animal Research Station in Cambridge using donor nuclei from early embryos in NT; the second began a decade later, in 1996, when Roslin workers described the birth of lambs from differentiated, cultured, embryo-derived cells and a year later from fetal and adult sheep cells. The successful cloning of sheep from an adult somatic cell and the birth of 'Dolly' was to capture the imagination of researchers and public alike around the world and marked a scientific breakthrough destined to play an important role in the development of new procedures for genetic engineering in farm animals (see Chapter 13).

Commercial use of embryo splitting

The technique employed to produce demi- and quarter-embryos in the late 1970s, although of interest to those researching in developmental biology, was too tedious and time-consuming to warrant consideration for possible commercial application. By the early 1980s, however, technology had moved to splitting embryos at the blastocyst stage of development; for cattle, this meant that embryos could be flushed non-surgically from a superovulated donor cow and, after splitting, used to increase the total number of transfers from a valuable donor animal. For those with micromanipulation equipment and the required skills, split embryos could mean valuable additional pregnancies; the same equipment could also be employed to provide biopsy cell samples for embryo sexing purposes (see Fig. 12.1).

Survival of demi-embryos

In general, survival of half-embryos or demi-embryos is such that more offspring can be produced per donor female by transferring demi-embryos than by transferring embryos intact. Survival of each demi-embryo, on the other hand, is lower than that of intact embryos; additional insults applied to demi-embryos (e.g. freezing and thawing) reduce demi-embryo survival even lower. Reduced survival of demi-embryos may result from an insufficient number of viable cells to sustain normal development or because the smaller embryonic mass provides a weaker anti-luteolytic signal, which is essential for the maintenance of pregnancy. Once produced, demi-embryos have either been transferred fresh to synchronized recipients of the same species or frozen in liquid nitrogen. In some circumstances, it may be useful to give some hormonal support to the recipient animal. In the USA, workers found evidence suggesting that some goat demi-embryos may provide inadequate signalling for maternal recognition of pregnancy but that such pregnancies could be maintained with

appropriate progestogen treatment. As to the normality of young derived from split embryos, which must run into countless thousands, all the evidence supported the view that they were no different from offspring born from whole embryos. Once the hurdle of initiating a pregnancy in the surrogate mother was passed, a matter largely of cell number, the rest of prenatal life was uneventful (see Fig. 12.2).

Cambridge in the 1980s

The pioneering work of Steen Willadsen in the mid-1980s in Cambridge was to

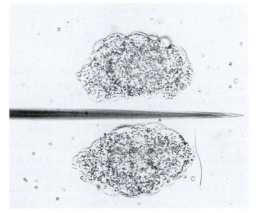

Fig. 12.2.   Demi-embryos result in normal young.

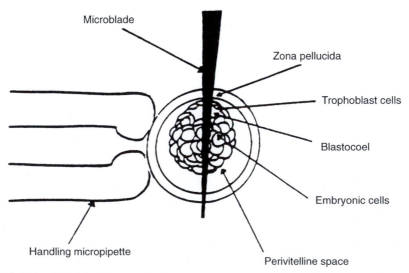

Fig. 12.1.   Splitting the bovine blastocyst.

demonstrate the possibility of producing sheep clones by fusing a whole nucleated blastomere from a donor sheep embryo with an enucleated recipient oocyte. This opened up an important new area of exploration in embryology and developmental biology. In Willadsen's procedure, a slit was first made in the zona pellucida of the recipient oocyte at a point close to the polar body. Enucleation was achieved by aspirating the polar body and some of the ooplasm. Using the same pipette, a nucleated blastomere was introduced into the perivitelline space of the oocyte and fused into the ooplasm. Fusion of the blastomere and the enucleated oocyte (cytoplast) was achieved in various ways, including the use of electrofusion; in this, the blastomere and cytoplast were aligned by alternating current and a pulse of direct current was used to effect fusion. The micromanipulation was carried out in the presence of a cytoskeletal inhibitor (cytochalasin B); use of the inhibitor allowed the removal of the metaphase II chromosomes without rupturing the ooplasm membrane (Fig. 12.3).

Willadsen's early efforts suffered from the need to develop the cloned embryo (embedded in agar) in the sheep oviduct, due to the absence at that time of a suitable *in vitro* culture system. It later became possible to replace the sheep oviduct with an *in vitro* culture system to grow the cloned embryos to the blastocyst stage. After moving to the USA, Willadsen continued his pioneering work by demonstrating the effectiveness of his technique in cattle. A paper published in 1991 dealt with several hundred transfers of cloned cattle embryos and the birth of more than 100 calves; the paper also recorded the birth of some cloned calves with exceptionally high birth-weights (up to 155 lb).

Studies elsewhere in the early 1990s also recorded that 20–30% of cloned calves were larger than usual (up to twice the normal birth weight); it was now evident to all that a new phenomenon, the 'large-offspring syndrome' (LOS), had emerged. Some proportion of cloned calves were not only overweight but could be showing evidence of defects in metabolic regulation. This was in contrast to the many thousands of cattle,

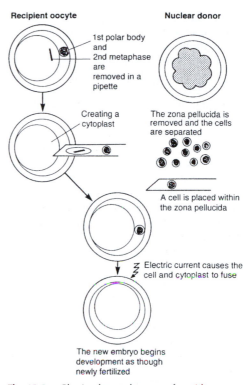

Recipient oocyte   **Nuclear donor**

1st polar body and 2nd metaphase are removed in a pipette

Creating a cytoplast

The zona pellucida is removed and the cells are separated

A cell is placed within the zona pellucida

Electric current causes the cell and cytoplast to fuse

The new embryo begins development as though newly fertilized

Fig. 12.3. Cloning by nuclear transfer with blastomeres.

sheep, pig and horse embryos that had been transferred to recipients in conventional embryo transfers without any concern about their health and normality.

Current cloning efficiency

Current cloning efficiency, as determined by the proportion of live offspring developed from all oocytes that received donor cell nuclei, is low, regardless of the cell type and animal species used. In all animals, the great majority of cloned embryos perish before reaching full term. Even in Japanese Black cattle, where some of the most encouraging results have been found, less than 20% of cloned embryos reach adulthood. Such low efficiency of cloning is believed to be largely due to faulty epigenetic reprogramming of donor cell nuclei after transfer into recipient oocytes; cloned embryos with major epigenetic errors die before or soon after implantation. Those with relatively minor epigenetic errors may survive birth and reach adulthood.

It is well known that cloned embryos contain fewer cells than normal embryos; recent reports show 9% fewer cells in cattle embryos, 19% in pig, 43% in rabbit and 55% in mice. As observed by Houdebine (2003), it is of interest to note that these proportions correlate with the difficulty of cloning these different species.

According to Yanagimachi (2002), although a trial and error approach may discover better cells for cloning, what is urgently required is a greater understanding of the molecular mechanisms of epigenetic nuclear programming and reprogramming to find the way to make cloning safer and more efficient. The same author suggests that the relatively high cloning success rate in Japanese Black cattle may provide clues to solving the problem of high mortality of cloned offspring. Questions have been raised as to how old cloned animals are in terms of their genetic age and how long they will be able to live and reproduce. In Japan, Miyashita et al. (2003) studied the telomere lengths of sperm in cloned bulls; they concluded from their data that the telomere lengths of sperm are maintained throughout ageing in normal cattle and that the same held good for bulls produced by cloning. The same workers produced calves from their somatic-cell-cloned bulls and found telomere length to be normal; they suggest that such bulls could be used as breeding sires.

Intergeneric nuclear transfer embryos

Studies reported in recent years have demonstrated the posssibility of intergeneric NT, with some researchers showing that bovine oocyte cytoplasm is capable of supporting development of transplanted somatic-cell nuclei of buffaloes, sheep, pigs, monkeys and rats. In Japan, Atabay et al. (2004) used vitrified bovine oocytes as recipient cytoplasts for buffalo somatic-cell nuclear transfer (SCNT); results from this work (Table 12.1) are encouraging to those who are trying to find new ways of producing buffalo embryos for embryo transfer purposes. However, it must be stressed that the birth of young by this approach has not been reported.

12.1.2. Human interest

There are those who believe that one of the greatest potential benefits of cloning technology may lie in therapeutic cloning as applied in human medicine, where SCNT is used to generate replacement tissues or organs. This form of cloning would avoid the risks of tissue rejection by providing a patient with new tissue of their own genetic type. The most widely quoted instance of animal cloning was the report from the Roslin Institute in Scotland by Wilmut and associates in 1997; 'Dolly' was to show the world that cloning with cells taken from the adult mammal was possible. This was after a period during which well-respected scientists, working with mice, had proclaimed that there was no possibility of such cloning.

Immortality of the soul is a view passionately held by a majority of people around the world, although with serious reservations by the minority. To some human eyes, cloning by way of an adult somatic cell held out the possibility of a form of immortality not previously considered; to others, cloning was a profound insult to the Creator

Table 12.1. Using cattle oocytes to produce buffalo embryos (from Atabay et al., 2004).

Origin of fibroblasts	No. of oocytes (replicates)	% of fused couplets	% of fused couplets that developed to		Blastocyst (cell no.)
			> 2 cells	Blastocysts	
Water buffalo	61 (3)	67.0 ± 3.8	60.1 ± 4.6	16.2 ± 4.6	85.7 ± 7.6 (7)
Domestic cattle	83 (3)	67.4 ± 2.3	67.6 ± 3.4	21.2 ± 3.2	118.2 ± 8.0 (12)

and undermined the uniqueness of the human being. There are, however, those less ready to see a horror in human cloning; if cloning could be developed without the risk of producing abnormal offspring, why should there be objections to the use of the technology in certain areas of human assisted reproduction? After all, it is not as if human clones are unheard of; the medical literature records a set of identical twins, and these are clones, once in every 250 births. Whatever the view taken, there has been huge human interest in cloning, the Roslin report sparking off banner headlines in the world press and debates in countries far and wide. However, for those in farming and those in biotechnology, the commercial development of farm-animal cloning is likely to be dependent on three factors: economics, the views of regulatory agencies and consumer acceptability.

Cloning for biomedical purposes

Commercial interest in cloning for biomedical purposes is now sharply focused on its use in the production of transgenic animals, a process not necessarily limited by current inefficiencies in the NT procedure; often, only a few cloned transgenic founder animals are required. There are those who believe that cloning can bring many benefits to society, becoming an efficient, fast and useful method of producing transgenic fetuses for cell therapies, adult animals for pharmaceutical protein production and organs for xenotransplantation.

12.1.3. Cattle

Advantages

Among the possible advantages of cloning in cattle, the following have been mentioned in various reports: (i) herds with homogeneous production, nutritional and health characteristics optimal for a given local environment could be created; (ii) identical replacement animals could be generated for such herds and production costs could be markedly reduced; (iii)

particular genotypes (e.g. casein milk producers) could be rapidly increased in number as required; (iv) the preservation of genetic material could take the form of frozen tissue samples rather than cryopreserved gametes; and (v) cloning would allow basic research to be focused on problems such as genetic imprinting, maternal effects and environment/production interactions.

Emphasis on cattle

Within 3 years of the birth of 'Dolly', several laboratories had proved the feasibility of this new technique in cattle. The transfer of a somatic nucleus into an enucleated recipient oocyte has now been confirmed in several mammalian species as a method which can give rise to live animals (Table 12.2). Despite its low efficiency, this approach has been shown to be a new means of reproducing mammals (reproductive cloning). Although initial commercial interest in cloning technology centred on the production of large numbers of genetically élite animals for agricultural purposes, this particular commercial avenue quickly lost its appeal. Several companies in the USA and elsewhere financed ambitious research and development programmes only to be faced with problems of low pregnancy rates, increased calf weights, neonatal anomalies and poor survivability of cloned calves. It was later to become clear that abnormal placental development, especially the allantois, is responsible for much of the

Table 12.2. Young after cloning by somatic-cell nuclear transfer.

Year	Species	Researchers
1997	Sheep	Wilmut *et al.*
1998	Mouse	Wakayama *et al.*
1999	Cattle	Wells *et al.*
2000	Pig	Polejaeva *et al.*
2001	Goat	Keefer *et al.*
2002	Rabbit	Chesne *et al.*
2002	Cat	Shin *et al.*
2003	Mule	Woods *et al.*
2003	Horse	Galli *et al.*
2003	Rat	Zhou *et al.*

severe fetal loss observed in the early months of pregnancy.

Élite cattle

There are two main ways in which NT is likely to be useful in cattle breeding: first, the possibility of producing multiple copies of élite cattle; and, secondly, NT from cultured bovine cell populations can be used to produce genetically modified progenitor cattle, to introduce genetic change into nucleus herds or to increase the rate of genetic progress in the general cattle population.

Commercial interest

Currently, there are probably more laboratories worldwide working on cattle cloning than on cloning in all other mammalian species combined. The success of several groups to clone cattle is a result not only of numerous research programmes focused on bovine NT, but also on the enormous base of knowledge that has been built up over the past 20 years in the application of assisted reproduction techniques in this species. Successful and repeatable procedures for *in vitro* oocyte maturation, *in vitro* fertilization and *in vitro* culture are now better established in cattle than in any other species; each of these procedures is important for successful cloning. In part, this is due to the ability to obtain large numbers of oocytes from abattoirs for use in research at relatively low cost. Exploitation of cloning opportunities in cattle is likely to depend largely upon optimizing procedures for NT: this can only be achieved through a greater understanding of how factors in the oocyte cytoplasm act upon the DNA of the transferred nucleus to regulate gene expression.

Speeding up genetic progress

Cloning has the potential to revolutionize commercial animal production by enabling the current method of genetic improvement, which involves breeding commercial females to high-merit males, to be replaced with the transfer, to them, of cloned embryos with the genetic make-up of the most outstanding animals from within an élite nucleus population. Using such a system, there would be an immediate increase in the genetic merit of the resulting commercial offspring, which might be five to ten times the current annual response. Cloning could offer opportunities to supply markets with products of uniform quality and the technique applied to hybrids could be used to disseminate the benefits of hybrid vigour. All this, of course, depends on a cloning method with a proven negligible risk of producing abnormal calves; whether that day arrives sooner or later depends on research.

Clones for research

Genetically identical animals are valuable models for research; over the years, considerable use has been made of monozygotic twins in dairy cattle research in several countries (Fig. 12.4). As natural monozygotic twin calves are rare (about one set in 2000 births), NT has offered the possibility of generating sets of animals with the same nuclear genome. In cattle, cloned animals are already used as models for nutrition; some studies have reported that the coefficient of variation of voluntary food intake, eating time and rumination periods between animals of the same clone was reduced by 60% compared with non-cloned animals. There have also been reports showing the growth pattern of cloned cattle to be more homogeneous than that of non-cloned animals. Work in France reported by Heyman *et al.* (2003a), the first systematic comparison of the developmental potential of oocytes from cloned cattle, indicated that cloned cattle yielded similar numbers of oocytes and embryos to those of non-genetically related control animals.

12.1.4. Horses

Advantages

Cloning of horses is currently in the early development stage, but its eventual

Fig. 12.4. Identical cattle twins by micromanipulation.

application to the needs of those in the horse industry may see it taken up by the owners of high-genetic-value animals. As a means of retrieving the genetics of a valuable animal that has died prematurely or is barren for one reason or another, cloning may be the only option. Many of the outstanding eventing horses are geldings and as such are certainly not in a position to reproduce themselves by the usual means; using donor cells from valuable geldings could be the way towards producing clones of that particular genotype. Using current cloning technology, there is concern about mitochondrial inheritance and epigenetic modification arising from NT. However, there is evidence that epigenetic changes in cloned offspring are not carried over into subsequent generations. The application of knowledge derived from the Equine Genome Sequencing Project may prove particularly rewarding; breeders will be able to make well-informed mating decisions, which may improve the health of horses by screening donor animals for heritable diseases.

There appear to be marked differences in cleavage rate after NT in the horse in comparison with cattle and other farm mammals. Several studies have investigated the basic conditions required for the production of horse embryos by the transfer of the nuclei of fetal and adult fibroblast cells to enucleated oocytes; results reported by Li *et al.* (2002) in the UK, for example, demonstrated a very limited potential for *in vitro* development of horse embryos after nuclear reprogramming following transfer of such fibroblasts. In further studies, Li *et al.* (2003) recorded half the cleavage rate in reconstructed equine oocytes (25–27%) in comparison with cattle data; these workers suggested that progression to the two-cell stage in reconstructed horse oocytes needed to be much greater before attempts to produce viable cloned embryos in the horse were likely to be successful. However, in Italy, the work of Galli *et al.* (2003b) was to add the horse to the list of mammals so far successfully cloned from an adult somatic cell. The authors of the report note that cloning might now enable gelding champions to contribute their genotype to future generations. The cloned foal, named 'Prometea', was derived from a cell taken in a skin biopsy from the recipient mare mother; the clone was therefore genetically identical to the mother that carried it to term.

12.1.5. Pigs

Although SCNT has been reported in cattle, sheep and goats, in the pig the viability of such embryos has proved to be particularly poor, with an extremely low rate of cloned piglet production. Some of the problems associated with successful cloning in pigs are species-specific; in the pig, there is the difficulty that several (at least four) good-quality embryos are required to maintain a pregnancy. Work in the USA by Polejaeva *et al.* (2000) led to the birth of a litter of five cloned pigs, by the use of a new NT method in which fertilized zygotes were used as cytoplast recipients. As developmentally competent embryos are rare with current NT technology, there is the need to transfer large numbers of reconstructed embryos into pig recipients. It is known that many factors are involved in the development of porcine SCNT embryos and that culture medium and recipient oocyte are important factors.

Fetal fibroblasts

The developmental competence of porcine oocytes/embryos derived *in vitro* is lower than those produced *in vivo*, which suggests that the current *in vitro* production system is suboptimal. In South Korea, studies by Hyun *et al.* (2003) showed that sow oocytes have a greater maturation rate and developmental competence than gilt oocytes when used for SCNT, regardless of the maturation medium used. Studies in pigs in the same country reported by Lee *et al.* (2003) demonstrated that the type of donor somatic cell is important for improving reconstructed embryo development after SCNT; they found fetal fibroblasts to be the most effective among the donor cells evaluated, using both *in vitro*- and *in vivo*-matured oocytes as recipients.

12.2. Developments in Cloning Technology

Somatic-cell cloned animals, produced by NT of cultured somatic cells into enucleated oocytes, have so far been produced from most farm-animal species as well as in mice and some other laboratory animals. There are two essential components of the NT process: the donor nuclear genome (karyoplast), which is the target for clonal replication, and the enucleated oocyte (cytoplast), whose cytoplasmic make-up is sufficiently competent to facilitate genome reprogramming and to support embryonic and fetal development to term.

12.2.1. Oocyte sources

The sources of oocytes used as recipients in cloning programmes are likely to vary according to the species. In the cow, the source is likely to be oocytes artificially matured after recovery from abattoir ovaries or by way of ovum pick-up (OPU) from live donor animals. In pigs, it may be a matter of recovering oocytes from the ovaries of pigs after predicting the time of ovulation by real-time ultrasonics; a report by King *et al.* (2001) in Scotland suggested that transcutaneous ultrasonography may be a reliable method of predicting the time of ovulation after hormone treatment. In small ruminants and pigs, surgery is usually required to obtain oocytes from the oviducts of live animals. In the cow, whether a matter of using ovaries from abattoirs or recovering oocytes from the live animal, the procedures do not bring welfare concerns.

Nutritional effects

Some studies in sheep have shown that the nutrition of the oocyte donor ewe can influence the success of somatic-cell cloning; such results make it all the more important for more research on the effect of nutrition on oocyte quality to be conducted. Quite apart from cloning considerations, it is of interest to understand the reasons for this in the context of fertility in sheep as a whole. Results presented by Peura *et al.* (2003) in Australia are among those showing that nutrition of the oocyte donor in cloning studies influences the outcome of NT, as reflected by a greater pregnancy rate in high-nutrition sheep compared with low-nutrition animals.

In vivo vs. in vitro *oocytes*

In pigs, Adams *et al.* (2003) compared the developmental potential of NT embryos derived from *in vitro*- as opposed to *in vivo*-matured oocytes; although the fusion, cleavage and blastocyst formation rates were similar, the quality of embryos (based on number of nuclei) was higher in those derived from *in vivo*-matured oocytes.

12.2.2. Nuclear transfer procedures

There are three procedures currently used to produce cloned animals. The cell-fusion method involves placing a donor cell in the perivitelline space of an enucleated recipient oocyte and fusing the donor and recipient cells with electrical pulses; the second method is one in which the donor nucleus is injected into enucleated oocytes by piezo-actuated microinjection; and the third is one involving a whole-cell injection into the oocyte (Lee *et al.*, 2003c). There are studies showing that direct nuclear injection using the piezo drill is an efficient method for NT in horse and cattle oocytes and that sperm extract can efficiently activate horse oocytes after NT (Choi *et al.*, 2002). With this method, however, prolonged manipulation of the donor cell is required to isolate the nucleus.

The third method, as reported by Lee *et al.* (2003c), is a technique involving direct injection of a whole cell into an enucleated oocyte, effectively bypassing both the fusion and the nucleus isolation process; using this new cloning procedure, it has proved possible to produce cloned piglets from the fibroblast cells of an adult transgenic sow. The whole-cell injection procedure resulted in a blastocyst development rate as high as 37%. This success rate was attributed to the fact that the method reduced the manipulation time of donor cells and recipient oocytes compared with the other two NT procedures and to the fact that whole-cell injection assured delivery of all cellular components to the enucleated oocytes.

When the nucleus from a fetal or adult cell is inserted into an oocyte, the genetic information in the nucleus is able to direct the development of the whole organism, but it is becoming increasingly clear that this direction may often be seriously faulty; the reactivation of genes that are silent in most adult tissues, but needed for early development, is defective. Regardless of the arguments for and against cloning, studies in this area will inevitably provide much insight into the biology of cell differentiation. In due course, it seems likely that such research will lead to more efficient and safer applications of reprogramming in agriculture and for biomedical purposes.

12.2.3. Reprogramming the nucleus

Incomplete reprogramming

In cattle and other farm species, the preparation of developmentally competent enucleated oocytes is a key factor that determines the overall success of cloning. However, the low efficiency of somatic-cell cloning is a major obstacle to the adoption of this technology, in the laboratory and on the farm. It is believed that incomplete nuclear reprogramming following the transfer of donor nuclei into recipient oocytes has been implicated as a primary reason for this low efficiency of the cloning procedure (see Miyoshi *et al.*, 2003); the mechanisms and factors involved in nuclear reprogramming have yet to be fully understood, but identification of these factors is likely to increase efficiency of cloning. Studies by Tani *et al.* (2003) in Japan have shown that reprogramming is not directly regulated by either maturation-promoting factor (MPF) or mitogen-activated protein kinase (MAPK) activity in bovine oocytes.

Of interest is the technique reported by Boiani *et al.* (2003), suggesting that the relatively simple procedure of embryo aggregation may markedly improve the outcome of cloning in mice; by aggregating two or three cloned embryos at the four-cell stage, they achieved an eightfold increase in the number of young born (Fig. 12.5). It is possible that the presence of additional blastomeres corrects various defects in the

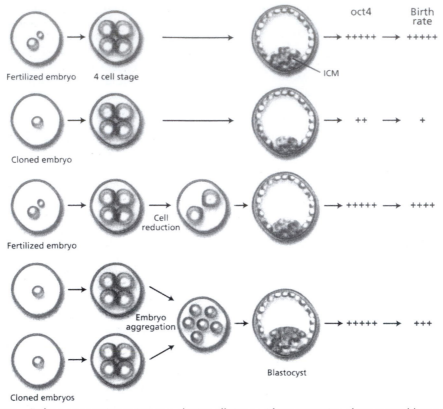

Fig. 12.5. Embryo aggregation to improve cloning efficiency. After aggregation, these mouse blastocysts contained more cells and resulted in higher birth rates (+++). It is possible that this effect was due to improved communication between the cells of the embryo, which favoured genome reprogramming during embryo development. ICM, inner cell mass. (From Houdebine, 2003.)

cloned embryos by acting as a source of key substances for reprogramming. Reviewing this new evidence, Houdebine (2003) suggests that embryo aggregation may be one way of improving cloning efficiency, and it will be of interest to see whether this approach can be extended to other species, particularly farm animals.

Normality of adult clones

It is evident from many studies that adult clones do not differ significantly from *in vivo*-derived animals, which is in marked contrast to the large variability that is usually found within cloned fetuses and newborn young. According to Cezar *et al.* (2003) such findings indicate that there is an epigenetic reprogramming threshold for normal development. Clones that have survived to adulthood have clearly shown an ability to overcome epigenetic challenges derived from their somatic-cell origin. An adequate understanding of how this threshold is achieved in some clones but not in others may suggest approaches that can improve cloning efficiency.

Source of donor cells

Early and late gestation losses in pregnancies from SCNT have been linked to incomplete reprogramming of the donor karyoplast. There is evidence showing that the incidence of fetal losses in bovine embryo (blastomere) NT is significantly less than in SCNT. In the USA, Du *et al.* (2002) reported data suggesting that embryonic and somatic nuclei require different recipient cytoplast environments for remodelling/reprogramming; this is probably due to the different cell-cycle stage and profiles of molecular differentiation of the transferred donor nuclei. For such reasons, Behboodi *et al.* (2003) used a recloning protocol in an effort to improve the reprogramming of SCNT; they used blastomeres from NT embryos derived from transfected somatic cells as donors in their transgenic goat founder programme. These workers found that the somatic-cell-generated NT embryos at the 16–32 cell stage could be used for embryo cloning in goats.

Incomplete nuclear reprogramming following the transfer of donor nuclei into recipient oocytes has been implicated as a primary reason for the low efficiency of the procedure (Miyoshi *et al.*, 2003); although many research groups are working in this area, the mechanisms and factors that affect the progression of the nuclear reprogramming process have yet to be fully elucidated. Some of the factors known to be important in determining the success of cloning efforts are detailed in Fig. 12.6.

Improving competence of donor cells

Studies in Korea showed it to be possible to improve the *in vitro* development of bovine cloned embryos by exposing donor somatic cells to β-mercaptethanol (ME) or haemoglobin (Hb). These studies were subsequently extended to show that the use of ME-treated donor cells and the culture of cloned embryos in ME + Hb

supplemented medium yielded the best results in promoting the production of better-quality blastocysts (Park *et al.*, 2004); treatment with ME + Hb apparently decreased blastomere apoptosis in the cloned embryos.

12.2.4. Gestational and perinatal losses of clones

Recently developed assisted reproductive technologies, such as *in vitro* embryo production, cloning and transgenic technology, have encountered perinatal morbidity and mortality of the offspring produced, which has severely limited the commercial application of these techniques. Various perinatal complications have been reported after the transfer of cloned embryos to surrogate mothers. These complications include increases in the duration of pregnancy, birth weight, incidence of parturient dystocia and susceptibility to neonatal infections. There have been studies, however, which have shown that such problems may be alleviated by appropriate management of animals at the time of delivery. It is believed that many of the parturient complications may be due to inadequate signalling between mother and fetus during the preparation for parturition; in sheep, for example, it appears that in some instances there may be problems with either the maturation of the fetal hypothalamic–anterior pituitary axis or the signals necessary to increase plasma cortisol levels.

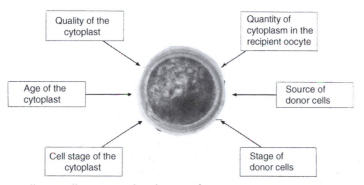

Fig. 12.6. Factors affecting effectiveness of nuclear transfer.

Facilitating delivery of clones

In an effort to minimize losses in clones at birth, induction protocols have been devised, using a combination of glucocorticoids and oestrogen injections, to ensure correct preparedness for delivery in foster-mother ewes, thus ensuring more adequate fetal pulmonary and glucogenic maturation and improved development of the mammary gland, which facilitates the proper adaptation of offspring after birth (Ptak *et al.*, 2002). In cattle, Funk (2002) noted that, historically, there has been a lack of spontaneous parturition in NT animals produced by SCNT; in an attempt to circumvent some of the problems experienced with calves at birth, the author described various measures adopted at a commercial centre (Trans-Ova Genetics) to induce recipients to calve 1 week earlier than full term; it was considered that this measure had a favourable effect on the survival of the calves.

12.2.5. The large-offspring syndrome (LOS)

It is evident from several studies that a proportion of genes in mammals (estimated at 0.1–1% of all genes) is repressed on one of the chromosomes and that this is dependent on the parental origin of the gene; this is known as genomic imprinting. In mammals, imprinted genes are clearly implicated in the regulation of fetal growth and in the development and function of the placenta. Imprinting is one of the best-known systems of epigenetic gene regulation and has been dealt with in various reviews (Reik *et al.*, 2003). The fact that embryo loss in *in vitro*-produced (IVP) embryos and NT embryos is apparently linked to compromised placental function has focused attention on placental development. There are many studies, particularly in sheep, showing a strong correlation between placental size, fetal birth weight and perinatal survival, which emphasize the importance of normal placental development. In the farm mammals, which have a chorio-allantoic placenta, the fate of the embryo and fetus

depends on the appropriate formation and vascularization of the allantois. The allantois is formed as an endoderm-derived diverticulum from the hind-gut and its vascularization develops from the splanchic mesoderm. The establishment and function of the chorio-allantoic placenta requires proper formation of the germ layers (endoderm, mesoderm and ectoderm) during the process of gastrulation. Although some features of gastrulation have been covered in reports over the years, the information available has remained limited.

It is known that, after shedding the zona pellucida, which occurs around days 8–10 in cattle, the embryo is transformed from a spherical to an ovoid form. This marks the onset of embryo elongation, which is observed between days 12 and 14. The first hypoblast cells (also known as the primitive endoderm) are formed from the inner cell mass (ICM) around day 8 and in a further 2 days these cells have established a confluent hypoblast lining the inside of the trophoblast. At about day 12, the ICM, now known as the epiblast, displaces the thin overlying trophoblast lining and establishes the embryonic disc. By day 14 to 16, formation of the mesoderm occurs and by day 20 the allantois buds into the extraembryonic coelom. Further information on development of the bovine embryo and its associated membranes during the 2 weeks following hatching has been provided by studies in Denmark (Maddox-Hyttel *et al.*, 2003).

12.2.6. Development of clones after birth

Rhind *et al.* (2003) have provided evidence based on detailed pathological studies of a group of lambs that were not viable after birth; such evidence may raise questions about the validity of statements that surviving clones are apparently normal and healthy. The authors point to the need for similar studies to be conducted by other groups working with cloned animals; it was concluded that looking harder at failures might provide the information necessary for cloning success.

12.2.7. Simplification of technology

Although bovine SCNT is being performed in many laboratories around the world, successful application of NT technology is a difficult task, requiring expensive equipment and highly skilled personnel. A paper by Vajta *et al.* (2003) was the first account describing SCNT carried out by hand, i.e. without micromanipulation. The paper dealt with a technique that was regarded as a viable alternative to the traditional methods of SCNT. The authors also dealt with work which they had conducted to optimize the chemical environment for oocyte bisection, fusion methodology, the medium used to culture the reconstructed cloned embryos and the transfer of cloned embryos, either fresh or vitrified, into recipient cattle. It was found that the cattle serum employed in their studies was superior to other protein sources for *in vitro* embryo development; a possible explanation of this effect was the marked mitogenic activity of the cattle serum compared with that of commercially available fetal calf serum.

In a report by Pedersen *et al.* (2003), pregnancies established after the transfer of embryos produced by Vajta's novel method of hand-made somatic-cell cloning were evaluated; the fate of these embryos was an almost complete tale of disaster, with the pregnancy rate falling from 35% at 28 days of gestation to 6% at 250 days. The pregnancies generally ended in early fetal death, hydrallantois, large placentomes and fetal overgrowth. In Germany, Bhojwani *et al.* (2003) used the 'zona-free' method of SCNT as described by Vajta and reported the first birth of a calf by this procedure; this success followed the transfer of 18 blastocysts to 18 recipients, of which only one gave birth. Clearly, while simplication of cloning procedures is a worthy goal, in the final analysis it is the ability of the technology to deliver a normal healthy calf that is the prime consideration.

13

Production of Transgenics

13.1. Practical Implications of Technology

The elucidation of the structure of DNA by Watson and Crick in Cambridge in 1953 heralded the start of an era in biochemical research in cellular biology that has steadily gathered momentum over the past five decades. The application of genetic engineering in plants and animals is certain to be of immense value in expanding the range of options available in the 21st century to meet global food requirements; it is also certain that the technology will bring immense benefits in many other ways, particularly in human medicine. Genetic engineering of livestock to produce animals with altered traits such as disease resistance, wool growth, body growth and milk composition is being researched in laboratories worldwide (Fig. 13.1). Quite apart from the possible economic benefits of genetic engineering, there is an urgent need to keep the public well informed about the possible benefits.

Although the uproar over genetically modified (GM) foods in countries of the European Union (EU) was in marked contrast to what occurred in the USA, consumer sensitivities to genetic engineering must be kept very much in mind. Dealing with the press and media generally requires professional skills of the highest order. It is not a matter of trying to tell anything other than the truth but it is the way in which that information is presented. During the past two

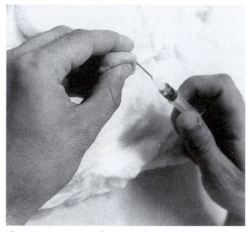

Fig. 13.1. Source of oocytes for use in transgenic studies.

decades, research aimed at the production of transgenic farm animals has been steadily progressing and many see future development in this area as being closely allied to cloning technology. In the early 1980s, transgenic farm animals of various species, but mainly pigs, suffered from the ways in which transgenes were expressed in their bodies. This led to serious criticism of the emerging technology, which is still carried in the minds of many who remember those early days. Current methods take due care to use more site-specific and targeted gene constructs to minimize the risk of inappropriate expression of genes in vulnerable tissues.

13.1.1. Historical

The term 'transgenic' was coined in the early 1980s to describe mammals that had new genes incorporated into their genome; several research groups at that time had reported success in gene transfer and the development of transgenic mice. Since those days, there has been a rapid advance in the application of genetic engineering technology to increasingly complex organisms, from bacteria and yeasts to fish, birds and mammals, including most farm mammals. For such reasons, quite apart from their possible use in commercial ventures, transgenic animals are among the most exciting research tools found in agricultural and biological science; there are literally hundreds of articles and several comprehensive books that have dealt with the production of transgenic animals. Some of the landmark events along the way towards the production of transgenic farm animals are noted in Table 13.1.

One of the earliest studies to influence the thinking of those working in animal agriculture was the work of Palmiter and Brinster in the early 1980s; their reports, featuring 'Supermouse', an animal growing to become 100% larger than its littermates, was a dramatic example of mice transgenic for growth hormone. Although farmers have, for centuries, manipulated the genetic constitution of their farm animals, usually in attempts to make them more productive, it is only in the last quarter-century that developments in recombinant DNA technology have enabled the molecular biologist to isolate and modify genes in such a way as to enable gene constructs to be introduced into the farm-animal genome. For those interested in cattle cloning and transgenic research, the fact that reasonably effective methods of maturing bovine oocytes and culturing embryos became available around 1990 made it possible to use cheap abattoir-derived ovaries and to substantially reduce the need for live cattle and the costly facilites normally associated with them.

Manipulating genes

The huge research commitment made to recombinant DNA technology in the 1970s and 1980s has brought developments in many fields, ranging from the food industry to farming and biomedicine. Gene constructs have been introduced into most species of food animal, including cattle, sheep, goats, pigs, rabbits, poultry and fish. Genetic modification techniques, developed initially in laboratory strains of selected bacteria and viruses, are essential tools for understanding the genomes of farm livestock. These tools enable researchers to isolate, sequence and characterize genes; identifying key genes affecting animal productivity is possible only by using such techniques. A review by Crawford (2003) described the techniques now in use and showed how molecular geneticists may eventually use these to help identify genetic alterations and to breed healthier and more productive animals. As with any new

Table 13.1. Landmark events in transgenic research.

Year	Event	Researchers
1971	First demonstration of sperm-mediated gene transfer	Brackett and associates
1977	mRNA transferred into *Xenopus* eggs	Gurdon
1980	mRNA transferred into mammalian embryos	Brinster and associates
1980–1981	Transgenic mice first recorded	Palmiter and associates
1981	Transgenic mice first documented	Six research groups
1982	Transgenic mice demonstrate an enhanced growth phenotype (Supermouse)	Palmiter and associates
1985	First transgenic pigs and sheep produced	Hammer and associates
1997	Production of transgenic sheep by nuclear transfer of nuclei from transfected fetal fibroblasts	Schnieke and associates

technology, a new language makes communication easier among participants in the field but more difficult for others to understand the technology.

Animal welfare considerations

A major concern of agricultural researchers must be in ensuring that applications of recombinant DNA technology do not compromise farm-animal welfare. Several unfortunate examples from early attempts to apply the technology have been widely quoted. Farm animals that are destined to produce novel animal products by way of genetic engineering are likely to raise questions far removed from technical problems in the laboratory. It is important that the public becomes aware of the implications of the work and that researchers should discuss animal welfare, food safety and other considerations with them. In the USA, the Food and Drug Administration (FDA) has advised companies developing somatic-cell nuclear clones for agricultural use that products from cloned animals or their progeny may not be entered into the human or animal food supply until its evaluation of the issues involved has been completed.

On the other hand, it may well be that the modification of disease resistance or disease susceptibility by gene transfer will eventually come to be seen as a major asset to animal welfare as well as to the economics of livestock production.

Commercialization of products of transgenic animals

As yet, no protein produced in a transgenic animal has been taken all the way through to regulatory approval and commercialization (Powell, 2003); for that reason, there are several uncertainties facing those engaged in this area of biotechnology. It is still not clear that producing human therapeutics in transgenic animals is a viable business strategy; this is likely to remain so until an FDA-approved drug finally

reaches the market-place. According to Powell (2003), some commercial companies may have tried to develop too many products at once or concentrated on pharmaceutical products which have no clear end-point, such as an obviously cured medical condition, when used in clinical trials (Table 13.2).

For cattle genetically engineered to produce high-quality therapeutic proteins in their milk, there may be caution about using them due to fears about bovine spongiform encephalopathy (BSE), which could lead regulatory authorities to be reluctant to approve such drugs for human use. In this context, it is of some relevance to note that recent work in Korea by cloning researcher Woo-Suk Hwang claims to have resulted in the birth of 'BSE-proof' cloned calves. More than 100 people have died, most of them in Britain, since BSE was found to have crossed from cattle to humans in the 1990s. Although the incidence of new variant Creutzfeldt–Jakob disease (CJD) in humans is currently decreasing rather than increasing, the nature of the disease in cattle is such as to warrant the most stringent human health precautions. The disease in cattle and humans is believed to be caused by prions, aberrant forms of a normal protein, known as PrP; prions convert other PrPs to their own misshapen form, causing a build-up in plaques of protein and the death of adjacent nerve cells. Although costs would rule out the use of the Korean cloned animals in conventional milking herds, they could prove valuable when used on a

Table 13.2. Transgenic products from animals currently in clinical development.

Company/ sponsor	System	Product and its possible use
GTC	Goats	Antithrombin III for antithrombin deficiency, rheumatoid arthritis and multiple sclerosis
PPL	Sheep	Alpha-1-antitrypsin against cystic fibrosis and oedema

limited scale for producing pharmaceutical proteins.

13.1.2. Cattle

Milking cows

Productivity traits in cattle and goats that are major targets for genetic engineering include the quality and quantity of caseins, lactose and butterfat. The milk of cows is currently an important food source, whether in its liquid form or as cheese and a variety of processed products. Approximately 80% of milk protein is casein, which makes this one of the most valuable components of milk because of its nutritional value and processing properties. For such reasons, an important research area in the production of transgenic cattle lies in attempts to improve the quality and quantity of casein in milk. Milk is also a unique source of certain other proteins, such as those in whey (α-lactalbumin and β-lactoglobulin).

Altering milk composition

Workers in New Zealand have shown that it is feasible to substantially alter a major component of milk in high-producing dairy cows by a transgenic approach, thereby improving the functional properties of dairy milk. Brophy *et al.* (2003) introduced additional copies of the genes encoding bovine β- and κ-casein into bovine fibroblasts and used nuclear transfer to produce transgenic calves; when they started milking, nine transgenic cows produced milk with 8–20% more β-casein and a twofold increase in κ-casein. The control of milk composition by genetic engineering could improve the processing characteristics in milk and have profound effects on the milk-processing industry.

High-casein milks

According to Brophy *et al.* (2003), it will take about 4 years to introduce the transgenes which they produced into the dairy cattle population on a large scale. Once a highly expressing founder line has been identified, it becomes possible to expand the number of homozygous animals within a year by way of conventional reproductive technologies. Larger-scale production herds would require the generation of homozygous transgenic bulls, which could then be used to transmit the novel genotype rapidly and at low cost by way of artificial insemination. Milks with increased casein content could be exploited in the manufacturing of milk-protein concentrates and casein products.

Edible casein is used in vitamin tablets, instant drinks and infant formulas; technical acid caseins are used for paper coatings, cosmetics, button-making, paints and textile fabrics. There is a considerable international trade in casein and the development of high-casein milks could have a substantial economic impact. It is likely that designer milks, speciality milks or humanized milks may be competing in the next decade to capture part of the global dairy product market worth US$400 billion annually.

13.1.3. Pigs

Hearts and valves

The transplantation of organs and tissues between animal species, or xenotransplantation, is the focus of a very active field of research, due primarily to the increasing shortage of allogeneic donor organs. Pigs have been used for many years as a model for investigating various human medical conditions and as a source of materials in dealing with various problems (e.g. heart valves); because of size and anatomical similarities, the pig is believed to be the most suitable non-human organ source. The use of the animal would be greatly enhanced by the ability to modify the pig genome. One of the major constraints to using pig organs for xenotransplantation is human antibody-mediated hyperacute rejection (HAR), and it has been necessary to devise appropriate strategies to deal with this problem.

As noted by Prather *et al.* (2003), clinical use of transplantation has become a major feature in dealing with many forms of terminal organ failure in human patients; the authors point to more than 80,000 patients waiting for organs in the USA in 2001, but with less than one-third receiving transplants. None of the current human possibilities offers a solution to this acute shortage of organs for transplantation. The pig, on the other hand, in which the physiology of most of the major organs is similar to those of humans, is available in large numbers (> 95 million used for food annually in the USA) and it possesses organs of the required size for transplantation to humans.

Islet transplantation

Juvenile-onset diabetes is a major health problem and exogenous insulin therapy is only partially successful in preventing its many complications. Although islet transplantation holds great promise for a cure, the number of potential human pancreas donors makes it extemely unlikely that there is enough islet tissue to provide for the needs of millions of human patients worldwide. The xenotransplantation of pig islets is currently being investigated as a potential alternative clinical therapy; however, the use of islet tissue, as well as hearts, must await clarification of certain health risks. Although genetic engineering may overcome the natural immune barrier, xenotransplantation to human patients may result in exposure to replication-competent porcine endogenous retrovirus. Obviously, extreme caution is the keyword in all attempts to make progress in this field of research.

Carcass quality considerations

Quite apart from the possible use of transgenic pigs in biomedical applications, modification of the pig genome could have several useful applications on the farm. It could provide animals with an altered carcass composition which would make them more acceptable for human health, alter growth rates to make animals more efficient and produce pig strains that are resistant to specific diseases.

13.1.4. Sheep and goats

Pharmaceutical proteins

Although the production of valuable recombinant pharmaceutical proteins in the mammary glands of transgenic livestock has progressed to the stage where it may become a commercial possibility, the technology of generating transgenic animals is currently inefficient, time-consuming and labour- and cost-intensive. In comparison with cattle, sheep and goats have the advantage of a much shorter generation interval and a greater number of offspring. At least five different pharmaceutical proteins have been produced in the mammary glands of sheep and the usual method of producing transgenic animals has been by way of the microinjection of DNA constructs into the pronuclei of zygotes. In goats, animals have also been developed carrying genes that encode for pharmaceutical, diagnostic and industrially important proteins targeted to the mammary gland for expression. Among such proteins, human serum albumin (HSA) is one which is regarded as especially valuable; the normal source of HSA is outdated donated blood. Among reports in recent years is one by workers in Israel dealing with the birth of transgenic kids carrying the HSA gene.

Developing industries

The production of recombinant proteins in the milk of transgenic animals has attracted the attention of several research groups in the past decade due to the outstanding protein synthesis capacity of the mammary gland. Transgenic animals have the ability to produce recombinant proteins in a more efficient way than traditional systems based on microorganisms or animal cells. In several countries, new industries are in the process of development based on such transgenic animals. In Canada, for example, Nexia Biotechnologies Inc. has developed an approach to goat transgenics designed to enable rapid progress to be made from construct design to production of recombinant protein in milk; product manufacturing by Nexia has resulted in the BELE (breed early, lactate early) transgenic goat system, featuring short generation times, early sexual maturity, multiple offspring, lack of seasonality and good milk yields. There would be those who say, at this stage of development, that only those who are willing to accept an absence of job security and considerable risk are likely to be attracted to such industries (Table 13.3).

Pathogen-free herds

The production of human recombinant proteins in the milk of transgenic small ruminants requires the use of specific pathogen-free (SPF) herds, in particular scrapie-free herds. For such reasons, goats of dairy breeds (Saanen and Toggenburg) have been imported from New Zealand into Canada. In goats, the method of choice for the generation of transgenic animals has

been by microinjection of the pronuclei of zygotes. In Canada, results reported by Baldassarre *et al.* (2003) indicated that efficient transgenesis rates can be obtained by DNA microinjection of *in vitro*-produced zygotes; the same workers also demonstrated that laparoscopic ovum pick-up is a reliable and effective technique for the recovery of goat oocytes for the *in vitro* production of these zygotes.

13.1.5. Cats and dogs

According to some authors, one promising market for transgenesis could be in the pet world (Long *et al.*, 2003). There is only the need to produce a relatively small population by genetic engineering and then to produce the offspring by natural mating at a much lower cost. One current commercial venture is directed towards the production of allergen-free cats by way of cloning and transgenic technology. Apparently, there are estimates suggesting that about one in four human pet owners are allergic to their cats; the market for such allergen-free cats could well be considerable, depending on the public's response. As well as a possible line in unique cats, cloning and genetic engineering could bring even more unique dogs, one line mentioned being a green fluorescent dog. Such an addition to the many breeds of dog already available, many of them excellent examples of endocrinological abnormalities in perpetuity, may find a welcome from some quarters. Public reaction in general to such novel possibilities in pets, in the light of views already expressed in some countries on GM crops and such, may prove somewhat less than favourable.

However, enormous amounts of money are currently spent on pets and pet foods and clearly this is an animal world expanding at a rapid rate. In the USA, the transgenic GloFish, a zebra fish with sea anemone genes inserted into its genome, is now on sale. It is said that in the Western world, with its much reduced family size, increasing attention and money are being directed

Table 13.3. Companies currently working on transgenic protein production.

Transgenic system	Company
Cattle	Gala Biotech (Sauk City, Wisconsin, USA)
	GTC (Framingham, Massachusetts, USA)
	Hematech (Westport, Connecticut, USA)
	Pharming (Leiden, The Netherlands)
Chickens	Avigenics (Athens, Georgia, USA)
	Vivalis (Nantes, France)
	TransXenoGen (Shrewsburg, Massachusetts, USA)
Goats	GTC
	Nexia Biotechnologies (Montreal, Quebec, Canada)
Rabbits	Pharming
	BioProtein Technology (Paris, France)
Sheep	GTC
	PPL Therapeutics (Edinburgh, UK)

towards pets, which in some ways are in danger of becoming substitutes for children.

13.2. Producing Transgenics

13.2.1. Improving the technology

Improvements in the technology employed in the production of transgenic farm animals are important, for both biotechnology and basic research. The main barrier in transgenic animal production remains the identification of more efficient systems of transgene delivery and more effective methods of regulating transgene expression levels. Fortunately, the production of somatic-cell clones derived from different tissue types of cultured cells has opened the way to new approaches. Much evidence is now available in cattle and other farm animals that, when transfected donor cells are properly selected, a high proportion of the young derived from nuclear transfer are transgenic.

13.2.2. DNA transfer

Until a decade or so ago, the standard method of inserting exogenous genes into an animal was pronuclear microinjection

(Fig. 13.2). Many transgenic animals were produced by this method, but the cost and effort required were often considerable, primarily because all the injected embryos needed to be transferred to foster-mothers and maintained to term until the small proportion of offspring could be identified. However, from 1998, somatic-cell transfection followed by nuclear transfer has been the method of choice in the production of transgenics. The advantage of somatic-cell nuclear transfer is that gene transfer and selection of transgenic cells can be done in culture. Using cultured cells, gene transfer can be done by various approaches, such as electroporation or lipofection. Selection of transgenic cells is done by incorporating an antibiotic resistance gene into the genetic construct and culturing transfected cells in medium containing the appropriate antibiotic; the surviving colonies are clones of single transgenic cells. Although the efficiency of development following nuclear transfer is low, all the offspring are transgenic.

Although the production of transgenic pigs has relied on the DNA injection of pronuclei for the past two decades, it is believed by some that major changes are now likely to be made in transgenesis technology. According to Nagashima *et al.* (2003), *in vitro*-matured oocytes will become dominant in porcine developmental technologies,

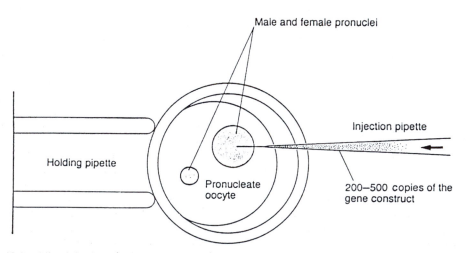

Fig. 13.2.   Microinjection of DNA into a pronucleate oocyte.

replacing *in vivo*-derived oocytes. Such a transition is probable because of the work done in cattle showing that artificially matured oocytes are suitable as recipients in nuclear-transfer programmes. The same Japanese authors note that a combined technique of sperm vectors and *in vitro*-matured porcine oocytes would make the production of transgenic pigs quite feasible.

13.2.3. Transfected cells for nuclear transfer

In pigs, it is now possible to combine the technology for homologous recombination in fetal somatic cells with that of nuclear transfer to create specific modifications to the pig genome. One example of such genome modification is the knocking out of the gene responsible for hyperacute rejection when pig organs are transferred to primates. It should be noted that the ability to generate mice with a targeted mutation in a particular gene has been one of the most important advances in understanding the function of gene products and has enabled numerous mouse models of human diseases to be developed (Tymms and Kola, 2001). The same type of technology may be valuable in producing pig models of human genetic disease. An example given by Prather *et al.* (2003) is in using the transgenic pig to allow basic research and drug development in the treatment of patients with cystic fibrosis. Although producing pigs with specific genetic modifications is now possible, improvements to cloning and transgenic technology are still required to make this approach more effective.

Gene targeting

For more than a decade, it has been possible to modify endogenous genes in the mouse by manipulating embryonic stem (ES) cells; researchers have been able to generate specific mutations and alter specfic gene sequences in ES cells. Such modifed cells retain their developmental potential and when inserted into a developing embryo can contribute to all its tissues, including sperm and oocytes. The ability to use cells to transfer a predetermined genetic modification to the whole animal would have obvious appeal in farm animals. Workers in Scotland have described efficient and reproducible gene targeting in fetal fibroblasts and demonstrated that viable sheep could be produced by using the fibroblasts in nuclear transfers. The Roslin work showed that nuclear transfer in transgenesis does not require ES cell- or primordial germ-cell (PGC)-derived cells and avoids the need to generate chimeric animals, which can be costly and time-consuming.

13.2.4. Sperm-mediated DNA transfer

A report by Lavitrano and colleagues in Italy in the late 1980s describing the production of transgenic mice using sperm as a vector of exogenous DNA into the oocyte attracted much attention and some degree of scepticism. However, compelling evidence of successful sperm-mediated gene transfer has been provided more recently by Lavitrano *et al.* (2002), who reported the birth of transgenic pigs for use in xeno-transplantation studies; they were able to show a 25-fold improvement in their success rate in comparison with pronuclear injection. According to the Italian team, it is necessary to free the sperm of seminal plasma because of its interferon-1 content, which usually prevents the sperm accepting the new DNA; the majority of the piglets born (up to 80%) in their latest work had the gene incorporated into the genome and most transcribed the gene (human decay-accelerating factor (hDAF)) in a stable manner, with the gene being transmitted to progeny. According to Wall (2002), it is likely that sperm-mediated gene transfer protocols will continue to be refined; if reliable gene expression can be achieved, then the method could become the method of choice in species, such as the pig, in which manipulation of oocytes and zygotes presents particular difficulties.

13.2.5. Retroviral infection of early embryos

According to Wall (2002), the advantages of retrovirus-mediated gene transfer include a high frequency of gene transfer across embryonic membranes, high integration into the oocyte/zygotic genome and minimal embryo manipulation. Disadvantages include the fact that the technique can probably handle relatively small amounts of genetic information (< 10 kb) in a field where increasing emphasis is on the use of increasingly longer gene sequences; there is also the complexity of introducing the transgenes into the retrovirus, a process involving many steps. A final difficulty may be that of public perception of this approach, bearing in mind the inevitable connections drawn between viruses and human disease conditions. Some of the other avenues explored in gene transfer are indicated in Fig. 13.3.

13.2.6. Identifying transgenic embryos

Various workers have noted the need for accurate methods for screening both GM donor cells and reconstructed embryos. There is also the fact that progress in transgenesis in farm animals may be limited by the high cost involved in maintaining large numbers of pregnant recipient animals

carrying non-transgenic fetuses. For such reasons, the ability to identify those embryos carrying gene inserts before they are transferred to recipients could be of considerable value. Various methods have been reported. Early work in the 1990s attempted to identify transgenic embryos by taking a biopsy sample of 10–30 trophoblast cells and analysing these by PCR for the presence of the transgene. In France, workers described a luminescent screening test to select transgenic cattle embryos; it was suggested that selecting luminescent blastocysts on the basis of signal intensity and distribution could markedly reduce the numbers of recipients required.

In Japan, a group reported the quick detection of firefly luciferase gene expression in early cattle embryos by photon-counting; in Poland, workers used a green fluorescent protein (GFP) gene reporter to select cattle embryos and found this to be a useful method for increasing the number of transgenic calves born (Duszewska *et al.*, 2003). Work in Canada demonstrated that GFP transgenic calves could be obtained from transfected somatic-cell nuclei transferred into either metaphase- or telophase-enucleated oocytes (Bordignon *et al.*, 2003); the same workers showed that the GFP transgene was transmissible in the germ line and could be used to select bovine embryos derived from a transgenic cloned bull.

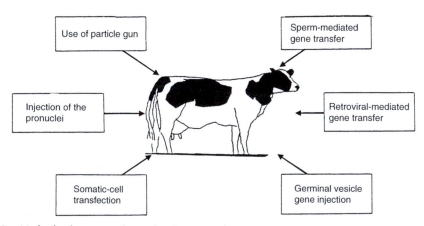

Fig. 13.3. Methods of gene transfer used in farm animals.

13.2.7. Future developments

Artificial chromosomes

Although most studies in the production of transgenic animals have involved the insertion of relatively short DNA sequences (5–30 kb) prepared in plasmid vectors, there are those interested in developing approaches to produce genetically modified farm animals with very long exogenous DNA sequences, incorporating large and complex genes. The target of one study, reported by Robl *et al.* (2003), was to produce a genetic line of cattle capable of producing high titres of human polyclonal antibodies for therapeutic applications; these workers believe that such polyclonal antibodies, produced in a bovine-based system, may have broad application, such as the treatment or prevention of human infectous disease, including antibiotic-resistant infections, autoimmune diseases and certain forms of cancer.

Controlled expression of sex phenotype

As noted earlier in dealing with the sexing of sperm, sex phenotype is an important production parameter in animal agriculture. It is anticipated that genetic engineering methods will eventually be developed for the controlled expression of genes involved in mammalian sex determination; such methods may involve pharmacological control of transgene expression by oral supplementation in water or feed, by injection of an appropriate agent during development or by crossing two lines of animals to induce the desired transgene expression.

14

Suppressing Reproductive Activity

The onset of puberty and gonadal function in male and female farm animals is regulated by a complex interaction of hypothalamic, pituitary and gonadal hormones. There may be some circumstances, particularly in cattle and horses, when it may be desirable to suppress normal reproductive activity by endocrine or other forms of intervention. It may not necessarily be a permanent suppression of reproductive function that is sought; some practical needs may be covered satisfactorily by temporary forms of suppression. One of the best-known means of inhibiting sexual activity and the ability to reproduce is by castration, practised by the ancients on man and beast over many centuries (Fig. 14.1). In cattle country, spaying, the surgical removal of the ovaries, was the female equivalent of castration; animal welfare considerations make the practice no longer acceptable.

14.1. Advantages of Technology

The usual reasons why farm livestock are castrated include: (i) management difficulties; (ii) to reduce aggressive and sexual behaviour; and (iii) to improve meat and carcass quality. Where surgical intervention or physical castration is questioned on welfare grounds, other options have been suggested, including immunocastration, where animals are immunized to control

Fig. 14.1. Castrating ram lambs by the rubber-ring method.

gonadotrophin-releasing hormone (GnRH) secretion, effectively suppressing gonadotrophin release and generating a temporary castration-like effect. It is not only on the

©I.R. Gordon 2004. *Reproductive Technologies in Farm Animals*
(I.R. Gordon)

farm that immunosuppressant approaches may prove useful; in the control of certain wild animal populations, they may have particular relevance.

14.1.1. Cattle

Bulls have an advantage over castrated steers in being capable of producing 15% more carcass gain and 15% more lean meat, while eating 10% less feed per unit gain. Having a bull showing such characteristics without its male aggression and sexual behaviour has obvious commercial attractions. Unfortunately, although encouraging at the research station, none of the vaccine treatments thus far devised appears to be commercially acceptable, at least not to the point where they can be considered suitable for extensive use. On the female side, again dealing with cattle, the expression of behavioural oestrus in postpubertal beef heifers may result in undesirable management difficulties; such problems may include pregnancies, an increase in the number of injuries due to riding behaviour and the potential adverse effect on carcass value if slaughtered during oestrus. There can be sound commercial justification for having a safe and reliable means of dealing with the problem of oestrous beef heifers.

Cryptorchid bulls

Artificial manipulation of the scrotum has been used in the USA and elsewhere as a means of sterilizing bulls. In this, the scrotum is artificially shortened to hold the testes close to the body, resulting in elevated testicular temperature and causing significantly reduced spermatogenesis. A bull subjected to such treatment is referred to as a 'short-scrotum' bull, and is effectively an artificial cryptorchid. The testes are artificially forced into the dorsal region of the scrotum by placing a rubber ring around the lower portion of the scrotum. In 3–4 weeks, the empty scrotal sac sloughs at the juncture of the rubber band because of restricted circulation. As might be expected, the weight of the testes in these bulls is less than in normal bulls, reaching only about one-half the weight of the intact animal. These males, although sterile, appear to maintain normal testosterone levels and a high rate of growth. Such animals have increased growth efficiency and have leaner carcasses in comparison with steers. This procedure shows the importance of testicular cooling for normal fertility and the importance of androgens for growth and leanness.

14.1.2. Sheep and goats

Cryptorchid lambs

As noted above for cattle, there are occasions when it may be desirable to render male sheep infertile, but without this involving conventional castration. In New Zealand, Tarbotton *et al.* (2002) conducted a survey of 400 sheep farmers to determine the incidence and perceptions of cryptorchid lambs in the year 2000. The main reason for farmers producing cryptorchids was to achieve good lamb growth rates and lean carcasses without having fertile male lambs; there was also the added benefit that cryptorchids were cleaner and more easily managed than ram lambs. The survey showed that, although the production of cryptorchids has declined in recent years, the majority of farmers who currently produce them have a preference for this method and were unwilling to change their practice.

Hormonally castrated goats

Studies in Australia tested the usefulness of immunizing goats against GnRH by way of a commercially available vaccine (Vaxtrate) which suppresses gonadotrophin secretion, steroidogenesis, sperm production, the production of male odour and agonistic behaviour between males; in 90% of the immunized bucks, the testes remained small for > 1 year after primary vaccination and a 2-week interval was as effective as a 4-week interval between primary and booster immunizations.

14.1.3. Horses

Stallions

Colts often become sexually aggressive with
the onset of puberty at 14–15 months of age.
Such aggressive sexual behaviour in domestic stallions is undesirable because it can
limit achievement of performance potential
and increase the risk of injury to other
horses or humans. Castration may not
necessarily be an option for behavioural
control in many individual stallions due to
the potential loss of valuable blood-lines; in
any event, castration may not always quell
sexual behaviour in horses.

Mares

A method that suppresses oestrous cyclicity
in mares is of considerable interest in
eventing and showjumping fillies; it would
enable more of these animals to take part in
competitions.

Melatonin treatment

Because melatonin secretion is an
important signal for the maintenance of
anoestrus, administration of melatonin may
provide a means for artificially inducing
anoestrus in horses during the summer
without the use of exogenous steroids. This
would allow for the suppression of oestrous
behaviour where mares are used for other
industry-related events (e.g. showing,
racing, working).

14.1.4. Pigs

Boar taint

In pigs, there is considerable interest in
methods of suppressing boar taint without
this involving the pain and stress of surgical
castration. Fattening boars rather than castrated pigs for pork production could be
one way of increasing the efficiency of pig
production. However, in most countries,
male pigs intended for meat production are
castrated at an early age to avoid potential

consumer dissatisfaction due to a sexual
odour in boar adipose tissue. This offensive
odour is caused mainly by the pheromone
androstenone, which is synthesized in the
testes.

14.1.5. Deer

Wild populations

There are occasions when research efforts
have been directed towards preventing
pregnancies in certain deer species rather
than trying to control or facilitate them.
One such instance, as noted earlier, is
in control of wild deer populations. In the
USA, for example, there have been attempts
to develop appropriate contraceptive techniques; the effectiveness of GnRH to disrupt
oestrous cycles has been one of the
approaches used.

14.2. Development and Application of Technology

14.2.1. Hormonal approach

Cattle

Work in the mid-1960s with melengestrol
acetate (MGA) administered orally was the
first to show that a progestogen can have a
growth-promoting action over and above its
oestrus-inhibiting ability; a comprehensive
evaluation of MGA at the time in beef heifers showed that the agent resulted in an
average 11% improvement in daily rate
of gain and a 7.6% improvement in feed-
conversion efficiency over untreated controls. It was believed that MGA exerted its
growth-promoting action indirectly by permitting substantial oestrogen production
in the ovaries of the oestrus-suppressed
animals; growth promotion was apparently
achieved in much the same way as
oestradiol exerts its action in castrated
steers.

The fact that the growth-promoting
effect is not apparent until heifers reach

puberty may explain some differences in the way that female cattle respond to the progestogen. In non-European Union (EU) countries, where beef heifers may be at pasture rather than in feedlots, implantation may be the method of choice. Work in Ireland at one time showed that a single MGA implant would hold heifers out of oestrus for about 4 months, with a useful growth-promoting effect being evident during this period (Fig. 14.2). Although the commercial availability of progestogens such as MGA is banned under current EU regulations, in countries that do permit their use, principally the USA, there can be obvious economic advantages.

GnRH agonist implants

Studies reported from Australia have examined the use of GnRH agonist implants for long-term suppression of fertility in heifers and cows. A paper by D'Occhio *et al.* (2002) showed that, in most heifers and cows treated with the GnRH agonist implants, ovarian follicular growth was restricted to early antral follicles (2–4 mm); it was concluded that such implants have considerable potential as a practical technology to suppress ovarian activity and control reproduction in cattle kept under extensive rangeland environments; the same technology may have wider applications in cattle production systems.

Applications in horses

For various reasons, there may be the need to delay the occurrence of oestrus in the cyclic mare that is to participate in racing or showjumping. As in the other farm mammals, endogenous progesterone from the corpus luteum inhibits oestrus and ovulation in the mare during the period of luteal activity. Based on this fact, many reports have appeared over the past 40 years on the use of progesterone or progestogens, administered in various ways.

Stallions

Some of those who have looked to alternatives to castration in horses have used a hormonal approach. It is known that testosterone, synthesized and secreted by the Leydig (interstitial) cells of the testes in response to blood-borne luteinizing hormone (LH), plays a role in aggressive and sexual behaviour in stallions. It is also known that short-term administration of

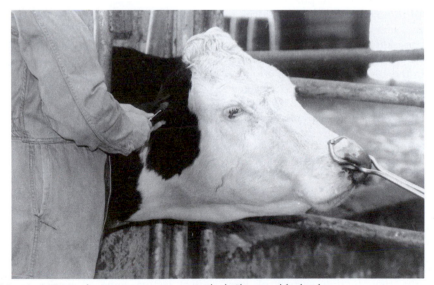

Fig. 14.2. An MGA implant to suppress oestrus in the heifer reared for beef.

exogenous progesterone to stallions or geldings may reduce pituitary LH secretion; on that basis, chronic administration of progesterone to stallions might suppress secretion of LH sufficiently to reduce testosterone levels and hence suppress sexual behaviour. Treatment of sexually mature stallions with the synthetic progestogen Regumate (altrenogest) has resulted in decreases of serum concentrations of LH and testosterone in some studies reported from the USA; such suppression appears to be partially reversible, although > 90 days are required before sperm morphology improves to pretreatment levels.

GnRH agonists

In horses, there have been studies to examine the ability of GnRH agonists to inhibit reproductive activity by virtue of the agent's ability to down-regulate ovarian activity. Californian researchers showed that ovarian and sexual activity could be suppressed in the mare, but it took up to a month of treatment before an effect was evident.

Non-pharmacological approach

In horses, Cambridge workers have described an effective non-pharmacological means of inducing prolonged dioestrus in the mare (Lefranc and Allen, 2003). Using manual pressure applied *per rectum*, they showed it was possible to terminate pregnancy in the 16–22-day period of gestation and induce a state of 'pseudopregnancy', which held for an average of 82 ± 13 days. During this time, oestrous behaviour was not observed, uterine tone stayed high and the cervix remained closed.

14.2.2. Immunological approach

Active immunization against GnRH was recognized as early as the 1970s, when this releasing hormone was first identified, and was suggested as a potential means by which the reproductive system of farm mammals might be suppressed for various practical and clinical reasons. Since that time, many reports have appeared as workers attempted to determine the applicability of the technique as an alternative to surgical removal of the testes. The possible advantages of such immunocastration include the improvement of meat and carcass characteristics for cattle, sheep, goats and pigs, improvement in feed efficiency relative to castrated animals in the same species, reduction in male aggressive behaviour and reduction in male-associated odours in goats and pigs. GnRH is the key hypothalamic hormone in the regulation of the pituitary–testicular axis and therefore exerts a profound effect on functions such as spermatogenesis and sexual activity.

Cattle

Various researchers have examined the possibility of devloping a vaccine against GnRH which might provide an alternative to surgical castration in the cattle industry. Some of the attempts to immunize bulls against GnRH used Freund's complete adjuvant, a potent immunological adjuvant; however, for commercial use, there are serious objections to Freund's adjuvant, particularly the fact that animals can react as false positives in tuberculosis testing.

Horses

In one recent study reported in Australia, workers used a water-soluble GnRH vaccine to suppress the ovarian activity of fillies at the peak of the breeding season; suppression of activity occurred for 25 and 30 weeks in fillies receiving 200 and 400 mg of the vaccine, respectively. It was found that the vaccine suppressed ovarian function and prevented oestrous behaviour; these effects were reversible and the subsequent fertility of the vaccinated fillies was normal. In stallions, a vaccine against GnRH was also used to suppress testosterone secretion and depressed testicular function in young horses; using an implant vaccine, it proved possible to produce an effect lasting some 30 weeks, which was reversible.

Pigs

In pigs, a study by Zeng *et al.* (2002) determined the optimal dose of a GnRH vaccine for immunocastration of Chinese male pigs, based on immune, endocrine and testicular responses; these authors concluded that, within the dose range examined, a 62.5 µg dose was optimal for GnRH immunization studies or practical applications in Chinese male pigs. Results presented by Thun *et al.* (2003) suggested that vaccination against GnRH (Improvac, CLS, Australia), carried out 8 and 4 weeks before slaughter, may be a practical and effective method to suppress boar taint; not only was boar taint suppressed in all animals, but the yield of lean meat was significantly improved by the procedure.

14.2.3. Wild animal populations

Horses

A problem of serious public concern in some countries is that posed by wild horses. In the USA, for example, the feral horse population has increased markedly in spite of attempts to control numbers. Such attempts have included adoption programmes, confinement in feedlots, vasectomy of the dominant stallion, intrauterine devices (IUDs), progestogen-releasing implants and remotely delivered immunization vaccines against zona pellucida proteins. In terms of research needs, there is certainly a case for developing a practical, economical, safe and reversible method of contraception for mares that would be effective for several years without compromising the horse's future fertility. An acceptable contraceptive method for use in wild horses should be not only effective but fully reversible.

Of all the large farm animals, the mare lends itself best to the use of IUDs. Workers in California in the mid-1990s provided evidence that IUDs can be used as an effective contraceptive in the mare; due to its size and the anatomy of the genital tract, placement and removal of the IUD is possible in the standing animal without the need for instrumentation.

Deer

It has been demonstrated that porcine zona pellucida (PZP) protein is highly effective as an immunocontraceptive in white-tailed deer (*Odocoileus virginianus*), but problems in this approach include batch variations in the immunogenicity of PZP and the potential hazards of extracts containing either viral or pathogenic material. In the USA, Miller and Killian (2002) attempted to select epitopes of PZP, which could be synthesized and used as a safe and consistently effective vaccine.

Camelids

To end on a historical note, it may be mentioned that the Arabs of many centuries ago are reputed to have used an IUD, in the shape of a small, smooth pebble, inserted into the uterus through the cervix of the camel, as a means of preventing pregnancy.

Bibliography

Abeydeera, L.R., Wang, W.H., Cantley, T.C., Rieke, A., Murphy, C.N., Prather, R.S. and Day, B.N. (2000) Development and viability of pig oocytes matured in a protein-free medium containing epidermal growth factor. *Theriogenology* 54, 787–797.

Adams, C.S., Martin, M.J., Thomas, D., Booth, M., Cottrill, F.N., Keirns, J. and Wiseman, B.S. (2003) A comparison of the developmental potential of porcine nuclear transfer (NT) embryos derived from *in vitro* as opposed to *in vivo* matured ova. *Biology of Reproduction* 68 (Suppl. 1), Abstract 216, p. 200.

Adoff, G., Skjennum, F.C. and Engelsen, R. (2002) Experience and prospects of Norwegian cod farming. *Bulletin of the Aquaculture Association of Canada* 102, 8–11.

Ahmad, Z., Anzar, M., Shahab, M., Ahmad, N. and Andrabi, S.M.H. (2003) Sephadex and sephadex ion-exchange filtration improves the quality and freezability of low-grade buffalo semen ejaculates. *Theriogenology* 59, 1189–1202.

Alabart, J.L., Folch, J., Fernandez-Arias, A., Ramon, J.P., Garbayo, J.M. and Cocero, M.J. (2003) Screening of some variables influencing the results of embryo transfer in the ewe. Part II: Two-day old embryos. *Theriogenology* 59, 1345–1356.

Al-Aghbari, A.M. and Menino, A.R. Jr (2002) Survival of oocytes recovered from vitrified sheep ovarian tissues. *Animal Reproduction Science* 71, 101–110.

Alan, M. and Tasal, I. (2002) Efficacy of prostaglandin F2-alpha and misoprostol in the induction of parturition in goats. *Veterinary Record* 150, 788–789.

Al-Eknah, M., Homeida, N. and Al-Haider, A. (2001) A new method for semen collection by artificial vagina from the dromedary camel. *Journal of Camel Practice and Research* 8, 127–130.

Allen, W.R. (1978) Control of oestrus and ovulation in the mare. In: Foxcroft, G.R., Haynes, B. and Lamming, G.E. (eds) *Control of Ovulation*. Butterworths, London, pp. 453-468.

Allen, W.R. and Antczak, D.F. (2000) Reproduction and modern breeding technologies in the mare. In: Bowling, A.T. and Ruvinsky, A. (eds) *The Genetics of the Horse*. CAB International, Wallingford, UK, pp. 307–341.

Aller, J.F., Rebuffi, G.E., Cancino, A.K. and Alberio, R.H. (2002) Successful transfer of vitrified llama (*Lama glama*) embryos. *Animal Reproduction Science* 73, 121–127.

Alm, H. and Torner, H. (2003) Increase of embryo developmental rate *in vitro* by selection of bovine oocytes before IVM using a staining test: preliminary results. In: *Proceedings 19th Meeting European Embryo Transfer Association (Rostock)*, p. 132.

Amorin, C.A., Goncalves, P.B.D. and Figueiredo, J.R. (2003) Cryopreservation of oocytes from preantral follicles. *Human Reproduction Update* 9, 119–129.

Andersson, M., Taponen, J., Koskinen, E. and Dahlbom, M. (2003) Effect of insemination with a dose of 2 or 15 million frozen–thawed spermatozoa and semen deposition on pregnancy rate in dairy cows. *Reproduction in Domestic Animals* 38, 336 (Abstract).

Anel, L., de Paz, P., Alvarez, M., Chamorro, C.A., Boixo, J.C., Manso, A., Gonzalez, M., Kaabi, M. and Anel, E. (2003) Field and *in vitro* assay of three methods for freezing ram semen. *Theriogenology* 60, 1293–1308.

Arav, A., Yavin, S., Zeron, Y., Natan, D., Dekel, I. and Gacitua, H. (2002) New trends in gamete's cryopreservation. *Molecular and Cellular Endocrinology* 187(1/2), 77–81.

Arnold, D.R., Bordignon, V. and Smith, L.C. (2003) Comparisons in trophoblast function and differentiation in d-17 *in vitro* produced and nuclear transfer bovine embryos. *Biology of Reproduction* 68 (Suppl. 1), Abstract 343, p. 253.

Atabay, E.C., Takahashi, Y., Katagiri, S., Nagano, M., Koga, A. and Kanai, Y. (2004) Vitrification of bovine oocytes and its application to intergeneric somatic cell nucleus transfer. *Theriogenology* 61, 15–23.

Azevedo, E.M.P., Cavalcanti, M.C.O., Guerra, M.M.P., Catanho, M.T.J.A., Souza, A.F.S., Loureiro, C.C., Bispo, C.A.S. and Barreto, M.B.P. (2002) Pregnancy diagnosis through the evaluation of serum estrone sulfate levels in dairy goats. *Revista Brasileira de Reproducao Animal* 26, 262–264.

Badi, A.M., O'Byrne, T.M. and Cunningham, E.P. (1981) An analysis of reproductive performance in Thoroughbred mares. *Irish Veterinary Journal* 35, 1–12.

Baker, R.D. and Coggins, E.G. (1968) Control of ovulation rate and fertilization in prepubertal gilts. *Journal of Animal Science* 27, 1607–1610.

Baldassarre, H., Wang, B., Kafidi, N., Gauthier, M., Neveu, N., Lapointe, J., Sneek, L., Leduc, M., Duguay, F., Zhou, J.F., Lazaris, A. and Karatzas, C.N. (2003) Production of transgenic goats by pronuclear microinjection of *in vitro* produced zygotes derived from oocytes recovered by laparoscopy. *Theriogenology* 59, 831–839.

Bareil, G., Casamitjana, P., Perrin, J. and Vallet, J.C. (1988) Embryo production, freezing and transfer in Angora, Alpine and Saanen goats. In: *Proceedings 4th Meeting European Embryo Transfer Association (Lyon)*, pp. 67–93.

Bark, M.C., Souza, C.J.H., MacDougall, C. and Telfer, E.E. (2002) The isolation and location of the oocyte-specific factor GDF-9 in the porcine ovary. *Reproduction, Abstract Series* No. 29, Abstract 1, p. 4.

Bartlewski, P.M., Beard, A.P. and Rawlings, N.C. (2000) Ultrasonographic study of ovarian function during early pregnancy and after parturition in the ewe. *Theriogenology* 53, 673–689.

Bartlewski, P.M., Beard, A.P. and Rawlings, N.C. (2003) Antral follicular development and FSH secretion during sexual maturation in the ewe lamb. *Reproduction, Abstract Series* No. 30, Abstract P46, pp. 65–66.

Baruselli, P.S., Madureira, E.H., Visintin, J.A., Porto-Filho, R., Carvalho, N.A.T., Campanile, G. and Zicarelli, L. (2000) Failure of oocyte entry into oviduct in superovulated buffalo. *Theriogenology* 53, 491.

Beebe, D.J., Wheeler, M., Zeriingue, H., Walters, E. and Raty, S. (2002) Microfluidic technology for assisted reproduction. *Theriogenology* 57, 125–135.

Behboodi, E., Memili, E., Ayres, S.L., Coin, M., Chen, L.H., Meade, H.M. and Echelard, Y. (2003) Application of embryo cloning to the generation of transgenic founder goats. *Theriogenology* 59, 237.

Benoit, M. and Veysset, P. (2003) Conversion of cattle and sheep suckler farming to organic farming: adaptation of the farming system and its economic consequences. *Livestock Production Science* 80, 141–152.

Berg, D.K. and Asher, G.W. (2003) New developments in reproductive technologies in deer. *Theriogenology* 59, 189–205.

Berg, D.K., Pugh, P.A., Thompson, J.G. and Asher, G.W. (2002) Development of *in vitro* embryo production systems for red deer (*Cervus elaphus*). Part 3. *In vitro* fertilization using sheep serum as a capacitating agent and the subsequent birth of calves. *Animal Reproduction Science* 70, 85–98.

Berthelot, F., Marinat-Botte, F., Locatelli, A. and Terqui, M. (2000) Vitrification of pig embryos aged between 5 to 6 days using an ultra-rapid freezing method: open pulled straw (OPS) method. *Journées de la Recherche Porcine en France* 32, 433–437.

Berthelot, F., Locatelli, A. and Martinat-Botte, F. (2003) *In vivo* development after surgical transfer of porcine embryos vitrified with the open pulled straw method. *Reproduction in Domestic Animals* 38, 320 (Abstract).

Bertolini, M., Beam, S.W., Shim, H., Bertolini, L.R., Moyer, A.L., Famula, T.R. and Anderson, G.B. (2002) Growth, development, and gene expression by *in vivo*- and *in vitro*-produced day 7 and 16 bovine embryos. *Molecular Reproduction and Development* 63, 318–328.

Betteridge, K.J. (2000) Reflections on the golden anniversary of the first embryo transfer to produce a calf. *Theriogenology* 53, 3–10.

Bhojwani, S., Vajta, G., Callesen, H., Alm, H., Torner, H., Roschlau, K., Kuwer, A., Becker, F., Klukas, H., Kanitz, W. and Poehland, R. (2003) Zona-free somatic cell nuclear transfer in cattle-first HMC calf in Europe. In: *Proceedings 19th Meeting European Embryo Transfer Association (Rostock)*, p. 136.

Bhowmick, S., Zhu, L., McGinnis, L., Lawitts, J., Nath, B.D., Toner, M. and Biggers, J. (2003) Desiccation tolerance of spermatozoa dried at ambient temperature: production of fetal mice. *Biology of Reproduction* 68, 1179–1786.

Biao, X. and Xiaorong, W. (2003) Organic agriculture in China. *Outlook on Agriculture* 32(3), 161–164.

Bilodeau, J.F., Blanchette, S., Gagnon, C. and Sirard, M.A. (2001) Thiols prevent H_2O_2-mediated loss of sperm motility in cryopreserved bull semen. *Theriogenology* 56, 275–286.

Boelling, D., Groen, A.F., Sorensen, P., Madsen, P. and Jensen, J. (2003) Genetic improvement of livestock for organic farming systems. *Livestock Production Science* 80, 79–88.

Boiani, M., Eckardt, S., Leu, N.A., Scholer, H.R and McLaughlin, K.J. (2003) Pluripotency deficit in clones overcome by clone–clone aggregation: epigenetic complementation? *EMBO Journal* 22, 5304–5312.

Boland, M.P. (2002) Impact of currrent management practices on embryo survival in the modern dairy cow. *Cattle Practice* 10, 337–347.

Bordignon, V., Keyston, R., Lazaris, A., Bilkodeau, A.S., Pontes, J.H.F., Arnoild, D., Fecteau, G., Keefer, C. and Smith, L.C. (2003) Transgene expression of greeen fluorescent protein and germ line transmission in cloned calves derived from *in vitro*-transfected somatic cells. *Biology of Reproduction* 68, 2013–2023.

Bormann, C.L., Moige Ongeri, E. and Krisher, R.L. (2003) The effect of vitamins during maturation of caprine oocytes on subsequent developmental potential *in vitro*. *Theriogenology* 59, 1373–1380.

Bourke, S., Diskin, M.G. and Sreenan, J.M. (1995) Field-scale test of IVF cattle embryo transfer. In: *Proceedings of the 11th Meeting European Embryo Transfer Association* (Hanover), p. 136.

Bousquet, D., Burnside, E.B. and Van Doormaal, B.J. (2003) Biotechnologies of reproduction applied to dairy cattle production: embryo transfer and IVF. *Canadian Journal of Animal Science* 83, 403–407.

Bretzlaff, K.N. and Romano, J.E. (2001) Advanced reproductive techniques in goats. *Veterinary Clinics of North America, Food Animal Practice* 17, 421–434.

Brophy, B., Smolenski, G., Wheeler, T., Wells, D., L'Huillier, P. and Laible, G. (2003) Cloned transgenic cattle produce milk with higher levels of beta-casein and k-casein. *Nature Biotechnology* 21, 157–162.

Buchanan, B.R., Seidel, G.E. Jr, McHue, P.M., Schenk, J.L., Herickhoff, L.A. and Squires, E.L. (2000) Insemination of mares with low numbers of either unsexed or sexed spermatozoa. *Theriogenology* 53, 1333–1344.

Burkin, H.R. and Miller, D.J. (2000) Zona pellucida protein binding ability of porcine sperm during epididymal maturation and the acrosome reaction. *Developmental Biology* 222, 99–109.

Byron, C.R., Embertson, R.M., Bernard, W.V., Hance, S.R., Bramlage, L.R. and Hopper, S.A. (2003) Dystocia in a referral hospital setting: approach and results. *Equine Veterinary Journal* 35, 82–85.

Cabianca, G., Rota, A., Barnini, C. and Vincenti, L. (2003) Current techniques for *in vitro* horse reproduction. *Ippologia* 14(1), 15–31.

Cacis, M. and Ivankovic, A. (2001) Body condition scoring with special emphasis on reproductive performance of mares. *Stocarstvo* 55, 461–472.

Campbell, B.K., Baird, D.T., Souza, C.J.H. and Webb, R. (2003) The Fec-B (Booroola) gene acts at the ovary: *in vivo* evidence. *Reproduction* 126, 101–111.

Carleton, C.L. and Threlfall, W.R. (1986) Induction of parturition in the mare. In: Morrow, D.A. (ed.) *Current Therapy in Theriogenology*. W.B. Saunders, Philadelphia, Pennsylvania, pp. 689–692.

Carvalho, N.A.T., Baruselli, P.S., Zicarelli, L., Madureira, E.H., Visintin, J.A. and D'Occhio, M.J. (2002) Control of ovulation with a GnRH agonist after superstimulation of follicular growth in buffalo: fertilization and embryo recovery. *Theriogenology* 58, 1641–1650.

Cassady, J.P., Johnson, R.K. and Ford, J.J. (2000) Comparison of plasma FSH concentration in boars and gilts from lines selected for ovulation rate and embryonal survival, and litter size and estimation of (co)variance components for FSH and ovulation rate. *Journal of Animal Science* 78, 1430–1435.

Cavestany, D., Meikle, A., Kindahl, H., Van Lier, E., Moreira, F., Thatcher, W.W. and Forsberg, M. (2003) Use of medroxyprogesterone acetate (MAP) in lactating Holstein cows within an Ovsynch protocol: follicular growth and hormonal patterns. *Theriogenology* 59, 1787–1798.

Cech, S., Havlicek, V., Lopatarova, M., Lorincova, L., Zahradnikova, J. and Dolezel, R. (2003) Repeated ovum pick-up in stimulated pregnant dairy cows. *Reproduction in Domestic Animals* 38, 349–350 (Abstract).

Cezar, G.G., Bartolomei, M.S., Forsberg, E.J., First, N.L., Bishop, M.D. and Eliertsen, K.J. (2003) Genome-wide epigenetic alterations in cloned bovine fetuses. *Biology of Reproduction* 68, 1009–1014.

Chastant-Maillard, S., Quinton, H., Lauffenburger, J., Cordonnier-Lefort, N., Richard, C., Marchal, J., Mormede, P. and Renard, P.J. (2003) Consequences of transvaginal follicular puncture on well-being in cows. *Reproduction* 125, 555–563.

Chemineau, P. and Malpaux, B. (1998) Melatonin and reproduction in farm livestock. *Comptes Rendus des Seances de la Societe de Biologie et de ses Filiales* 192, 669–682.

Cheng, H., Althouse, G.C. and Hsu, W.H. (2001) Prostaglandin F2-alpha added to extended boar semen at processing elicits *in vitro* myometrial contractility after 72 hours of storage. *Theriogenology* 55, 1901–1906.

Choi, Y.H., Love, C.C., Chung, Y.G., Varner, D.D., Westhusin, M.E., Burghardt, R.C. and Hinrichs, K. (2002) Production of nuclear transfer horse embryos by Piezo-driven injection of somatic cell nuclei and activation with stallion sperm cytosolic extract. *Biology of Reproduction* 67, 561–567.

Christensen, C.R., Redmond, M.J. and Laarveld, B. (2000) Vaccination against follistatin affects reproductive potential in cycling gilts. *Canadian Journal of Animal Science* 80, 337–342.

Christenson, R.K. and Leymaster, K.A. (2003) Effects of selection for ovulation rate or uterine capacity on number and weight of pigs at birth and weaning. *Biology of Reproduction* 68 (Suppl. 1), Abstract 321, p. 244.

Christie, W.B., Mullan, J.S. and Harding, D.A. (2002) Current status of OP/IVF in cattle. *Cattle Practice* 10, 349–354.

Clark, S.G. and Knox, R.V. (2000) Utilizing real-time ultrasound to optimize swine reproduction. *Embryo Transfer Newsletter* 18(4), 16–22.

Cognie, Y., Baril, G., Poulin, N. and Mermillod, P. (2003) Current status of embryo technologies in sheep and goat. *Theriogenology* 59, 171–188.

Colenbrander, B., Gadella, B.M. and Stout, T.A.E. (2003) The predictive value of semen analysis in the evalutaion of stallion fertility. *Reproduction in Domestic Animals* 38, 305–311.

Comizzoli, P., Mermillod, P., Legendre, X., Chai, N. and Mauget, R. (2000) *In vitro* production of embryos in the red deer (*Cervus elaphus*) and the sika deer (*Cervus nippon*). *Theriogenology* 53, 327.

Comizzoli, P., Mauget, R. and Mermillod, P. (2001) Assessment of *in vitro* fertility of deer spermatozoa by heterologous IVF with zona-free bovine oocytes. *Theriogenology* 56, 109–120.

Comizzoli, P., Urner, F., Sakkas, D. and Renard, J.P. (2003) Up-regulation of glucose metabolism during male pronucleus formation determines the early onset of the S phase in bovine zygotes. *Biology of Reproduction* 68, 1934–1940.

Corcoran, D., Fair, T., Rizos, D., Wade, M., Boland, M.P. and Lonergan, P. (2003) Identification of differentially expressed genes in bovine embryos using suppressive subtractive hybridization. In: *Proceedings 19th Meeting European Embryo Transfer Association (Rostock)*, p. 150.

Cordoba, M.C. and Fricke, P.M. (2002) Initiation of the breeding season in a grazing-based dairy by synchronization of ovulation. *Journal of Dairy Science* 85, 1752–1763.

Cordoba, R., Pkiyach, S., Roole, J.A., Edwards, S.A., Penny, P.C. and Pike, I. (2000) The effect of feeding salmon oil during pregnancy on causes of piglet deaths prior to weaning. In: *Proceedings British Society of Animal Science (Winter Meeting)*, p. 105.

Cormier, N. and Bailey, J.E. (2003) A differential mechanism is involved during heparin- and cryopreservation-induced capacitation of bovine spermatozoa. *Biology of Reproduction* 69, 177–185.

Crawford, A.M. (2003) The use of genetic modification technologies in the discovery of genes affecting production traits and disease resistance in animals. *New Zealand Veterinary Journal* 51, 52–57.

Cuello, C., Berthelot, F., Martinat-Bote, F., Venturi, E., Vazquez, J.M., Roca, J. and Martinez, E.A. (2002) Pregnancy rate after non-transfer of vitrified pig embryos. *Reproduction, Abstract Series* No. 29, Abstract 52, pp. 20–21.

Dalbies-Tran, R. and Mermillod, P. (2003) Use of heterologous complementary DNA array screening to analyze bovine oocyte transciptome and its evolution during *in vitro* maturation. *Biology of Reproduction* 68, 252–261.

Davis, K.L. and Macmillan, K.L. (2002) Predicting the onset of parturition during late gestation in dairy cows using udder scoring and hormonal profiling. *Proceedings New Zealand Society of Animal Production* 62, 345–347.

Day, B.N. (2000) Reproductive biotechnologies: current status in porcine reproduction. *Animal Reproduction Science* 60/61, 161–172.

Day, F.T. (1940) Clinical and experimental observations on reproduction in the mare. *Journal of Agricultural Science (Cambridge)* 30, 244–246.

Dell'Aquila, M.E., Albrizio, M., Maritato, F., Minoia, P. and Hinrichs, K. (2003) Meiotic competence of equine oocytes and pronucleus formation after intracytoplasmic sperm injection (ICSI) as related to granulosa cell apoptosis. *Biology of Reproduction* 68, 2065–2072.

De Pauw, I.M.C., Van Soom, A., Mintiens, K., Verberckmoes, S. and de Kruif, A. (2003a) *In vitro* survival pf bovine spermatozoa stored at room temperature under epididymal conditions. *Theriogenology* 59, 1093–1107.

De Pauw, I.M.C., Van Soom, A., Maes, D., Verberckmoes, S. and de Kruif, A. (2003b) Effect of sperm coating on the survival and penetrating ability of *in vitro* stored bovine spermatozoa. *Theriogenology* 59, 1109–1122.

De Rensis, F. and Scaramuzzi, R.J. (2003) Heat stress and seasonal effects on reproduction in the dairy cow – a review. *Theriogenology* 60, 1139–1151.

De Rensis, F., Benedetti, S., Silva, P. and Kirkwood, R.N. (2003) Fertility of sows following artificial insemination at a gonadotrophin-induced oestrus coincident with weaning. *Animal Reproduction Science* 76, 245–250.

Dinnyes, A., Kikuchi, K., Watanabe, A., Fuchimoto, D., Iwamoto, M., Kaneko, H., Noguchi, J., Somfai, T., Onishi, A. and Nagai, T. (2003) Successful cryopreservation of *in vitro* produced pig embryos by the solid surface vitrification (SSV) method. *Theriogenology* 59, 299.

Diskin, M.G. and Larkin, J. (2003) Teagasc. Athenry, Ireland.

Diskin, M.G. and Sreenan, J.M. (2003) Oestrous expression and detection in cattle. *Reproduction in Domestic Animals* 38, 322 (Abstract).

Diskin, M.G., Austin, E.J. and Roche, J.F. (2002) Exogenous hormonal manipulation of ovarian activity in cattle. *Domestic Animal Endocrinology* 23, 211–228.

Dobrinsky, J.R., Pursel, V.G., Long, C.R. and Johnson, L.A. (2000) Birth of piglets after transfer of embryos cryopreserved by cytoskeletal stabilization and vitrification. *Biology of Reproduction* 62, 564–570.

D'Occhio, M.J., Fordyce, G., Whyte, T.R., Jubb, T.F., Fitzpatrick, L.A., Cooper, N.J., Aspden, W.J., Bolam, M.J. and Trigg, T.E. (2002) Use of GnRH agonist implants for long-terrm suppression of fertility in extensively managed heifers and cows. *Animal Reproduction Science* 74, 151–162.

Donadeu, F.X. and Ginther, O.J. (2003) Interactions of follicular factors and season in the regulation of circulating concentrations of gonadotrophins in mares. *Reproduction* 125, 743–750.

Donovan, G.A., Bennett, F.L. and Springer, F.S. (2003) Factors associated with first service conception in artificially inseminated nulliparous Holstein heifers. *Theriogenology* 60, 67–75.

Douglas, C.G.B., Perkins, N.R., Stafford, K.J. and Hedderley, D.I. (2002) Prediction of foaling using mammary secretion constituents. *New Zealand Veterinary Journal* 50, 99–103.

Du, F.L., Sung, L.Y., Tian, X.C. and Yang, X.Z. (2002) Differential cytoplast requirement for embryonic and somatic cell nuclear transfer in cattle. *Molecular Reproduction and Development* 63, 183–191.

Dufort, I., Foret, A., Bousquet, D. and Sirard, M.A. (2003) Changes in the relative mRNA transcripts abundance of genes involved in lipid pathways from preimplantation bovine embryos produced in different culture systems. *Biology of Reproduction* 68 (Suppl. 1), Abstract 114, p. 159.

Duggavathi, R., Bartlewski, P.M., Barrett, D.M.W. and Rawlings, N.C. (2003) Use of high-resolution transrectal ultrasonography to assess changes in numbers of small ovarian antral follicles and their relationships to the emergence of follicular waves in cyclic ewes. *Theriogenology* 60, 495–510.

Duque, P., Diez, C., Royo, L., Lorenzo, P.L., Carneiro, G., Hidalgo, C.O., Facal, N. and Gomez, E. (2002) Enhancement of developmental capacity of meiotically inhibited bovine oocytes by retinoic acid. *Human Reproduction* 17, 2706–2714.

Duszewska, A.M., Kozikova, L., Szydlik, H., Cybulska, M., Korwin-Kossakowski, M., Was, B., Poloszynowicz, J., Wicinska, K. and Rosochacki, S.J. (2003) The use of green fluorescent protein (GFP) to select bovine embryos. *Journal of Animal and Feed Sciences* 12, 71–81.

Dwyer, C.M., Lawrence, A.B., Bishop, S.C. and Lewis, M. (2003) Ewe–lamb bonding behaviours at birth are affected by maternal undernutrition in pregnancy. *British Journal of Nutrition* 89, 123–136.

Edwards, J.L., Coy, P., Romar, R., Payton, R.R., Dunlap, J. and Saxton, A.M. (2003) Culture of bovine oocytes in roscovitine for 24 or 48 hours before oocyte maturation: effects on nuclear and cytoplasmic maturation, zona pellucida hardening and fertilization. *Biology of Reproduction* 68 (Suppl. 1), Abstract 346, p. 254.

Elli, M., Gaffuri, B., Frigerio, A., Zanardelli, M., Covini, D., Candiani, M. and Vignali, M. (2001) Effect of a single dose of ibuprofen lysinate before embryo transfer on pregnancy rates in cows. *Reproduction* 121, 151–154.

Esaki, R., Yamashita, C., Yoshioka, H., Ushijima, H., Kuwayama, M. and Nagashima, H. (2003) Successful cryopreservation of porcine IVM-derived embryos. *Theriogenology* 59, 300.

Estienne, M.J. and Harper, A.F. (2000) PGF2-alpha facilitates the training of sexually active boars for semen collection. *Theriogenology* 54, 1087–1092.

Evans, A.C.O. (2003) Characteristics of ovarian follicle development in domestic animals. *Reproduction in Domestic Animals* 38, 240–246.

Evans, A.C.O. and O'Doherty, J.V. (2001) Endocrine changes and management factors affecting puberty in gilts. *Livestock Production Science* 68, 1–12.

Evans, A.C.O., Duffy, P., Hynes, N. and Boland, M.P. (2000) Waves of follicle development during the estrous cycle in sheep. *Theriogenology* 53, 699–715.

Fabbrocini, A., Sorbo, C.D., Fasano, G. and Sansone, G. (2000) Effect of differential addition of glycerol and pyruvate to extender on cryopreservation of Mediterranean buffalo (*B. bubalis*) spermatozoa. *Theriogenology* 54, 193–207.

Faber, D.C., Molina, J.A., Ohlrichs, C.L., Vander Zwaag, D.F. and Ferre, L.B. (2003) Commercialization of animal biotechnology. *Theriogenology* 59, 125–138.

Fair, T., Gutierrez-Adan, A., Murphy, M., Rizos, D., Martin, F., Boland, M.P. and Lonergan, P. (2004) Search for the bovine homolog of the murine Ped gene and characterization of its messenger RNA expression during bovine preimplantation development. *Biology of Reproduction* 70, 488–494.

Ferguson, E.M., Ashworth, C.J., Edwards, S.A., Hawkins, N., Hepburn, N. and Hunter, M.G. (2003) Effect of different nutritional regimens before ovulation on plasma concentrations of metabolic and reproductive hormones and oocyte maturation in gilts. *Reproduction* 126, 61–71.

Fernandes, C.A.C., Viana, J.H.M., Ferreira, A.M. and Sa, W.F. (2002) Fertility of heifers after abortion induced by cloprostenol. *Arquivo Brasileiro de Medicina Veterinaria e Zootecnia* 54, 279–282.

Firk, R., Stamer, E., Junge, W. and Krieter, J. (2002) Automation of oestrus detection in dairy cows: a review. *Livestock Production Science* 75, 219–232.

Fischer-Brown, A.E., Northey, D.L., Parrish, J.J. and Rutledge, J.J. (2003) Post-transfer growth of *in vitro* produced bovine embryos. *Biology of Reproduction* 68 (Suppl. 1), Abstract 340, p. 252.

Fisher, M. (2003) New Zealand farmer narratives of the benefits of reduced human intervention during lambing in extensive farming systems. *Journal of Agricultural and Environmental Ethics* 16, 77–90.

Flint, A.P.F., Wall, E., Coffey, M., Simm, G., Brotherstone, S., Stott, A.W., Santarossa, J., Royal, M.D. and Wooliams, J.A. (2002) Introducing a UK fertility index. *Cattle Practice* 10, 373–378.

Flint, A., Brink, Z. and Seidel, G.E. Jr (2003) Use of heterospermic insemination and genotyping embryos to compare fertility of flow-sorted sperm from individual bulls. *Theriogenology* 59, 507.

Foote, R.H. (1974) Artificial insemination. In: Hafez, E.E.S. (ed.) *Reproduction in Farm Animals*, 3rd edn. Lea and Febiger, Philadelphia, Pennsylvania, pp. 409–431.

Foote, R.H., Brockett, C.C. and Kaproth, M.T. (2002) Motility and fertility of bull sperm in whole milk extender containing antioxidants. *Animal Reproduction Science* 71, 13–23.

Froman, D. (2003) Deduction of a model for sperm storage in the oviduct of the domestic fowl (*Gallus domesticus*). *Biology of Reproduction* 69, 248–253.

Funk, D.J. (2002) Induced parturition in recipient cattle carrying nuclear transfer calves. *Archiv für Tierzucht* 45, 443–449.

Gade, P.B. (2002) Welfare of animal production in intensive and organic systems with special reference to Danish organic pig production. *Meat Science* 62, 353–358.

Galli, C. and Lazzari, G. (2003) *in vitro* production of embryos in farm animals. In: *Proceedings 19th Meeting European Embryo Transfer Association (Rostock)*, p. 93.

Galli, C., Duchi, R., Crotti, G., Turini, P., Ponderato, N., Coilleoni, S., Lagutina, I. and Lazzari, G. (2003a) Bovine embryo technologies. *Theriogenology* 59, 599–616.

Galli, C., Lagutina, I., Crotti, G., Colleoni, S., Turini, P., Ponderato, N., Duchi, R. and Lazzari, G. (2003b) A cloned horse born to its dam twin. *Nature* 424, 635.

Galli, C., Vassiliev, I., Lagutina, I., Galli, A. and Lazzari, G. (2003c) Bovine embryo development following ICSI: effect of activation, sperm capacitation and pre-treatment with dithiothreitol. *Theriogenology* 60, 1467–1480.

Garcia, A.J., Landete-Castilklejos, T., Gomez, J.A., Albinana, B., Garde, J.J. and Gallego, L. (2001) Increasing the prolificacy in Iberian deer (*Cervus elaphus hispanicus*) by hormone treatment. *Extra* 22, 751–753.

George, F., Simonis, I., Feugang, J.M., Massip, A., Verhoeye, F., Schneider, Y.J. and Donnay, I. (2002) Vegetal peptones as a substitute for animal proteins in embryo freezing medium. *Reproduction, Abstract Series* No. 29, Abstract 51, p. 20.

Ginther, O.J., Beg, M.A., Bergfelt, D.R. and Kot, K. (2002) Activin A, estradiol, and free insulin-like growth factor I in follicular fluid preceding the experimental assumption of follicle dominance in cattle. *Biology of Reproduction* 67, 14–19.

Giussani, D.A., Forhead, A.J., Gardner, D.S., Fletcher, A.J.W., Allen, W.R. and Fowden, A.L. (2003) Postnatal cardiovascular function after manipulation of fetal growth by embryo transfer in the horse. *Journal of Physiology* 547, 67–76.

Glossop, C.E. (1991) Pig artificial insemination re-assessed. *In Practice* 13(5), 191–195.

Goff, A.K. (2002) Embryonic signals and survival. *Reproduction in Domestic Animals* 37, 133–139.

Gomes, L.G., Gastal, E.L., Gastal, M.O. and Godoi, D.B. (2003) Effect of the season of birth on the onset of puberty in fillies. *Biology of Reproduction*, 68 (Suppl. 1), Abstract 436, p. 291.

Goncalves, R.F., Chapman, D.A. and Killian, G.J. (2003) Effect of osteopontin on *in vitro* bovine embryo development. *Biology of Reproduction* 68 (Suppl. 1), Abstract 545, pp. 336–337.

Gonzalez-Bulnes, A., Carrizosa, J.A., Diaz-Delfa, C., Garcia-Garcia, R.M., Urrutia, B., Santiago-Moreno, J., Cocero, M.J. and Lopez-Sebastian, A. (2003) Effects of ovarian follicular status on superovulatory response of dairy goats to FSH treatment. *Small Ruminant Research* 48, 9–14.

Gorecka, A. and Jezierski, T. (2003) Effect of single vs. muliple ovulations on oestrous behaviour and pregnancy rate in Thoroughbred mares. *Animal Science Papers and Reports* 21, 27–33.

Graham, J.K. and Purdy, P.H. (2002) Altering sperm membranes to improve cryosurvival. *Journal of Animal Science* 80 (Suppl. 2), Abstract 209, p. 83.

Grazul-Bilska, A.T., Choi, J.T., Bilski, J.J., Weigl, R.M., Kirch, J.D., Kraft, K.C., Reynolds, L.P. and Redmer, D.A. (2003) Effects of epidermal growth factor on early embryonic development after *in vitro* fertilization of oocytes collected from ewes treated with follicle stimulating hormone. *Theriogenology* 59, 1449–1457.

Gualtieri, R. and Talevi, R. (2003) Selection of highly fertilization-competent bovine spermatozoa through adhesion to the Fallopian tube epithelium *in vitro*. *Reproduction* 125, 251–258.

Guthrie, H.D., Liu, J. and Critser, J.K. (2002) Osmotic tolerance limits and effects of cryoprotectants on motility of bovine spermatozoa. *Biology of Reproduction* 67, 1811–1816.

Hafez, E.S.E. and Gordon, I. (1962) Female reproductive organs of farm mammals. In: Hafez, E.S.E. (ed.) *Reproduction of Farm Animals*. Lea and Febiger, Philadelphia, Pennsylvania, p. 79.

Hafez, E.S.E. and Hafez, B. (2001) Reproductive parameters of male dromedary and Bactrian camels. *Archives of Andrology* 46, 85–98.

Hammond, J. Jr, Bowman, J.C. and Robinson, T.J. (1983) *Hammond's Farm Animals*, 5th edn. Edward Arnold, London, 305 pp.

Hansel, W., Concannon, P.W. and Lukaszewska, P.H. (1973) Estrous cycle of the pig. *Biology of Reproduction* 8, 222.

Hasler, J.F. (2002) The freezing, thawing and transfer of cattle embryos. In: Fields, M.J., Sands, R.S. and Yelich, J.V. (eds) *Factors Affecting Calf Crop: Biotechnology of Reproduction*. CRC Press, Boca Raton, Florida, pp. 119–130.

Hazeleger, W., Bouwman, E.G., Noorhuizen, J.P.T.M. and Kemp, B. (2000a) Effect of superovulation induction on embryonic development on day 5 and subsequent development and survival after nonsurgical embryo transfer in pigs. *Theriogenology* 53, 1063–1070.

Hazeleger, W., Noordhuizen, J.P.T.M. and Kemp, B. (2000b) Effect of asynchronous non-surgical transfer of porcine embryos on pregnancy rate and embryonic survival. *Livestock Production Science* 64, 281–284.

Hemsworth, P.H. (2003) Human–animal interactions in livestock production. *Applied Animal Behaviour* 81, 185–198.

Henderson, K.M., Franchimont, P., Lecomte-Yerna, M.J., Charlet-Renard, C., Hudson, N., Ball, K. and McNatty, K.P. (1984) Ovarian inhibin: a hormone with potential to increase ovulation rate in sheep. *Proceedings of the New Zealand Society of Animal Production* 49, 97–101.

Henig, R.M. (2003) Pandora's baby. *Scientific American* 288(6), 51–55.

Hermansen, J.E. (2003) Organic livestock production systems and appropriate development in relation to public expectations. *Livestock Production Science* 80, 3–15.

Heyman, Y., Tamassia, M., Richard, C., Renard, J.P. and Chastant-Maillard, S. (2003a) Preliminary results on variability in oocyte recovery and developmental competence in cattle derived from embryonic cloning: work in progress. *Theriogenology* 60, 891–900.

Heyman, Y., Richard, C., Lavergney, Y., Menezo, Y. and Vignon, X. (2003b) Outcome of pregnancies after transfer of bovine IVP embryos cultured in the serum free sequential media, ISM1/ISM2. *Proceedings 19th Meeting European Embryo Transfer Association (Rostock)*, p. 160.

Hodges, J. (2002) Conservation of farm animal biodiversity: history and prospects. *Animal Genetic Resources Information* 32, 1–12.

Hodges, J. (2003) Livestock, ethics and quality of life. *Journal of Animal Science* 81, 2887–2894.

Hoffmann, B. and Schuler, G. (2002) The bovine placenta: a source and target of steroid hormones: observations during the second half of gestation. *Domestic Animal Endocrinology* 23, 309–320.

Hopkins, S.M. and Evans, L.E. (2003) Artificial insemination. In: Pineda, M.H. and Dooley, M.P. (eds) *McDonald's Veterinary Endocrinology and Reproduction*. Iowa State University Press, Ames, Iowa, pp. 341–375.

Horvat, G. and Bilkei, G. (2003) Exogenous prostaglandin F2-alpha at time of ovulation improves reproductive efficiency in repeat breeder sows. *Theriogenology* 59, 1479–1484.

Hoshi, H. (2003) *In vitro* production of bovine embryos and their application for embryo transfer. *Theriogenology* 59, 675–685.

Houdebine, L.M. (2003) Cloning by numbers. *Nature Biotechnology* 21, 1451–1452.

Hovi, M., Sundrum, A. and Thamsborg, S.M. (2003) Animal health and welfare in organic livestock production in Europe: current state and future challenges. *Livestock Production Science* 80, 41–53.

Hughes, A.M.E., Allcock, J.G. and Richardson, J.S. (2000) Strategic use of gonadotrophins in first litter sows after weaning. *Veterinary Record* 146, 164–165.

Hughes, P.E. (1998) Effects of parity, season and boar contact on the reproductive performance of weaned sows. *Livestock Production Science* 54, 151–157.

Hunter, R.H.F. (2002) Vital aspects of Fallopian tube physiology in pigs. *Reproduction in Domestic Animals* 37, 186–190.

Hyun, S., Lee, G., Kim, D., Kim, H., Lee, S., Nam, D., Jeong, Y., Kim, S., Yeom, S., Kang, S., Han, J., Lee, B. and Hwang, W. (2003) Production of nuclear transfer-derived piglets using porcine fetal fibroblasts transfected with the enhanced green fluorescent protein. *Biology of Reproduction* 69, 1060–1068.

Irvine, C.H.G., Alexander, S.L. and McKinnon, A.O. (2000) Reproductive hormone profiles in mares during the autumn transition as determined by collection of jugular blood at 6 h intervals throughout ovulatory and anovulatory cycles. *Journal of Reproduction and Fertility* 118, 101–109.

Johansson, I. and Hansson, A. (1943) The sex ratio and multiple births in sheep. *Annals of the Agriculture College of Sweden* 11, 145–171.

Johnson, L.A., Weitze, K.F., Fiser, P. and Maxwell, W.M.C. (2000) Storage of boar semen. *Animal Reproduction Science* 62, 143–172.

Jones, C.J.P., Abd-Elnaeim, M., Bevilacqua, E., Oliveira, L.V. and Leiser, R. (2002) Comparison of uteroplacental glycosylation in the camel (*Camelus dromedarius*) and alpaca (*Laca pacos*). *Reproduction* 123, 115–126.

Jopson, N.B., Davis, G.H., Farquhar, P.A. and Bain, W.E. (2002) Effects of mid-pregnancy nutrition and shearing on ewe body reserves and foetal growth. *Proceedings New Zealand Society of Animal Production* 62, 49–52.

Joyce, M.M., Burghardt, R.C., Bazer, F.W., Zaunbrecher, G.M. and Johnson, G.A. (2003) Interferon-stimulated genes (ISGs) are induced in the endometrium of pregnant but not pseudopregnant pigs. *Biology of Reproduction* 68 (Suppl. 1), Abstract 230, p. 206.

Ju, J.C., Tsay, C. and Ruan, C.W. (2003) Alterations and reversibility of the chromatin, cytoskeleton and development of pig oocytes treated with roscovitine. *Molecular Reproduction and Development* 64, 482–491.

Kabanov, V. (2002) Biological foundations of increasing the intensity of pig breeding. *Svinovodstvo (Moska)* 2, 27–28.

Kaeoket, K. (2003) The use of real time B-mode ultrasound to detect ovulation and pregnancy in swine. *Thai Journal of Veterinary Medicine* 33, 15–24.

Kagami, H. (2003) Sex reversal in chicken. *World's Poultry Science Association* 59, 15–18.

Kaneko, H., Kikuchi, K., Noguchi, J., Hosoe, M. and Akita, T. (2003) Maturation and fertilization of porcine oocytes from primordial follicles by a combination of xenografting and *in vitro* culture. *Biology of Reproduction* 69, 1488–1493.

Kanitz, W., Unger, C., Nurnberg, G., Hoppen, H.O., Becker, F. and Schneider, F. (2003) Pregnancy rates and endocrine events after treatment of mares with the gonadotrophin releasing hormone agonist buserelin during luteal phase. *Reproduction in Domestic Animals* 38, 355–356 (Abstract).

Karadjov, T. (2001) Duration of pregnancy in mares of Arabian breed and factors influenced on it. *Zhivotnovdni Nauki* 38, 39–41.

Karen, A., Beckers, J.F., Sulon, J., Melo de Sousa, N., Szabados, K., Reczigel, J. and Szenci, O. (2003) Early pregnancy diagnosis in sheep by progesterone and pregnancy-associated glycoprotein tests. *Theriogenology* 59, 1941–1949.

Karlsen, A., Klemetsdal, G. and Ruane, J. (2000) Twinning in cattle. *Animal Breeding Abstracts* 68, 1–8.

Katila, T. (2001) Sperm-uterine interactions: a review. *Animal Reproduction Science* 68, 267–272.

Kelley, T. (2003) Will A.I. be going deep? *Pork* 23(2), 24–25.

Kenyon, P.R., Morris, S.T. and McCutheon, S.N. (2002) Does an increase in lamb birth weight through mid-pregnancy shearing necessarily mean an increase in lamb survival rates to weaning? *Proceedings New Zealand Society of Animal Production* 62, 53–56.

Keskintepe, L., Pacholczyk, G., Machnicka, A., Norris, K., Curuk, M.A., Khan, I. and Brackett, B.G. (2002) Bovine blastocysts development from oocytes injected with freeze-dried spermatozoa. *Biology of Reproduction* 67, 409–415.

Kesler, D.J., Wherley, N.R. and Faulkner, D.B. (2002) Synchronization of estrus in beef heifers with

MGA and PGF. *Journal of Animal Science* 80 (Suppl. 2), Abstract 214, p. 84.

Kidson, A., Schoevers, E., Langendijk, P., Verheijden, J., Colenbrander, B. and Bevers, M. (2003) The effect of oviductal epithelial cell co-culture during *in vitro* maturation on sow oocyte morphology, fertilization and embryo development. *Theriogenology* 59, 1889–1903.

Kindahl, H., Kornmatitsuk, B., Konigsson, K. and Gustafsson, H. (2002) Endocrine changes in late bovine pregnancy with special emphasis on fetal well-being. *Domestic Animal Endocrinology* 23, 321–328.

King, T.J., Dobrinsky, J.R., Bracken, J., McCorquodale, C. and Wilmut, I. (2001) Ovulated pig oocyte production using trans-cutaneous ultrasonography to determine ovulation time. *Veterinary Record* 149, 362–363.

Klocek, C., Szewczyk, A. and Nowicki, J. (2000) Behaviour of periparturient sows allowed freedom of movement in pens. *Annals of Animal Science – Roczniki Naukowe Zootechniki* 27, 161–171.

Knight, T.W. (1983) Ram induced stimulation of ovarian and oestrous activity in anoestrous ewes – a review. *Proceedings of the New Zealand Society of Animal Production* 43, 7–11.

Knijn, H.M., Gjorret, J.O., Vos, P.L.A.M., Hendriksen, P.J.M., van der Weijden, B.C., Maddox-Hyttel, P. and Dieleman, S.J. (2003a) Consequences of *in vivo* development and subsequent culture on apoptosis, cell number and blastocyst formation in bovine embryos. *Biology of Reproduction* 69, 1371–1378.

Knijn, H.M., Wrenzycki, C., Zeinstra, E.C., Vos, P.L.A.M., van der Weijden, G.C., Niemann, H. and Dieleman, S.J. (2003b) Glucose transporter expression in bovine expanded blastocysts cultured until day 7 post ovulation after collection at different times of *in vivo* development. In: *Proceedings 19th Meeting European Embryo Transfer Association (Rostock)*, p. 174.

Knox, R.V., Vatzias, G., Naber, C.H. and Zimmerman, D.R. (2003) Plasma gonadotropins and ovarian hormones during the estrous cycle in high compared to low ovulation rate gilts. *Journal of Animal Science* 81, 249–260.

Kobayashi, S., Kano, M., Takei, M. and Tajima, S. (2003) Viability and telomerase activity of porcine embryos vitrified by ultra-rapid cooling method using the open pulled straw. *Theriogenology* 59, 304.

Koeman, J., Keefer, C.L., Baldassarre, H. and Downey, B.R. (2003) Developmental competence of prepubertal and adult goat oocytes cultured in semi-defined media following laparoscopic recovery. *Theriogenology* 60, 879–889.

Kolle, S., Stojkovic, M., Boie, G., Wolf, E. and Sinowatz, F. (2003) Growth hormone-related effects on apoptosis, mitosis, and expression of Connexin 43 in bovine *in vitro* maturation cumulus–oocyte-complexes. *Biology of Reproduction* 68, 1584–1589.

Kramer, J.M., Ridriguez-Sallaberry, C.J. and Moore, K. (2003) Aberrant expression of the insulin-like growth factor family genes in day 25 nuclear transfer derived bovine conceptuses. *Biology of Reproduction* 68 (Suppl. 1), Abstract 342, pp. 252–253.

Kreider, D., Rorie, R., Brown, D., Miller, F. and Wright, S. (2000) Enhancement of ovulation rate and litter size in swine. *Research Series – Arkansas Agricultural Experiment Station* No. 478, 138–142.

Kruip, T.A.M., Pieterse, M.C., van Beneden, T.A.M., Vos, P.L.A.M., Wurth, Y.A. and Taverne, M.A.M. (1991) A new method for bovine embryo production: a potential alternative to superovulation. *Veterinary Record* 128, 208–210.

Kunavongkrit, A. and Heard, T.W. (2000) Pig reproduction in south east Asia. *Animal Reproduction Science* 60/61, 527–533.

Kurykin, J., Jaakma, U., Majas, L., Jalakas, M., Aidnik, M., Waldmann, A. and Padrik, P. (2003) Fixed time deep intracornual insemination of heifers at synchronized estrus. *Theriogenology* 60, 1261–1268.

Kuzmina, T., Pozdnyakova, T., Alm, H., Torner, H. and Kanitz, W. (2003) Effect of somatotropin on developmental capacity of *in vitro* matured bovine oocyte from animals of different age. In: *Proceedings 19th Meeting European Embryo Transfer Association (Rostock)*, p. 176.

Kvasnicki, A.V. (1951) Interbreed ova transplantation. *Sovetsk. Zooteh.*, pp. 36–42. *Animal Breeding Abstracts* 19, 224.

Kwon, H., Spencer, T.E., Bazer, F.W. and Wu, G. (2003) Developmental changes of amino acids in ovine fetal fluids. *Biology of Reproduction* 68, 1813–1820.

Lamb, G.C. and Dahlen, C.R. (2002) Past, present and future impact of ultrasound technology on beef cattle reproductive research and management strategies. *Journal of Animal Science* 80 (Suppl. 2), Abstract 218, p. 85.

Lamberson, W.R. and Safranski, T.J. (2000) A model for economic comparison of swine insemination programs. *Theriogenology* 54, 799–807.

Lane, M., Gardner, D.K., Hasler, M.J. and Hasler, J.F. (2003) Use of G1.2/G2.2 media for commercial bovine embryo culture: equivalent development and pregnancy rates compared to co-culture. *Theriogenology* 60, 407–419.

Laven, R.A., Biggadike, H.J. and Allison, R.D. (2002) The effect of pasture nitrate concentration and concentrate intake after turnout on embryo growth and viability in the lactating dairy cow. *Reproduction in Domestic Animals* 37, 111–115.

Lavitrano, M., Bacci, M.L., Forni, M., Lazzereschi, D., Di Stefano, C., Fioretti, D., Giancotti, P., Marfe, G., Pucci, L., Renzi, L., Wang, H., Stoppacciaroi, A., Stassi, G., Sargiacomo, M., Sinibaldi, P., Turchi, V., Giovannoni, R., Della Casa, G., Seren, E. and Rossi, G. (2002) Efficient production by sperm-mediated gene transfer of human decay accelerating factor (hDAF) transgenic pigs for xenotransplantation. *Proceedings of the National Academy of Sciences USA* 99, 14230–14235.

Lazzari, G., Wrenzycki, C., Herrmann, D., Duchi, R., Kruip, T., Niemann, H. and Galli, C. (2002) Cellular and molecular deviations in bovine *in vitro*-produced embryos are related to the large offspring syndrome. *Biology of Reproduction* 67, 767–775.

Lee, G., Hyun, S., Kim, H., Kim, D., Lee, S., Lim, J., Lee, B. and Hwang, W. (2003) Improvement of a porcine somatic cell nuclear transfer technique by optimizing donor cell and recipient oocyte preparations. *Theriogenology* 59, 1949–1957.

Lee, J.W., Dinnyes, A. and Yang, X. (2003a) Development of transgenic embryos following injection of freeze-dried sperm in pigs. *Theriogenology* 59, 305.

Lee, J.W., Tian, X.C. and Yang, X. (2003b) Failure of male pronucleus formation is the major cause of lack of fertilization and embryo development in pig oocytes subjected to intracytoplasmic sperm injection. *Biology of Reproduction* 68, 1341–1347.

Lee, J.W., Wu, S.C., Tian, X.C., Barber, M., Hoagland, T., Riesen, J., Lee, K.H., Tu, C.F., Cheng, W.T.K. and Yang, X. (2003c) Production of cloned pigs by whole cell intracytoplasmic microinjection. *Biology of Reproduction* 69, 995–1001.

Lefranc, A.-C. and Allen, W.R. (2003) Non-pharmacological suppression of oestrous cyclicity in the mare. *Reproduction in Domestic Animals* 38, 320–321 (Abstract).

Leiding, C. (2002) The current position and future for artificial insemination in Germany. *Zuchtwahl und Besamung* 147, 62–67.

Leroy, J.L.M.R., Genicot, G., Opsomer, G., Donnay, I. and Van Soom, A. (2003) A comparison of the lipid content of immature and mature bovine oocytes and of morulae after staining with Nile Red. In: *Proceedings 19th Meeting European Embryo Transfer Association (Rostock)*, p. 180.

Li, X., Morris, L.H.A. and Allen, W.R. (2002) *In vitro* development of horse oocytes reconstructed with the nuclei of fetal and adult. *Biology of Reproduction* 66, 1288–1292.

Li, X., Tremoleda, J.L. and Allen, W.R. (2003) Effect of number of passages of fetal and adult fibroblasts on nuclear remodelling and first embryonic division in reconstructed horse oocytes after nuclear transfer. *Reproduction* 125, 535–542.

Liebermann, J., Nawroth, F., Isachenko, V., Isachenko, E., Rahimi, G. and Tucker, M.J. (2002) Potential importance of vitrification in reproductive medicine. *Biology of Reproduction* 67, 1671–1680.

Lindeberg, H., Karjalainen, H., Koskinen, E. and Katila, T. (1999) Quality of stallion semen obtained by a new semen collection phantom (Equidame) versus a Missouri artificial vagina. *Theriogenology* 51, 1157–1173.

Lonergan, P. (2003) Overall bovine embryo transfer activity in Europe in 2002. In: *Proceedings 19th Meeting European Embryo Transfer Association (Rostock)*, p. 77.

Lonergan, P., Rizos, D., Kanka, J., Nemcova, L., Mbaye, A.M., Kingston, M., Wade, M., Duffy, P. and Boland, M.P. (2003a) Temporal sensitivity of bovine embryos to culture environment after fertilization and the implications for blastocyst quality. *Reproduction* 126, 337–346.

Lonergan, P., Rizos, D., Gutierrez-Adan, A., Moreira, P.M., Pintado, B., de la Fuente, J. and Boland, M.P. (2003b) Temporal divergence in the pattern of messenger RNA expression in bovine embryos cultured from the zygote to blastocyst stage *in vitro* or *in vivo*. *Biology of Reproduction* 69, 1424–1431.

Lonergan, P., Faerge, I., Hyttel, P., Boland, M. and Fair, T. (2003c) Ultrastructural modifications in bovine oocytes maintained in meiotic arrest using roscovitine or butyrolactone. *Molecular Reproduction and Development* 64, 369–378.

Long, C.R., Walker, S.C., Tang, R.T. and Westhusin, M.E. (2003) New commercial opportunities for advanced reproductive technologies in horses, wildlife and companion animals. *Theriogenology* 59, 139–149.

Lopatarova, M., Holy, L., Jindra, M., Krontorad, P. and Cech, S. (2003) Sex determination in split bovine embryos by polymerase chain reaction under farm conditions. *Reproduction in Domestic Animals* 38, 349 (Abstract).

Lopes, A.S., Lovendahl, P. and Callesen, H. (2003) Assessment of number of oocytes retrieved from individual cows can be based on few collections. In: *Proceedings 19th Meeting European Embryo Transfer Association (Rostock)*, p. 182.

Lopez-Gatius, F. (2003) Is fertility declining in dairy cattle? A retrospective study in northeastern Spain. *Theriogenology* 60, 89–99.

Lopez-Gatius, F., Yaniz, J. and Madriles-Helm, D. (2003) Effects of body condition score and scorer change on the reproductive performance of dairy cows: a meta-analysis. *Theriogenology* 59, 801–812.

Luo, H.L., Kimura, K., Aoki, M. and Hirako, M. (2002) Effect of vascular endothelial growth factor on maturation, fertilization and developmental competence of bovine oocytes. *Journal of Veterinary Medical Science* 64, 803–806.

Lymberopoulos, A.G., Boscos, C.M., Dellis, S., Papia, A. and Belibasaki, S. (2002) Oestrous synchronization under range conditions in dairy goats treated with different PGF2-alpha doses during the transitional period in Greece. *Animal Science* 75, 289–294.

McCauley, T.C., Mazza, M.R., Didion, B.A., Mao, J., Wu, G., Coppola, G., Coppola, G.F., Di Berardino, D. and Day, B.N. (2003) Chromosomal abnormalities in day-6, *in vitro* produced pig embryos. *Theriogenology* 60, 1569–1580.

MacDonnell, H.F., Mullins, S. and Gordon, I. (1993) Foetal progress and onset of parturition monitored by plasma oestrogen levels in the cow. *Irish Journal of Medical Science* 162(3), 108–109.

McEachern, M. and Tregear, A. (2000) Farm animal welfare in the UK: a comparison of assurance schemes. *Farm Management* 10, 685–708.

McGrath, M., Mee, J.F. and O'Callaghan, D. (2003) Onset of puberty and reproductive performance in different strains of dairy heifers. *Reproduction in Domestic Animals* 38, 362 (Abstract).

McGraw, S., Robert, C., Massicotte, L. and Sirard, M.A. (2003) Quantification of histone acetyltransferase and histone deacetylase transcripts during early bovine embryo development. *Biology of Reproduction* 68, 383–389.

McInerney, J.P. (2002) Animal welfare: ethics, economics and productivity. *Proceedings New Zealand Society of Animal Production* 62, 340–347.

Maclellan, L.J., Bass, L.D., McCue, P.M. and Squires, E.L. (2003) Effect of cooling large and small equine embryos prior to cryopreservation on pregnancy rates after transfer. *Theriogenology* 59, 306.

Maddox-Hyttel, P., Alexopoulos, N.I., Vajta, G., Lewis, I., Rogers, P., Cann, L., Callesen, H., Tveden-Nyborg, P. and Trounson, A. (2003) Immunohistochemical and ultrastructural characterization of the initial post-hatching development development of bovine embryos. *Reproduction* 125, 607–623.

Maes, D.G.D., Mateusen, B., Rijsselaere, T., De Vlieghher, S., Van Soom, A. and de Kruif, A. (2003) Motility characteristics of boar spermatozoa after addition of prostaglandin F2-alpha. *Theriogenology* 60, 1435–1443.

Mapletoft, R.J., Steward, K.B. and Adams, G.P. (2002) Superovulation in perspective. In: *Proceedings 18th Meeting European Embryo Transfer Association (Rolduc)*, pp. 119–127.

Marchant, J.N., Broom, D.M. and Cornoing, S. (2001) The influence of sow behaviour on piglet mortality due to crushing in an open farrowing system. *Animal Science* 72, 19–28.

Martin, P.A. (1986) Embryo transfer in swine. In: Morrow, D.A. (ed.) *Current Therapy in Theriogenology*. W.B. Saunders, Philadelphia, Pennsylvania, pp. 66–69.

Martinat-Botte, F., Bariteau, F., Badouard, B. and Terqui, M. (1985) Control of pig reproduction in a breeding programme. *Journal of Reproduction and Fertility* (Supplement) 33, 211–228.

Martinez, E.A., Vazquez, J.M., Roca, J., Lucas, X., Gil, M.A., Parrilla, I., Vazquez, J.L. and Day, B.N. (2001) Successful non-surgical deep intrauterine insemination with small numbers of spermatozoa in sows. *Reproduction* 122, 289–296.

Martinez, E.A., Vazquez, J.M. and Roca, J. (2003) Deep intrauterine embryo transfer in non-sedated gilts and sows. *Reproduction in Domestic Animals* 38, 320 (Abstract).

Martins, A. Jr, Silva, R.B., Zanon, J.E.O., Calegari, R.S. and Verona, D. (2003) Beneficial effect of catalase on development of bovine embryos. *Biology of Reproduction* 68 (Suppl. 1), Abstract 550, pp. 338–339.

Matas, C., Coy, P., Romar, R., Marco, M., Gadea, J. and Ruiz, S. (2003) Effect of sperm preparation method on *in vitro* fertilization in pigs. *Reproduction* 125, 133–141.

Matta, C.G.F., Silva, J.F.S., Da Silva, C.L., Soares, S.G., Van Tilburg, M.F., Barrabas, N., Fagundes, B, Souza, G.V., Souza, V.R. and Matta, M.F.R. (2001) Immuno-sexing of bovine spermatozoa using monoclonal antibody against a male-specific protein. *Revista Brasiliera de Reproducao Animal* 25, 402–404.

Mavrides, A. and Morroll, D. (2002) Cryopreservation of bovine oocytes: is cryoloop vitrification the future to preserving the female gamete? *Reproduction, Nutrition, Development* 42, 73–80.

Maxwell, W.M.C. and Evans, G. (2000) Recent developments in artificial insemination of sheep and goats with semen stored in chilled liquid or frozen state. In: *Proceedings 14th International Congress on Animal Reproduction (Stockholm)*, vol. 1, p. 268.

Medan, M.S., Watanabe, G., Sasaki, K., Nagura, Y., Sakaime, H., Fujita, M., Sharawy, S. and Taya, K. (2003) Effects of passive immunization of goats against inhibin on follicular development, hormone profile and ovulation rate. *Reproduction* 125, 751–757.

Mee, J.F. (2003) Trends in reproductive performance in Irish dairy cow herds. *Reproduction in Domestic Animals* 38, 362 (Abstract).

Mejia Silva, W., Cruz Arambulo, R., Calatayud Marques, D., Leon, G. and Quintero-Moreno, A. (2001) Use of real-time ultrasound for early pregnancy diagnosis in the sow. *Revista Científica, Facultad de Ciencias Veterinarias, Universidad del Zulia* 11, 418–422.

Melendez, P., Bartolome, J., Archbald, L.F. and Donovan, A. (2003) The association between lameness, ovarian cysts and fertility in lactating dairy cows. *Theriogenology* 59, 927–937.

Menchaca, A., Pinczak, A. and Rubianes, E. (2002) Follicular recruitment and ovulatory response to FSH treatment initiated on Day 0 or Day 3 post ovulation in goats. *Theriogenology* 58, 1713–1721.

Mendes, J.O.B. Jr, Burns, P.D., De la Torre-Sanchez, J.F. and Seidel, G.E. Jr (2003) Effect of heparin on cleavage rates and embryo production with four bovine sperm preparation protocols. *Theriogenology* 60, 331–340.

Merks, J.W.M., Cucro-Steverink, D.W.B. and Feitsma, H. (2000) Management and genetic factors affecting fertility in sows. *Reproduction in Domestic Animals* 35, 261–266.

Merton, J.S., de Roos, A.P.W., Mullaart, E., de Ruigh, L., Kaal, L., Vos, P.L.A.M. and Dieleman, S.J. (2003) Factors affecting oocyte quality and quantity in commercial application of embryo technologies in the cattle breeding industry. *Theriogenology* 59, 651–674.

Mialot, J.P., Constant, F., Dezaux, P., Grimard, B., Deletang, F. and Ponter, A.A. (2003) Estrus synchronization in beef cows: comparison between GnRH + PGF2-alpha + GnRH and PRID + PGF2-alpha + ECG. *Theriogenology* 60, 319–330.

Miller, D.J., Eckert, J.J., Lazzari, G., Duranthon-Richoux, V., Sreenan, J., Morris, D., Galli, C., Renard, J.P. and Fleming, T.P. (2003) Tight junction messenger RNA expression levels in bovine embryos are dependent upon the ability to compact and *in vitro* culture methods. *Biology of Reproduction* 68, 1394–1402.

Miller, L.A. and Killian, G.J. (2002) In search of the active PZP epitope in white-tailed deer immunocontraception. *Vaccine* 20, 2735–2742.

Misra, A.K. and Pant, H.C. (2003) Estrus induction following PGF2-alpha treatment in the superovulated buffalo (*Bubalus bubalis*). *Theriogenology* 59, 1203–1207.

Misumi, K., Suzuki, M., Sato, S. and Saito, N. (2003) Successful production of piglets derived from vitrified morulae and early blastocysts using a microdroplet method. *Theriogenology* 60, 253–260.

Miyano, T. (2003) Bringing up small oocytes to eggs in pigs and cows. *Theriogenology* 59, 61–72.

Miyashita, N., Shiga, K., Fujita, T., Umeki, H., Sato, W., Suzuki, T. and Nagai, T. (2003) Normal telomere lengths of spermatozoa in somatic cell-cloned bulls. *Theriogenology* 59, 1557–1565.

Miyoshi, K., Rzucidlo, S.J., Pratt, S.L. and Stice, S.L. (2003) Improvements in cloning efficiencies may be possible by increasing uniformity in recipient oocytes and donor cells. *Biology of Reproduction* 68, 1079–1086.

Mohan, M., Hurst, A.G. and Malayer, J.R. (2003) Comparative analysis of gene expression between *in vivo*-derived and *in vitro* produced day-7 bovine blastocysts. *Biology of Reproduction* 68 (Suppl. 1), Abstract 109, p. 157.

Moloney, A.P. (2002) The fat content of meat and meat products. In: Kerry, J., Kerry, J. and Ledward, D. (eds) *Meat Processing: Improving Quality*. Woodhead Publishing, Cambridge, UK, pp. 137–153.

Morel, M.C.G.D., Newcombe, J.R. and Holland, S.J. (2002) Factors affecting gestation length in the Thoroughbred mare. *Animal Reproduction Science* 74, 175–185.

Morris, L.H.A. and Allen, W.R. (2002) Reproductive efficiency of intensively managed Thoroughbred mares in Newmarket. *Equine Veterinary Journal* 34, 51–60.

Morris, L.H.A., Hunter, R.H.F. and Allen, W.R. (2000) Hysteroscopic insemination of small numbers of spermatozoa at the uterotubal junction of preovulatory mares. *Journal of Reproduction and Fertility* 118, 95–100.

Morton, K.M., Catt, S.L., Maxwell, W.M.C. and Evans, G. (2003) The effects of hormone stimulation and lamb age on *in vitro* production of embryos from prepubertal lambs. *Reproduction, Abstract Series* No. 30, Abstract P18, p. 56.

Moussa, M., Duchamo, G., Mahla, R., Bruyas, J.F. and Daels, P.F. (2003) *In vitro* and *in vivo* comparison of Ham's F-10, Emcare holding solution and Vigro holding plus for the cooled storage of equine embryos. *Theriogenology* 59, 1615–1625.

Mwanza, A.M., Englund, P., Kindahl, H., Lundeheim, N. and Einarsson, S. (2000) Effects of post-ovulatory food deprivation on the hormonal profiles, activity of the oviduct and ova transport in sows. *Animal Reproduction Science* 59, 185–199.

Mysterud, A., Steinheim, G., Yoccoz, N.G., Holand, O. and Stenseth, N.C. (2002) Early onset of reproductive senescence in domestic sheep, *Ovis aries*. *Oikoas* 97, 177–183.

Nagashima, H., Fujimura, T., Takahagi, Y., Kurome, M., Wako, N., Ochiai, T., Esaki, R., Kano, K., Saito, S., Okabe, M. and Murakami, H. (2003) Development of efficient strategies for the production of genetically modified pigs. *Theriogenology* 59, 95–106.

Nakai, M., Kashiwazaski, N., Takizawa, A., Hayashi, Y., Nakatsukasa, E., Fuchimoto, D., Noguchi, J., Kaneko, H., Shino, M. and Kikuchi, K. (2003) Viable piglets generated from porcine oocytes matured *in vitro* and fertilized by intracytoplasmic sperm head injection. *Biology of Reproduction* 68, 1003–1008.

Neglia, G., Gasparrini, B., di Brienza, V.C., Di Palo, R., Campanile, G., Presicce, G.A. and Zicarelli, L. (2003) Bovine and buffalo *in vitro* embryo production using oocytes derived from abattoir ovaries or collected by transvaginal follicle aspiration. *Theriogenology* 59, 1123–1130.

Nie, G.J., Johnson, K.E., Wenzel, J.G.W. and Braden, T.D. (2003) Effect of administering oxytocin or cloprostenol in the periovulatory period on pregnancy outcome and luteal function in mares. *Theriogenology* 60, 1111–1118.

Niemann, H., Rath, D. and Wrenzycki, C. (2003) Advances in biotechnologies: new tools in future pig production for agriculture and biomedicine. *Reproduction in Domestic Animals* 38, 82–89.

Niswender, G.D., Juengel, J.L., McGuire, W.J., Belifore, C.J. and Wiltbank, M.C. (1994) Luteal function: the estrous cycle and early pregnancy. *Biology of Reproduction* 50, 239–247.

O'Brien, M.J., Pendola, J.K. and Eppig, J.J. (2003) A revised protocol for *in vitro* development of mouse oocytes from primordial follicles dramatically improves their developmental competence. *Biology of Reproduction* 68, 1682–1686.

O'Farrell, K., Mee, J., Murphy, J. and Reitsma, P. (1991) Induced twinning in dairy cows. *Farm and Food Research* 21(2), 25–27.

Okuda, K., Miyamoto, Y. and Skarzynski, D.J. (2002) Regulation of endometrial prostaglandin F2-alpha synthesis during luteolysis and early pregnancy in cattle. *Domestic Animal Endocrinology* 23, 255–264.

Oliveira, M.A.L., Andrade, J.C.O., Lima, P.F., Santos Filho, A.S., Guido, S.I., Cavalcanti Neto, C.C., Tenorio Filho, F. and Oliveira, L.R.S. (2002) Influence of age on the superovulatory response

of Nelore donors: preliminary results. *Revista Brasileira de Reproducao Animal* 26, 246–248.

Orgeur, P., Le Dividich, J., Colson, V. and Meunier-Salaun, M.C. (2002) Mother–young relationships in pigs: from birth to weaning. *INRA, Productions Animales* 15, 185–198.

Palasz, A.T., Adams, G.P., Brogliatti, G.M. and Mapletoft, R.J. (2000) Effect of day of collection and of permeating cryoprotectants on llama (*Lama glama*) embryos and trophoblastic vesicles. *Theriogenology* 53, 341.

Panarace, M., Medina, M., Cattaneo, L., Caballero, J., Cerrate, H., Dalla Lasta, M. and Kaiser, G. (2003) Embryo production using sexed semen in superovulated cows and heifers. *Theriogenology* 59, 513.

Park, E.S., Hwang, W.S., Kang, S.K., Lee, B.C., Han, J.Y. and Lim, J.M. (2004) Improved embryo development with decreased apoptosis in blastomeres after the treatment of cloned bovine embryos with beta-mercaptoethanol and hemoglobin. *Molecular Reproduction and Development* 67, 200–206.

Parks, J.E., Lee, D.R., Huang, S. and Kaproth, M.T. (2003) Prospects for spermatogenesis *in vitro*. *Theriogenology* 59, 73–86.

Patterson, D.J. (2002) A review of methods to synchronize estrous cycles of postpartum suckled beef cows with the oral progestin, melengestrol acetate. *Journal of Animal Science* 80 (Suppl. 1), Abstract 215, pp. 84–85.

Paulenz, H., Kommisrud, E. and Hofmo, P.O. (2000) Effect of long-term storage at different temperatures on the quality of liquid boar semen. *Reproduction in Domestic Animals* 35, 83–87.

Pedersen, H.G., Schmidt, M., Strobech, L., Vajta, G., Callesen, H. and Greve, T. (2003) Evaluation of pregnancies following transfer of embryos produced by hand-made cloning. *Biology of Reproduction* 68 (Suppl. 1), Abstract 413, p. 282.

Pelletier, J., Sinoret, J.P., Cahill, L., Cognie, Y., Thimonier, J. and Ortavant, R. (1977) Physiological processes in oestrus, ovulation and fertility in sheep. In: *Symposium on Management of Reproduction in Sheep and Goats (Madison)*, Sheep Industry Development Program, pp. 1–14.

Peltoniemi, O.A.T., Tast, A. and Love, R.J. (2000) Factors affecting reproduction in the pig: seasonal effects and restricted feeding of the pregnant gilt and sow. *Animal Reproduction Science* 60/61, 173–184.

Perez-Hernandez, P., Garcia-Winder, M. and Gallegos-Sanchez, J. (2002) Postpartum anoestrus is reduced by increasing the within-day milking to suckling interval in dual purpose cows. *Animal Reproduction Science* 73, 159–168.

Perry, J.S. (1971) *The Ovarian Cycle of Mammals.* Oliver and Boyd, Edinburgh, 219 pp.

Peters, A.R., Dwyer, L., Canham, P.A. and Mackinnon, J.D. (2000) Effect of gonadotrophin-releasing hormone on the fertility of sows kept outdoors. *Veterinary Record* 147, 649–652.

Peterson, A.J. and Lee, R.S.F. (2003) Improving successful pregnancies after embryo transfer. *Theriogenology* 59, 687–697.

Peura, T.T., Kleeman, D.O., Rudiger, S.R., Nattrass, G.S., McLaughlan, C.J. and Walker, S.K. (2003) Effect of nutrition of oocyte donor on the outcomes of somatic cell nuclear transfer in the sheep. *Biology of Reproduction* 68, 45–50.

Pinto, M.M.V., Ramalho, S., Perestrelo-Vielra, R. and Rodrigues, J. (2000) Microbiological profile of pure semen and seminal doses of boars used in artificial insemination in Portugal. *Veterinaria Tecnica* 10(3), 24–28.

Pocar, P., Brevini, T.A.L., Fischer, B. and Gandolfi, F. (2003) The impact of endocrine disruptors on oocyte competence. *Reproduction* 125, 313–325.

Polejaeva, I.A., Chen, S.H., Vaught, T.D., Page, R.L., Mullins, J., Ball, S., Dal, Y., Boone, J., Walker, S., Ayares, D.L., Colman, A. and Campbell, K.H.S. (2000) Cloned pigs produced by nuclear transfer from adult somatic cells. *Nature* 407, 86–90.

Powell, K. (2003) Barnyard biotech – lame duck or golden goose? *Nature Biotechnology* 21, 965–967.

Prado, V., Orihuela, A., Lozano, S. and Perez-Leon, I. (2002) Management of the female stimulus during semen collection and its association with libido re-restablishment and semen characteristics of goats. *Journal of Animal Science* 80, 1520–1523.

Prado, V., Orihuela, A., Lozano, S. and Perez-Leon, I. (2003) Effect on ejaculatory performance and semen parameters of sexually-satiated male goats (*Capra hircus*) after changing the stimulus female. *Theriogenology* 60, 261–267.

Prather, R.S., Hawley, R.J., Carter, D.B., Lai, L. and Greenstein, J.L. (2003) Transgenic swine for biomedicine and agriculture. *Theriogenology* 59, 115–123.

Prescott, N.B., Wathes, C.M. and Jarvis, J.R. (2003) Light, vision and the welfare of poultry. *Animal Welfare* 12, 269–288.

Presicce, G.A., Parmeggiani, A., Senatore, E.M., Stecco, R., Barile, V.L., De Mauro, G.J., De Santis, G. and Terzano, G.M. (2003) Hormonal dynamics and follicular turnover in prepuberal Mediterranean Italian buffaloes (*Bubalus bubalis*). *Theriogenology* 60, 485–493.

Prunier, A. and Quesnel, H. (2000) Influence of the nutritional status on ovarian development in female pigs. *Animal Reproduction Science* 60/61, 185–197.

Ptak, G., Clinton, M., Tischner, M., Barboni, B., Mattioli, M. and Loi, P. (2002) Improving delivery and offspring viability of *in vitro*-produced and cloned sheep embryos. *Biology of Reproduction* 67, 1719–1725.

Quesnel, H. and Prunier, A. (1995) Endocrine bases of lactation anoestrus in the sow. *Reproduction, Nutrition, Development* 35, 395–414.

Rath, D. (2002) Low dose insemination in the sow – a review. *Reproduction in Domestic Animals* 37, 201–205.

Rath, D., Sieg, B., Leigh, J., Klinc, P., Besseling, M., Kruger, C., Wolken, A., Frenzel, A., Westermann, A., Probst, S., Grossfeld, R., Hadeler, K.G. and Ehling, C. (2003a) Current perspectives of sperm sorting in domestic farm animals. In: *Proceedings 19th Meeting European Embryo Transfer Association (Rostock)*, pp. 125–128.

Rath, D., Ruiz, S. and Sieg, B. (2003b) Birth of female piglets following intrauterine insemination of a sow using flow cytometrically sexed boar semen. *Veterinary Record* 152, 400–401.

Reik, W., Santos, F. and Dean, W. (2003) Mammalian epigenomics: reprogramming the genome for development and therapy. *Theriogenology* 59, 21–32.

Rejduch, B., Slota, E. and Gustavsson, I. (2000) 60, XY/60,XX chimerism in the germ cell line of mature bulls born in heterosexual twinning. *Theriogenology* 54, 621–627.

Revell, D.K., Morris, S.P., Cottam, Y.H., Hanna, J.E., Thomas, D.G., Brown, S. and McCutcheon, S.N. (2002) Shearing ewes at mid-pregnancy is associated with changes in fetal growth and development. *Australian Journal of Agricultural Research* 53, 697–705.

Rhind, S.M., King, T.J., Harkness, L.M., Bellamy, C., Wallace, W., DeSousa, P. and Wilmut, I. (2003) Cloned lambs – lessons from pathology. *Nature Biotechnology* 21, 744–745.

Riha, J. and Petelikova, J. (1990) Some experiences with induced twinning in the Czech Republic. *Nas Chov* 50, 489–491.

Rizos, D., Fair, T., Papadopoulos, S., Boland, M.P. and Lonergan, P. (2002) Developmental, qualitative and ultrastructural differences between ovine and bovine embryos produced *in vivo* or *in vitro*. *Molecular Reproduction and Development* 62, 320–327.

Rizos, D., Gutierrez-Adan, A., Perez-Garnelo, S., de la Fuente, J., Boland, M.P. and Lonergan, P. (2003a) Bovine embryo culture in the presence or absence of serum: implications for blastocyst development, cryotolerance and messenger RNA expression. *Biology of Reproduction* 68, 236–243.

Rizos, D., Burke, L., Duffy, P., Wade, M., O'Farrell, K., Mee, J.F., Macsiurtain, M., Boland, M.P. and Lonergan, P. (2003b) Developmental competence of oocytes from Holstein–Friesian nulliparous heifers and lactating cows in the early post partum period. In: *Proceedings 19th Meeting European Embryo Transfer Association (Rostock)*, p. 198.

Roberts, R.M., Kimura, K. and Larson, M. (2002) Sexual dimorphism among blastocysts may provide for sex adjustment in the bovine. *Journal of Animal Science* 80 (Suppl. 2), Abstract 212, p. 84.

Roberts, R.M., Green, M.P., Spate, L.D. and Kimura, K. (2003) The effects of high versus low oxygen and of the presence of glucose and fructose during culture on the sex ratio of bovine blastocysts. *Biology of Reproduction* 68 (Suppl. 1), Abstract 337, pp. 250–251.

Robinson, T.J. (1957) Pregnancy. In: Hammond, J. (ed.) *Progress in the Physiology of Farm Animals*, Vol. 3. Butterworths, London, pp. 793–904.

Robl, J.M., Kasinathan, P., Sullivan, E., Kuroiwa, Y., Tomizuka, K. and Ishida, I. (2003) Artificial chromosome vectors and expression of complex proteins in transgenic animals. *Theriogenology* 59, 107–113.

Roca, J., Carvajal, G., Lucas, X., Vazquez, J.M. and Martinez, E.A. (2003) Fertility of weaned sows after deep intrauterine insemination with a reduced number of frozen–thawed spermatozoa. *Theriogenology* 60, 77–87.

Roche, J.F. and Diskin, M.G. (2001) Resumption of reproductive activity in the early postpartum period of cows. In: Diskin, M.G. (ed.) *Fertility in the High-Producing Dairy Cow*. Occasional Publication No. 26, British Society of Animal Science, pp. 31–42.

Rodriguez, N., Gonzalez, F., Cabrera, F., Batista, M., Alamo, D., Sulon, J., Beckers, J.F. and Gracia, A. (2003) Comparison of transrectal ultrasound scanning, progesterone and pregnancy-associated glycoprotein assays in plasma samples for early pregnancy diagnosis in the goat. In: *Proceedings 19th Meeting European Embryo Transfer Association (Rostock)*, p. 202.

Romano, J.E. (2002) Does in proestrus–estrus hasten estrus onset in does estrous synchronized during breeding season. *Applied Animal Behaviour Science* 77, 329–334.

Rooke, J.A., Shao, C.C. and Speake, B.K. (2001) Effects of feeding tuna oil on the lipid composition of pig spermatozoa and *in vitro* characteristics of semen. *Reproduction* 121, 315–322.

Rorie, R.W., Bilby, T.R. and Lester, T.D. (2002) Application of electronic estrus detection technologies to reproductive management of cattle. *Theriogenology* 57, 137–148.

Rosa, H.J.D. and Bryant, M.J. (2002) The ram effect as a way of modifying the reproductive activity in the ewe. *Small Ruminant Research* 45, 1–16.

Rosa, H.J.D. and Bryant, M.J. (2003) Seasonality of reproduction in sheep. *Small Ruminant Research* 48, 155–171.

Rosenkrans, K.S. and Hardin, D.K. (2003) Repeatability and accuracy of reproductive tract scoring to determine pubertal status in beef heifers. *Theriogenology* 59, 1087–1092.

Roth, Z., Arav, A., Braw-Tal, R., Bor, A. and Wolfenson, D. (2002) Effect of treatment with follicle-stimulating hormone or bovine somatotropin on the quality of oocytes aspirated in the autumn from previously heat-stressed cows. *Journal of Dairy Science* 85, 1398–1405.

Roudebush, W.E. and Diehl, J.R. (2001) Platelet-activating factor content in boar spermatozoa correlates with fertility. *Theriogenology* 55, 1633–1638.

Rushen, J. (2003) Changing concepts of farm animal welfare: bridging the gap between applied and basic research. *Applied Animal Behaviour* 81, 199–214.

Sakaguchi, M., Geshi, M., Hamano, S., Yonai, M. and Nagai, T. (2002) Embryonic and calving losses in bovine mixed-twins induced by transfer of *in vitro*-produced embryos to bred recipients. *Animal Reproduction Science* 72, 209–221.

Sartori, R., Suarez-Fernandez, C.A., Monson, R.L., Guenther, J.N., Rosa, G.J.M. and Wiltbank, M.C. (2003) Improvement in recovery of embryos/ova using a shallow uterine horn flushing technique in superovulated Holstein heifers. *Theriogenology* 60, 1319–1330.

Schoevers, E.J., Kidson, A., Verheijden, J.H.M. and Bevers, M.M. (2003) Effect of follicle-stimulating hormone on nuclear and cytoplasmic maturation of sow oocytes *in vitro*. *Theriogenology* 59, 2017–2028.

Schwarz, T. and Wierzchos, E. (2000) Relationship between FSH and ovarian follicular dynamics in goats during the estrous cycle. *Theriogenology* 53, 381.

Seidel, G.E. Jr (2003) Economics of selecting for sex: the most important genetic trait. *Theriogenology* 59, 585–598.

Selles, E., Gadea, J., Romar, R., Matas, C. and Ruiz, S. (2003) Analysis of *in vitro* fertilizing capacity to evaluate the freezing procedures of boar semen and to predict the subsequent fertility. *Reproduction in Domestic Animals* 38, 66–72.

Sellier, P. (2002) Prospective in animal breeding. In: *28ème Journée de la Recherche Equine (Paris)*, pp. 199–209.

Senger, P.L. (1997) *Pathways to Pregnancy and Parturition*. Current Conceptions, Pullman, Washington.

Shaw, D.W. and Good, T.E. (2000) Recovery rates and embryo quality following dominant follicle ablation in superovulated cattle. *Theriogenology* 53, 1521–1528.

Shimada, M., Nishobori, M., Isobe, N., Kawano, N. and Terada, T. (2003) Luteinizing hormone receptor formation in cumulus cells surrounding porcine oocytes and its role during meiotic maturation of porcine oocytes. *Biology of Reproduction* 68, 1142–1149.

Shirasuna, K., Asaoka, H., Acosta, T.J., Hayashi, M., Tanaka, J., Ohtani, M. and Miyamoto, A. (2003) Evidence that the corpus luteum releases PGF2-alpha during spontanteous luteolysis in the cow. *Biology of Reproduction* 68 (Suppl. 1), Abstract 82, p. 146.

Signoret, J.P., Baldwin, B.A., Fraser, D. and Hafez, E.S.E. (1975) The behaviour of swine. In: Hafez, E.S.E. (ed.) *The Behaviour of Domestic Animals*, 3rd edn. Bailliere Tindall, London, pp. 295–329.

Silke, V., Diskin, M.G., Kenny, D.A., Boland, M.P., Dillon, P., Mee, J.F. and Sreenan, J.M. (2002) Extent, pattern and factors associated with late embryonic loss in dairy cows. *Animal Reproduction Science* 71, 1–12.

Silva, R.B., Martins, A. Jr, Zanon, J.E.O., Verona, D., Calegari, R.S. and Pereira, R.N. (2003) Effect of penicillamine on *in vitro* maturation of bovine oocytes. *Biology of Reproduction* 68 (Suppl. 1), Abstract 565, p. 341.

Sinclair, K.D., Molle, G., Revilla, R., Roche, J.F., Quintans, G., Marongiu, L., Sanz, A., Mackey, D.R. and Diskin, M.G. (2002) Ovulation of the first dominant follicle arising after day 21 postpartum in suckling beef cows. *Animal Science* 75, 115–126.

Smeaton, D.C., Harris, B.L., Xu, Z.Z. and Vivanco, W.H. (2003) Factors affecting commercial application of embryo technologies in New Zealand: a modelling approach. *Theriogenology* 59, 617–634.

Smith, R.F. and Dobson, H. (2002) Hormonal interactions within the hypothalamus and pituitary with respect to stress and reproduction in sheep. *Domestic Animal Endocrinology* 23, 75–85.

Soler, A.J., Astore, V., Sestelo, A., Rivolta, M., Jacome, L.N. and Garde, J.J. (2003) Effect of thawing procedure on cryosurvival of deer spermatozoa: work in progress. *Theriogenology* 60, 511–520.

Somfai, T., Bodo, S., Nagy, S., Papp, A.B., Ivancsics, J., Baranyai, B., Gocza, E. and Kovacs, A. (2002) Effect of swim-up and Percoll treatment on viability and acrosome integrity of frozen–thawed bull spermatozoa. *Reproduction in Domestic Animals* 37, 285–290.

Squires, E.L., Carnevale, E.M., McCue, P.M. and Bruemmer, J.E. (2003) Embryo technologies in the horse. *Theriogenology* 59, 151–170.

Steven, D. and Morris, G. (1975) Development of the foetal membrane. In: Steven, D. (ed.) *Comparative Placentation*. Academic Press, London, pp. 82–100.

Stocker, P. (2001) Principles of organic livestock production. In: Younie, D. and Wilkinson, J.M. (eds) *Organic Livestock Farming*. Chalcombe Publications, Lincoln, Nebraska, pp. 25–31.

Stout, T.A.E. and Allen, W.R. (2002) Prostaglandin E2 and F2-alpha production by equine conceptuses and concentrations in conceptus fluids and uterine flushings recovered from early pregnant and dioestrous mares. *Reproduction* 123, 261–268.

Stronge, A.J., Morris, D.G., Mee, J.F., Boland, M.P. and Sreenan, J.M. (2003) Post-insemination milk progesterone and early embryo survival in dairy cows. *Reproduction in Domestic Animals* 38, 357 (Abstract).

Suarez, S.S. (2002) Formation of a reservoir of sperm in the oviduct. *Reproduction in Domestic Animals* 37, 140–143.

Sugiyama, S., McGowan, M., Kafi, M., Phillips, N. and Young, M. (2003) Effects of increased ambient temperature on the development of *in vitro* derived bovine zygotes. *Theriogenology* 60, 1039–1047.

Suh, R.S., Phadke, N., Ohl, D.A., Takayama, S. and Smith, G.D. (2003) Rethinking gamete/embryo isolation and culture with microfluidics. *Human Reproduction Update* 9, 451–461.

Sundrum, A. (2001) Organic livestock farming: a critical review. *Livestock Production Science* 67, 207–215.

Tan, J., Wu, Y., Han, D., Zhou, J., Luo, M. and Chang, Z. (2003) Inhibition of meiotic resumption of goat oocytes by butyrolactone-I (BL-I), roscovitine (ROS) and cycloheximide (CHX). *Biology of Reproduction* 68 (Suppl. 1), Abstract 357, p. 259.

Tani, T., Kato, Y. and Tsunoda, Y. (2003) Reprogramming of bovine somatic cell nuclei is not directly regulated by maturation promoting factor or mitogen-activated protein kinase activity. *Biology of Reproduction* 69, 1890–1894.

Tantasuparuk, W., Lundeheim, N., Dalin, A.M., Kunavongkrit, A. and Einarsson, S. (2000) Reproductive performance of purebred Landrace and Yorkshire sows in Thailand with special reference to seasonal influence and parity number. *Theriogenology* 54, 481–496.

Tarbotton, I.S., Bray, A.R. and Wilson, J.A. (2002) Incidence and perceptions of cryptorchid lambs in 2000. *Proceedings New Zealand Society of Animal Production* 62, 334–336.

Taverne, M.A.M., Breeveld-Dwarkasing, V.N.A., Van Dissel-Emiliani, F.M.F., Bevers, M.M., De Jonng, R. and Van Der Weijden, G.C. (2002) Between prepartum luteolysis and onset of expulsion. *Domestic Animal Endocrinology* 23, 329–337.

Taylor, V.J., Beever, D.E., Bryant, M.J. and Waithes, D.C. (2003) Metabolic profiles and progesterone cycles in first lactation dairy cows. *Theriogenology* 59, 1661–1677.

Techakumphu, M., Tummaruk, P., Tantasuparuk, W. and Kunavongkrit, A. (2002) Progress of reproductive biotechnology in pig. *Thai Journal of Veterinary Medicine* 32 (Suppl.), 7–20.

Thibault, C. and Levasseur, M.C. (1974) Reproductive life cycle. In: Hafez, E.S.E. (ed.) *Reproduction in Farm Animals*, 3rd edn. Lea and Febiger, Philadelphia, Pennsylvania, pp. 82–100.

Thibier, M. and Stringfellow, D.A. (2003) Health and Safety Committee (HASAC) of the International Embryo Transfer Society (IETS) has managed critical challenges for two decades. *Theriogenology* 59, 1067–1078.

Thibier, M. and Wagner, H.G. (2000) World statistics for artificial insemination in cattle. In: *Proceedings 14th International Congress on Animal Reproduction*, Vol. 2, p. 15.2.

Thun, R., Jaros, P., Claus, R. and Hennessy, D.P. (2003) Immunocastration as an alternative to surgical castration of male piglets in Switzerland. *Reproduction in Domestic Animals* 38, 321 (Abstract).

Thurston, L.M., Holt, W.V. and Watson, P.F. (2003) Post-thaw functional status of boar spermatozoa cryopreserved using three controlled rate freezers: a comparison. *Theriogenology* 60, 101–113.

Tremoleda, J.L., Stout, T.A.E., Lagutina, I., Lazzari, G., Bevers, M.M., Colenbrander, B. and Galli, C. (2003) Effects of *in vitro* production on horse embryo morphology, cytoskeletal characteristics, and blastocyst capsule formation. *Biology of Reproduction* 69, 1895–1906.

Trewavas, A. (2001) Urban myths of organic farming. *Nature* 410, 409.

Tuchscherer, M., Puppe, B., Tuchscherer, A. and Tiemann, U. (2000) Early identification of neonates at risk: traits of newborn piglets with respect to survival. *Theriogenology* 54, 371–388.

Tullio, D.M. and Kozicki, L.H. (2001) Sonographic evaluation of Thoroughbred mares in the last month of pregnancy to determine the fetal biophysical profile. *Revista Brasileira de Reproducao Animal* 25, 367–369.

Tymms, M.J. and Kola, I. (2001) *Gene Knockout Protocols*. Humana Press, Totowa, New Jersey, 431 pp.

Ungerfeld, R. and Rubianes, E. (2002) Short term primings with different progestogen intravaginal devices (MAP, FGA and CIDR) for eCG-oestrous induction in anestrus ewes. *Small Ruminant Research* 46, 63–66.

Ungerfeld, R., Suarez, G., Carbajal, B., Silva, L., Laca, M., Forsberg, M. and Rubianes, E. (2003) Medroxyprogesterone priming and response to the ram effect in Corriedale ewes during the nonbreeding season. *Theriogenology* 60, 35–45.

Vaarst, M., Roderick, S., Lund, V. and Lockeretz, W. (2003) *Animal Health and Welfare in Organic Agriculture*. CAB International, Wallingford, UK, 448 pp.

Vajta, G., Lewis, I.M., Trounson, A.O., Purup, S., Maddox-Hyttel, P., Schmidt, M., Pedersen, I.G., Greve, T. and Callesen, H. (2003) Handmade somatic cell cloning in cattle: analysis of factors contributing to high efficiency *in vitro*. *Biology of Reproduction* 68, 571–578.

Van Arendonk, J.A.M. and Bijma, P. (2003) Factors affecting commercial application of embryo technologies in dairy cattle in Europe – a modelling approach. *Theriogenology* 59, 635–649.

Van Arendonk, J.A.M. and Liinamo, A.E. (2003) Dairy cattle production in Europe. *Theriogenology* 59, 563–569.

Van der Elst, J. (2003) Oocyte freezing: here to stay? *Human Reproduction Update* 9, 463–470.

Van Der Lende, T. and Van Rens, B.T.T.M. (2003) Critical periods for foetal mortality in gilts identified by analysing the length distribution of mummified foetuses and frequency of non-fresh stillborn piglets. *Animal Reproduction Science* 75, 141–150.

Van Eerdenburg, F.J.C.M. (2003a) Caring for ewe and lamb. *Dier en Arts* 18, 109–119.

Van Eerdenburg, F.J.C.M. (2003b) Oestrus detection in dairy cattle: how to beat the bull. *Reproduction in Domestic Animals* 38, 322 (Abstract).

Van Eerdenburg, F.J.C.M., Karthaus, D., Taverne, M.A.M., Merics, I. and Szenci, O. (2002) The relationship between estrous behavioral score and time of ovulation in dairy cattle. *Journal of Dairy Science* 85, 1150–1156.

Van Rens, B.T.T.M., Hazeleger, W. and van der Lende, T. (2000) Periovulatory hormone profiles and components of litter size in gilts with different estrogen receptor (ESR) genotypes. *Theriogenology* 53, 1375–1387.

Vandeplassche, M.M. (1986) Delayed embryonic development and prolonged pregnancy in mares. In: Morrow, D.A. (ed.) *Current Therapy in*

Theriogenology. W.B. Saunders, Philadelphia, Pennsylvania, pp. 685–689.

Vassena, R., Mapletoft, R.J., Allodi, S., Singth, J. and Adams, G.P. (2003) Morphology and developmental competence of bovine oocytes relative to follicular status.*Theriogenology* 60, 923–932.

Vazquez, J.M., Martinez, E.A., Parrilla, I., Roca, J., Gil, M.A. and Vazquez, J.L. (2003) Birth of piglets after deep intrauterine insemination with flow cytometrically sorted boar spermatozoa. *Theriogenology* 59, 1605–1614.

Verberckmoes, S., De Pauw, I., Van Soom, A., Dewulf, J. and de Kruif, A. (2003) Low dose insemination in cattle with the Ghent device. *Reproduction in Domestic Animals* 38, 335 (Abstract).

Verkerk, G. (2003) Pasture-based dairying: challenges and rewards for New Zealand producers. *Theriogenology* 59, 553–561.

Viana, J.H.M., Ferreira, A.M., Camargo, L.S.A., Sa, W.F., Rosa e Silva, A.A.M., Marques, A.P., de Leite, E.G. and de Fora, J. (2003) Follicular divergence in the presence of low or high progestagen concentrations in zebu cattle. *Biology of Reproduction* 68 (Suppl. 1), Abstract 521, pp. 326–327.

Vishwanath, R. (2003) Artificial insemination: the state of the art. *Theriogenology* 59, 571–584.

Wahner, M. and Huhn, U. (2003) Improving management for the benefit of people and pigs – synchronized farrowing in the pig. *Biotechnology in Animal Husbandry* 19, 1–9.

Wall, R.J. (2002) New gene transfer methods. *Theriogenology* 57, 189–201.

Wallenhorst, S. and Holtz, W. (2002) Embryo collection in prepubertal gilts and attempts to develop an improved embryo transfer technique. *Veterinary Record* 150, 749–751.

Walters, A.H., Bailey, T.L., Pearson, R.E. and Gwazdauskas, F.C. (2002) Parity-related changes in bovine follicle and oocyte populations, oocyte quality, and hormones to 90 days postpartum. *Journal of Dairy Science* 85, 824–832.

Wang, B., Baldassarre, H., Pierson, J., Cote, F., Rao, K.M. and Karatzas, C.N. (2003) The *in vitro* and *in vivo* development of goat embryos produced by intracytoplasmic sperm injection using tail-cut spermatozoa. *Zygote* 11, 219–227.

Ward, F., Rizos, D., Boland, M.P. and Lonergan, P. (2003) Effect of reducing sperm concentrations during IVF on the ability to distinguish between bulls of high and low field fertility: work in progress. *Theriogenology* 59, 1575–1584.

Webb, A.J. (1995) Future challenges in pig genetics. *Animal Breeding Abstracts* 63(10), 731–736.

Wei, H. and Fukui, Y. (2002) Birth of calves derived from embryos produced by intracytoplasmic

sperm injection without exogenous oocyte activation. *Zygote* 10, 149–153.

Wells, N.H., Hallford, D.M. and Hernandez, J.A. (2003) Serum thyroid hormones and reproductive characteristics of Rambouillet ewe lambs treated with propylthiouracil before puberty. *Theriogenology* 59, 1403–1413.

Wheaton, J.E. and Godfrey, R.W. (2003) Plasma LH, FSH, testosterone and age at puberty in ram lambs actively immunized against an inhibin alpha-subunit peptide. *Theriogenology* 60, 933–941.

White, F.J. and Wettemann, R.P. (2000) Season alters estrous behavior but not time of ovulation in beef cows. *Animal Science Research Report, Agricultural Experiment Station, Oklahoma State University*, No. P-980, 176–181.

Willard, S., Gandy, S., Bowers, S., Graves, K., Elias, A. and Whisnant, C. (2003) The effects of GnRH administration postinsemination on serum concentrations of progesterone and pregnancy rates in dairy cattle exposed to mild summer heat stress. *Theriogenology* 59, 1799–1810.

Willard, S.T., Neuendorff, D.A., Lewis, A.W. and Randel, R.D. (2002) A comparison of transvaginal artificial insemination procedures for use in commercially farmed deer. *Small Ruminant Research* 44, 135–140.

Williams, G.L., Amstalden, M., Garcia, M.R., Stanko, R.L., Nizielski, S.E., Morrison, C.D. and Keisler, D.H. (2002) Leptin and its role in the central regulation of reproduction in cattle. *Domestic Animal Endocrinology* 23, 339–349.

Wilmut, I. (2003) Livestock cloning: past, present and future. In: *Proceedings 19th Meeting European Embryo Transfer Association (Rostock)*, pp. 15–21.

Wilsher, S. and Allen, W.R. (2003) A novel method for non-surgical embryo transfer in the mare. *Reproduction in Domestic Animals* 38, 320 (Abstract).

Winston, R. (2003) No gain without pain: are we ignoring the risks of ART when pursuing its benefits? *Biology of Reproduction* 68 (Suppl. 1), Abstract K1, pp. 84–85.

Wood, S.L., Kucy, M.C., Smith, M.F. and Patterson, D.J. (2001) Improved synchrony of estrus and ovulation with the addition of GnRH to a melengestrol acetate–prostaglandin F2-alpha synchronization treatment in beef heifers. *Journal of Animal Science* 79, 2210–2216.

Wrenzycki, C., Herrmann, D. and Niemann, H. (2003) Timing of blastocyst expansion affects spatial messenger RNA expression patterns of genes in bovine blatocysts produced *in vitro*. *Biology of Reproduction* 68, 2073–2080.

Wright, C., Evans, A.C.O., Evans, N.P., Duffy, P., Fox, J., Boland, M.P., Roche, J.F. and Sweeney, T. (2002) Effect of maternal exposure to the environmental estrogen, octylphenol during fetal and/or postnatal life on onset of puberty, endocrine status, and ovarian follicular dynamics in ewe lambs. *Biology of Reproduction* 67, 1734–1740.

Wu, J., Carrell, D.T. and Wilcox, A.L. (2001) Development of *in vitro*-matured oocytes from porcine preantral follicles following intracytoplasmic sperm injection. *Biology of Reproduction* 65, 1579–1585.

Xu, J., Jiang, S. and Du, F. (2003) Comparison of the developmental competence of *in vitro* fertilized bovine oocytes after maturation in M199 medium supplemented with growth factor and gonadotropins. *Theriogenology* 59, 505.

Yanagimachi, R. (2002) Cloning: experience from the mouse and other animals. *Molecular and Cellular Endocrinology* 187, 241–248.

Yaniz, J., Santolaria, P. and Lopez-Gatius, F. (2003) Relationship between fertility and the walking activity of cows at oestrus. *Veterinary Record* 152, 239–240.

Yates, J.H., Chandler, J.E., Canal, A.L. and Paul, J.B. (2003) The effect of nocturnal sampling on semen quality and the efficiency of collection in bovine species. *Theriogenology* 60, 1665–1677.

Yavas, Y. and Walton, J.S. (2000) Postpartum acyclicity in suckled beef cows: a review. *Theriogenology* 54, 25–55.

Yoshioka, K. and Rodriguez-Martinez, H. (2003) Effects of embryo density on the development of porcine embryos cultured in a chemically defined medium. *Reproduction in Domestic Animals* 38, 352 (Abstract 106).

Young, I. (2003) Bioveterinary science: development of a discipline. *Nature Biotechnology* 21, 339–340.

Younie, D. (2001) Organic grassland: the foundation stone of organic livestock farming. In: Younie, D. and Wilkinson, J.M. (eds) *Organic Livestock Farming*. Chalcombe Publications, Lincoln, Nebraska, pp. 75–102.

Zeng, X.Y., Turkstra, J.A., Meloen, R.H., Liu, X.Y., Chen, F.Q., Schaaper, W.M.M., Oonk, H.B., Guo, D.Z. and Van de Wiel, D.F.M. (2002) Active immunization against gonadotrophin-releasing hormone in Chinese male pigs: effects of dose on antibody titer, hormone levels and sexual development. *Animal Reproduction Science* 70, 223–233.

Zhang, M., Lu, K.H. and Seidel, G.E. Jr (2003) Development of bovine embryos after *in vitro* fertilization of oocytes with flow cytometrically sorted, stained and unsorted sperm from different bulls. *Theriogenology* 60, 1657–1663.

Zhu, Z.N. and Ji, X.S. (2002) Organic food and its system of production. *Acta Agriculturae Shanghai* 18, 92–96.

Zi, X.D. (2003) Reproduction in female yaks (*Bos grunniens*) and opportunities for improvement. *Theriogenology* 59, 1303–1312.

Zicarelli, L., Boni, R., Campanile, G., Palo, R.D., Esposito, L., Roviello, S., Varricchio, E. and Langella, M. (1996) Superovulation, ovum pick-up and *in vitro* embryo production in Mediterranean buffaloes bred in Italy. *Bulgarian Journal of Agricultural Science* 2, 115–120.

Ziecik, A.J. (2002) Old, new and the newest concepts of inhibition of luteolysis during early pregnancy in pig. *Domestic Animal Endocrinology* 23, 256–275.

Appendix A

Books on Farm Animals (Breeding and Reproduction) Published in the Past Decade (1994–2004)

Arthur, G.H., Noakes, D.E., Pearson, H. and Parkinson, T.J. (eds) (1996) *Veterinary Reproduction and Obstetrics*. W.B. Saunders, London, UK.

Bowling, A.T. and Ruvinsky, A. (eds) (2000) *The Genetics of the Horse*. CAB International, Wallingford, UK.

Clark, A.J. (ed.) (1998) *Animal Breeding: Technology for the 21st Century*. Harwood Academic Publishers, Amsterdam, The Netherlands.

Dale, B. and Elder, K. (1997) In Vitro *Fertilization*. Cambridge University Press, Cambridge, UK.

Etches, R.J. (1996) *Reproduction in Poultry*. CAB International, Wallingford, UK.

Fahmy, M.H. (ed.) (1996) *Prolific Sheep*. CAB International, Wallingford, UK.

Fields, M.J., Sand, R.S. and Yelich, J.V. (eds) (2002) *Factors affecting Calf Crop: Biotechnology of Reproduction*. CRC Press, Boca Raton, Florida.

Foote, R.H. (1999) *Artificial Insemination to Cloning: Tracing 50 Years of Research*. Cornell University, Ithaca, New York.

Fries, R. and Ruvinsky, A. (eds) (1999) *The Genetics of Cattle*. CAB International, Wallingford, UK.

Gordon, I. (1996) *Controlled Reproduction in Cattle and Buffaloes*. CAB International, Wallingford, UK.

Gordon, I. (1997) *Controlled Reproduction in Sheep and Goats*. CAB International, Wallingford, UK.

Gordon, I. (1997) *Controlled Reproduction in Pigs*. CAB International, Wallingford, UK.

Gordon, I. (1997) *Controlled Reproduction in Horses, Deer and Camelids*. CAB International, Wallingford, UK.

Gordon, I. (2003) *Laboratory Production of Cattle Embryos*, 2nd edn. CAB International, Wallingford, UK.

Hafez, E.S.E. and Hafez, B. (eds) (2000) *Reproduction in Farm Animals*, 7th edn. Lippincott Williams and Wilkins, Baltimore, Maryland.

Houdebine, L.M. (ed.) (1997) *Transgenic Animals: Generation and Use*. Harwood Academic Publishers, Amsterdam, The Netherlands.

Knobil, E. and Neill, J. (eds) (1994) *The Physiology of Reproduction*. Raven Press, New York.

Morel, M.C.G.D. (2003) *Equine Reproductive Physiology, Breeding and Stud Management*, 2nd edn. CAB International, Wallingford, UK.

Muir, W.M. and Aggrey, S.A. (eds) (2003) *Poultry Genetics, Breeding and Biotechnology*. CAB International, Wallingford, UK.

Murray, J.D., Anderson, G.B., Oberbauer, A.M. and McGloughlin, M.M. (eds) (1999) *Transgenic Animals in Agriculture*. CAB International, Wallingford, UK.

Peters, A.R. and Ball, P.J.H. (1995) *Reproduction in Cattle*. Blackwell Science, Oxford, UK.

Phillips, C.J.C. (ed.) (1996) *Progress in Dairy Science*. CAB International, Wallingford, UK.

Pineda, M.H. and Dooley, M.P. (eds) (2003) *McDonald's Veterinary Endocrinology and Reproduction*. Iowa State University Press, Ames, Iowa.

Piper, L. and Ruvinsky, A. (eds) (1997) *The Genetics of Sheep*. CAB International, Wallingford, UK.

Renaville, R. and Burny, A. (eds) (2001) *Biotechnology in Animal Husbandry*. Kluwer Academic Publishers, Dordrecht, The Netherlands.

Rothschild, M.F. and Ruvinsky, A. (eds) (1998) *The Genetics of the Pig*. CAB International, Wallingford, UK.

Senger, P.L. (2003) *Pathways to Pregnancy and Parturition*, 2nd edn. Current Conceptions, Pullman, Washington.

Simm, G. (1998) *Genetic Improvement of Cattle and Sheep*. CAB International, Wallingford, UK.

Squires, E.J. (2003) *Applied Animal Endocrinology*. CAB International, Wallingford, UK.

Stringfellow, D.A. and Seidel, S.M. (eds) (1998) *Manual of the International Embryo Transfer Society*, 3rd edn. International Embryo Transfer Society, Savoy, Illinois.

Thibault, C., Levasseur, M.C. and Hunter, R.H.F. (eds) (1993) *Reproduction in Mammals and Man*. Ellipses, Paris, France.

Appendix B

Journals Dealing with Breeding and Reproduction in Farm Animals

AgBiotech News and Information. CAB International, Wallingford, UK.

Animal Breeding Abstracts. CAB International, Wallingford, UK.

Animal Reproduction Science. Elsevier Science BV, Amsterdam, The Netherlands.

Animal Science. British Society of Animal Science, Penicuik, UK.

Biology of Reproduction. Society for the Study of Reproduction, Champaign, Illinois.

British Veterinary Journal. Baillière Tindall, London, UK.

Canadian Journal of Animal Science. Agricultural Institute of Canada, Ottawa, Canada.

Cloning and Stem Cells. Mary Ann Liebert Inc., Larchmont, New York.

Embryo Transfer Newsletter. International Embryo Transfer Society, Champaign, Illinois.

Journal of Agricultural Science (Cambridge). Cambridge University Press, Cambridge, UK.

Journal of Animal Science. American Society of Animal Science, Champaign, Illinois.

Journal of Assisted Reproduction and Genetics. Kluwer Academic/Plenum Publishers, New York.

Journal of Dairy Science. American Dairy Science Association, Champaign, Illinois.

Journal of Reproduction and Development. Japanese Society of Animal Reproduction, Tokyo, Japan.

Livestock Production Science. Elsevier Science BV, Amsterdam, The Netherlands.

Molecular Reproduction and Development. John Wiley and Sons, Somerset, New Jersey.

Nature. Nature Publishing Group, London, UK.

Nature Biotechnology. Nature Publishing Group, London, UK.

Reproduction. Journal of Reproduction and Fertility, Cambridge, UK.

Reproduction in Domestic Animals. Blackwell Wissenschafts-Verlag GmbH, Berlin, Germany.

Reproduction, Nutrition, Development. Institut National de la Recherche Agronomique, Paris, France.

Science. American Association for the Advancement of Science, Washington, DC.

South African Journal of Animal Science. South African Society of Animal Science, Hatfield, South Africa.

Theriogenology. Elsevier Science, Oxford, UK.

Veterinary Record. British Veterinary Association, London, UK.

Zygote. Cambridge University Press, Cambridge, UK.

Index